STO

ACPL ITEM
DISCARDED

D0787512

AUG 10 '72

DYNAMIC RESPONSE

OF

STRUCTURES

DYNAMIC RESPONSE

OF

STRUCTURES

PROCEEDINGS OF A SYMPOSIUM

HELD AT

STANFORD UNIVERSITY, CALIFORNIA

JUNE 28 AND 29, 1971

Edited

by

GEORGE HERRMANN

and

NICHOLAS PERRONE

PERGAMON PRESS INC.

NEW YORK • TORONTO • OXFORD • SYDNEY • BRAUNSCHWEIG

PERGAMON PRESS INC.
Maxwell House, Fairview Park, Elmsford, N.Y. 10523

PERGAMON OF CANADA LTD.
207 Queen's Quay West, Toronto 117, Ontario

PERGAMON PRESS LTD.
Headington Hill Hall, Oxford

PERGAMON PRESS (AUST.) PTY. LTD.
Rushcutters Bay, Sydney, N.S.W.

VIEWEG & SOHN GmbH
Burgplatz 1, Braunschweig

Copyright © 1972, Pergamon Press Inc.
Library of Congress Catalog Card No. 75-171021

Printed in the United States of America

All Rights Reserved. No part of this publication may be reproduced, stored in a retrieval system or transmitted in any form, or by any means, electronic, mechanical, photocopying, recording or otherwise, without prior permission of Pergamon Press, Inc.

08 016850 7

1687179

CONTENTS

PREFACE

The dynamic behavior and response of structures continues to exercise a considerable and oftentimes decisive influence in the analysis or design and understanding of man-made and natural systems. To quote from an imaginary list of timely examples, mention could be made of problems associated with the dynamic response of transportation structures in all speed ranges, including aerospace, ground, and water vehicles, as well as various dynamic phenomena on the surface of the earth, e.g. earthquakes, tides, tornadoes, in their interaction with man-made structures.

In order to create a forum for an interchange of ideas and for communication of some new methodologies and results, it was deemed desirable to bring together workers in several diverse areas of dynamic response of structures for a two-day symposium. The principal aim of this meeting was to highlight some recent developments, without attempts at providing any degree of completeness of coverage. The Symposium was held on June 28 and 29, 1971, at Stanford University, Stanford, California, and the technical presentations were grouped in four sessions under the titles: I Structural Elements, II Structural Systems, III Aircraft Crashworthiness, and IV Materials and Solids.

The Symposium was made possible through the joint support of the National Science Foundation, the Office of Naval Research, the Army Research Office, and the Air Force Office of Scientific Research. The Symposium Committee responsible for the planning and execution consisted of the following: B. A. Boley (Cornell University), J. M. Crowley (Office of Naval Reserach), M. P. Gaus (National Science Foundation), G. Herrmann (Stanford University), J. Kempner (Polytechnic Institute of Brooklyn), H. Liebowitz (George Washington University), J. Mayers (Stanford University), J. J. Murray (Army Research Office), N. Perrone (Office of Naval Research), J. Pomerantz (Air Force Office of Scientific Research), M. Rogers (Air Force Office of Scientific Research), and R. B. Testa (Columbia University).

The complete program for the Symposium was as follows:

MONDAY, JUNE 28

Session I, 9:15 a.m., STRUCTURAL ELEMENTS
Chairman: R. L. Bisplinghoff*, National Science Foundation
Vice Chairman: James M. Gere, Stanford University
  N. Jones, J. W. Dumas, J. G. Giannotti, and K. E. Grassit, Massachusetts Institute of Technology
  The Dynamic Plastic Behavior of Shells
  G. R. Abrahamson and H. E. Lindberg, Stanford Research Institute
  Peak Load-Impulse Characterization of Critical Pulse Loads in Structural Dynamics
  S. H. Crandall and L. E. Wittig, Massachusetts Institute of Technology
  Chladni's Patterns for Random Vibration of a Plate
  J. B. Martin, Brown University
  On the Application of the Bounding Theorems of Plasticity to Impulsively Loaded Structures

Session II, 2:00 p.m., STRUCTURAL SYSTEMS
Chairman: Ronald Smelt, Lockheed Aircraft Corporation
Vice Chairman: Holt Ashley, Stanford University
  M. J. Turner and J. B. Bartley, The Boeing Company
  Flutter Prevention in Design of the SST

---

*Due to urgent Government business, Dr. Bisplinghoff was prevented from participating. He was replaced by a member of the Committee, Dr. J. Kempner.

vii

H. L. Runyan and R. C. Goetz, NASA Langley Research Center
Space Shuttle — A New Arena for the Structural Dynamicist
T. L. Geers and L. H. Sobel, Lockheed Palo Alto Research Laboratories
Analysis of the Transient Response of Shell Structures by Numerical Methods
D. E. Hudson, California Institute of Technology
Dynamic Properties of Full-Scale Structures Determined from Natural Excitations

TUESDAY, JUNE 29

Session III, 9:00 a.m., AIRCRAFT CRASHWORTHINESS
Chairman: Nicholas Perrone Office of Naval Research
Vice Chairman: John R. Manning Stanford University
   G. T. Singley III, U.S. Army Air Mobility Research & Development Laboratory
   A Survey of Rotary-Wing Aircraft Crashworthiness
   A. A. Ezra and R. J. Fay, University of Denver
   An Assessment of Energy Absorbing Devices for Prospective Use in Aircraft Impact Situations
   A. I. King, Wayne State University
   Human Tolerance Limitations Related to Aircraft Crashworthiness
   J. A. Collins and J. W. Turnbow, Arizona State University
   Response of a Seat-Passenger System

Session IV, 1:30 p.m., MATERIALS AND SOLIDS
Chairman: Robert Plunkett, University of Minnesota
Vice Chairman: E. H. Lee, Stanford University
   L. Knopoff and D. D. Jackson
   The Analysis of Underdetermined and Overdetermined Systems
   J. W. Dally*, Illinois Institute of Technology
   Applications of Photoelasticity to Elastodynamics
   H. Kolsky, Brown University
   Some Recent Experimental Investigations in Stress-Wave Propagation and Fracture
   T. Mura, Northwestern University
   Dynamic Response of Dislocations in Solids

Coincidental with the Symposium, a testimonial banquet was held on Monday, June 28, at the Faculty Club of Stanford University in honor of Dr. Nicholas J. Hoff, Chairman, Department of Aeronautics and Astronautics, Stanford University, who became Professor Emeritus at the end of the academic year 1970/71. At this banquet a copy of a bound volume *Aerostructures — Selected Papers of Nicholas J. Hoff,* which was edited by R. B. Testa and published by Pergamon Press, was presented to Professor Hoff, together with a bound volume of testimonial letters from his former students, his many friends, and his colleagues.

The undersigned have the pleasure of acknowledging the support and advice of numerous persons who contributed greatly to the success of the Symposium. Welcoming the Symposium participants were Dr. William R. Rambo, Associate Dean, on behalf of the School of Engineering and Dr. William F. Miller, Vice President and Provost, who spoke for the University. Local arrangements for technical sessions were under the supervision of Professor C. R. Steele. Mrs. LaVerne Faulmann took care of advance registration and numerous other organizational matters. The registration desk during the symposium was staffed by Mrs. Donna Loubsky.

George Herrmann
Nicholas Perrone

---

*Professor Dally is now affiliated with University of Maryland.

THE DYNAMIC PLASTIC BEHAVIOR OF SHELLS

NORMAN JONES, J. W. DUMAS, J. G. GIANNOTTI and K. E. GRASSIT

Department of Ocean Engineering, Massachusetts Institute of Technology

Abstract—A survey has been made of the literature published previously on the inelastic behavior of shells subjected to dynamic loads. An experimental investigation has also been undertaken to examine the behavior of various cylindrical shell panels which were loaded with an impulse on the inner surface. The panels were fully clamped along the two longitudinal edges and free on the other two. The initial kinetic energy of the dynamic loads was sufficiently large to cause inelastic behavior and to produce maximum permanent transverse deflections up to nearly twice the corresponding panel thickness. Tests were conducted on mild steel and aluminum 6061 T6 panels which had various thicknesses and an included angle of 90° approximately.

NOTATION

a	circumferential width of mid-plane of panel loaded with explosive sheet
b	axial length of explosive sheet
e_1, e_2	percentage elongation at fracture of tensile specimens cut across and parallel to the direction of rolling, respectively.
t	time
x	axial coordinate measured from the center of a panel as shown in Figure 2
E	Young's modulus
E_r	energy ratio defined in appendix
H	mean thickness of a cylindrical panel
I	Total impulse
I_s	specific impulse of Detasheet explosive
2L	axial length of a cylindrical panel as shown in Figure 2
M	mass of a timepiece or a cylindrical panel directly underneath a layer of sheet explosive
R	mean radius of a cylindrical panel
T	thickness of a foam attenuator
V	mean velocity of a timepiece during each interval
V_o	magnitude of the impulsive velocity applied to a cylindrical panel or a timepiece
W	permanent transverse deflection of a cylindrical panel with a positive sense as indicated in Figure 2

1

W_e	weight of sheet explosive
W_o	maximum permanent transverse deflection
2α	included angle of a panel as shown in Figure 2
θ	angle shown in Figure 2
λ	$\dfrac{\rho V_o^2 R^2}{\sigma_o H^2}$
ρ	density
σ_o	average yield stress in a tensile test
$\sigma_{y1}\ \sigma_{y2}$	yield stress of tensile specimens cut across and parallel to the direction of rolling, respectively
$\sigma_{u1}\ \sigma_{u2}$	ultimate stress of tensile specimens which are cut across and parallel to the direction of rolling, respectively

INTRODUCTION

Theoretical investigations into the dynamic inelastic response of engineering structures are often simplified by disregarding the influence of material elasticity. These so called rigid-plastic methods sometimes give rather good estimates of the behavior of structures which are subjected to dynamic loads with a magnitude sufficient to produce large permanent deformations. However, further studies are necessary in this general area in order to permit reliable predictions of structural damage to automobiles, trains, aircraft and high speed marine craft which is sustained during collisions, the safety of pressure vessels which contain nuclear reactors, the slamming damage of ships, and to develop energy absorbing devices for various applications.

In 1952 Lee and Symonds [1] demonstrated how rigid, perfectly plastic methods could be used in order to predict the inelastic response of beams subjected to dynamic loads. This theoretical work has been further developed by a number of authors in order to describe the behavior of beams with various boundary conditions and external dynamic loads. These theoretical studies and concomitant experimental investigations on beams have been summarized in a comprehensive report by Symonds [2] in which a number of important questions concerning the validity, or otherwise, of rigid-plastic procedures are also examined in detail.

It appears reasonable to disregard material elasticity when the dynamic energy is much larger than the maximum amount of strain energy which can be absorbed by a beam in a wholly elastic manner (larger than about three times [3]) and when the pulse duration is short compared with the corresponding natural period [2]. Material strain hardening is not important for moderate permanent deformations unless a material hardens significantly for small strains. Moreover, the influence of transverse shear forces and shear deformations may be disregarded in dynamic analyses of beams with rectangular or other compact sections [2, 4]. However, it is necessary to recognize the influence of strain-rate sensitivity when a beam is made from a strain-rate sensitive material and to retain the influence of finite-deflections when a beam is restrained axially. A few theoretical predictions which retain the influence of finite-deflections and material strain-rate sensitivity are compared in reference [5] with some experimental results recorded on axially restrained fully clamped beams loaded impulsively.

Hopkins and Prager showed in reference [6] how plasticity theory may be used in order to examine the response of a circular plate subjected to an axisymmetric dynamic load. The plate is assumed to be made from a rigid, perfectly plastic material which flows according to the piecewise linear Tresca yield criterion. The influence of different boundary conditions and various axisymmetric dynamic loads on the behavior of circular plates has been further studied in references [7-12, etc.]. More recently, Florence [13] conducted an experimental investigation into the behavior of circular metal plates when subjected to a uniformly distributed impulse. The circular plates were simply supported around the outer boundary and subjected to initial kinetic energies which caused maximum permanent deflections up to several plate thicknesses. It was observed that the experimental results for the strain rate insensitive plates and the corresponding rigid-plastic predictions [8] were similar for maximum permanent transverse deflections less than the plate thickness. The discrepancy between the experimental results and the theoretical rigid-plastic predictions increases with increase in the magnitude of the permanent deflections. It was shown in references [14] and [15] that this difference is due primarily to the influence of finite-deflections or geometry changes which introduce membrane forces into a plate during the response. It appears [16-19] that at low impulses, strain-rate strengthening could dominate for highly rate sensitive materials because of the initial rapid rise of the yield stress with increasing strain rate [2] and since the deflections are still small enough to keep membrane strengthening small. However, at larger impulses, membrane strengthening dominates because the flow stress of most metals increases only slightly with increasing strain-rate.

To the authors' knowledge the only "exact" rigid-plastic solution of a non-axisymmetric (external loading or geometry) flat plate loaded dynamically is the special case of a simply supported square plate subjected to a uniformly distributed dynamic load [20]. The inclusion of finite-deflections would further complicate any theoretical analyses of this class of structures since membrane forces in addition to the bending moments which are present in an infinitesimal analysis would control plastic yielding. In order to avoid some of these difficulties, an approximate rigid-plastic method which retains the influence of finite-deflections on the dynamic response of beams and plates is presented in reference [21]. The predictions according to this procedure agree reasonably well with the corresponding maximum permanent transverse deflections observed in some experiments on fully clamped strain-rate insensitive beams and rectangular plates reported in references [22] and [23].

Hodge [24-27] and other authors [28-32] have studied the response of thin-walled rigid, perfectly plastic cylindrical shells subjected to various axisymmetric dynamic loads. Pabjanek [33] has utilized the general "state of comparison" procedure developed by Wierzbicki [34] in order to examine the dynamic response of a simply supported cylindrical shell made from a rigid linear viscoplastic material. Duffey and Krieg [35] investigated the influence of material strain hardening and a linear form of strain-rate sensitivity on the dynamic behavior of an elastic-plastic cylindrical shell. They noted for a particular example that the permanent deflections calculated for a rigid plastic material were within 20% of the corresponding elastic-plastic values when the ratio of plastic to elastic energy was greater than three. In reference [36] Perrone presented a theoretical analysis for a tube loaded impulsively which avoids the important restrictions of linearized strain-rate sensitivity introduced in references [33] and [35]. This was accomplished by recognizing that the highly non-linear nature of the constitutive relations converts the initial kinetic energy into plastic work before the stress strain-rate point has moved appreciably from its initial value [37]. Thus the dynamic stresses associated with the initial strain-rates were assumed to remain constant throughout the subsequent motion. A simple analysis which capitalizes on this approximation was within a few percent of a solution which permits the variation of the dynamic stress throughout the response. It was observed that material strain-rate sensitivity is more important than strain hardening for small deflections and vice versa for large deflections. The theoretical solutions which were presented in references [24-33], [35] and [36] are all restricted to the inelastic response of cylindrical shells which undergo infinitesimal deflections and are subjected to axisymmetric dynamic loads. No rigid-plastic solutions appear to have been published for the dynamic behavior of cylindrical panels or cylindrical shells loaded non-axisymmetrically. An approximate rigid-plastic analysis which retains the influence of geometry changes or finite-deflections on the behavior of a cylindrical shell loaded with a uniformly distributed internal impulse is presented in reference [38]. It appears that the axial membrane forces which are introduced during the response of an axially restrained cylindrical shell reduce significantly the magnitude of the permanent deflections predicted by an infinitesimal analysis [24] when the transverse deflections exceed about one-half the shell thickness.

Sankaranarayanan [39, 40] has investigated the behavior of rigid, perfectly plastic spherical caps subjected to uniformly distributed dynamic pressure pulses. However, the static admissibility of these theoretical solutions are dependent on the satisfaction of certain inequalities which severely restrict the magnitude of the permissible dynamic pressure pulses. Baker presented an analysis in reference [41] which describes the elastic-plastic behavior of a complete spherical shell subjected to a suddenly applied pressure pulse. In reference [42], Wierzbicki studied a similar problem but disregarded elastic and strain hardening effects and concentrated on the influence of material strain-rate sensitivity which was represented by various constitutive relations. Recently, Duffey [43] has extended the earlier work of Baker [41] in order to examine the influence of material strain-rate sensitivity which is represented by a linear constitutive relation. It was observed that the influence of a rate-dependent tangent modulus can generally be disregarded in structural calculations. Moreover, a "size effect" was observed so that physically smaller spherical shells experience a greater sensitivity to strain-rate effects than larger ones. This observation has also been discussed for other structures in references [44] and [45]. It should be remarked that the theoretical solutions presented in references [41] to [43] are only valid for complete spherical shells loaded with a spherically symmetric pressure pulse.

In reference [46] an approximate rigid, perfectly plastic method, which is similar to that developed in reference [21] for flat plates, has been used to estimate the inelastic behavior of various shell intersections subjected to dynamic loads. In particular the two cases of cylindrical nozzles intersecting either spherical or cylindrical shells are studied in detail. However, these predictions should be viewed as tentative since to the authors' knowledge no other theoretical, numerical or experimental results have been published on the dynamic inelastic behavior of shell intersections.

It is evident from the previous comments that few "exact" theoretical solutions can be obtained which describe the dynamic inelastic behavior of structures even when the influence of material elasticity, strain hardening, strain-rate sensitivity and finite-deflections are disregarded. However, Martin [47] has derived a theorem which gives an upper bound to the permanent deflections of a rigid, perfectly plastic continuum subjected to an initial impulsive velocity field. A corresponding lower bound displacement theorem has recently been derived by Morales and Nevill [48]. If either the maximum permanent displacement or the displacement at a particular location in a structure is required then a statically admissible solution for a concentrated load which acts on the structure is necessary for the upper bound method in reference [47]. Unfortunately, as far as the authors are aware, and except for the axisymmetric cases considered in references [49, 50], no statically admissible rigid-plastic solutions are available for shells of any geometry or boundary conditions subjected to concentrated static loads. It would therefore be necessary to use linear elastic solutions which for most shells would probably give rather poor upper bounds. A number of authors [51-53, etc.] have extended Martin's theorems to examine the influence of dynamic pressure pulse duration, material viscoplasticity and elastic deformations.

In order to improve the accuracy of Martin's theorems, Martin and Symonds [54] developed a rational mode approximation procedure for evaluating the permanent deformation of impulsively loaded structures. Again, however, this approximate method is useful only when a statically admissible mode solution is available for the same structure loaded dynamically. The authors of references [44, 55-58] have utilized this mode approximation method to examine the viscoplastic behavior of beams and frames. However, the methods of references [47, 48, 51-54 and 56-58] neglect the influence of finite-deflections and therefore would overestimate the actual maximum permanent deflections of plates [23] and axially restrained cylindrical shells and beams particularly when the final transverse deflections exceed the corresponding structural thickness. Wierzbicki [59] has extended Martin's upper bound theorem in order to accommodate the influence of finite-deflections on a structural response. However, this method is only useful when the corresponding static load carrying capacity of a structure with finite-deflections is known. Unfortunately, few such theoretical solutions are available and almost all those that exist are approximate.

Rigid-plastic methods of analysis are frequently useful during the preliminary stages of a design. These procedures are also helpful for the planning and interpretation of experimental investigations, the selection of meaningful parametric quantities in numerical methods, the checking of computer programs and often provide increased understanding of the complex behavior of structures loaded dynamically. Nevertheless, once the approximate dimensions of a final design have been selected with the aid of a rigid-plastic method, it is sometimes necessary to have more accurate predictions and to obtain information on certain aspects of the response which may have been disregarded in an approximate method. Clearly, in order to obtain this additional information, the values of the principal parameters predicted by a rigid-plastic method should be used as input to a numerical procedure such as one of those developed in references [60-68, etc.]. It might be remarked here that due to the complexity of these numerical methods it is currently too expensive to use them for the preliminary stages of certain designs. Moreover, the predictions of both numerical and analytical methods must be considered somewhat tentative until additional experimental results become available on the constitutive relations of metals which are subjected to multi-dimensional dynamic stresses. Most of the multi-dimensional constitutive relations which are used currently in various numerical and analytical methods are based on uniaxial tensile or compressive data [69-70, etc.]. Dynamic tests have been conducted on beams in pure bending [71, 72] and tubes in pure torsion [73-75, etc.] but no combined stress states appear to have been systematically examined except for the studies in references [76, 77]. It was shown in references [44, 72] that the simplified uniaxial constitutive relation of Cowper and Symonds [69] may be used to provide a reasonable approximation to the experimentally observed moment-curvature behavior of a beam which is subjected to a dynamic pure bending moment.

The previous discussion has been concerned exclusively with the inelastic behavior of structures which have a "stable" response when subjected to dynamic loads. However, dynamic plastic buckling or "unstable" behavior can occur when certain structures which are not too thin are acted on by very large external dynamic loads [78]. Although this particular aspect of behavior is important in the design of protective systems for automobiles [79, 80] and aircraft, etc., it appears that scant attention has been directed towards the general area. Theoretical and experimental studies on dynamic plastic buckling have been undertaken on struts [78], frames [81], cylindrical shells [82-90, etc.] and hemispherical shells [60, 91]. Postlethwaite and Mills [92] have performed some dynamic tests on an idealized frontal structure of an automobile and various other sheet-metal structures. The numerical procedures of references [60-64, etc.] may also be used for dynamic plastic buckling studies.

An experimental investigation has been undertaken into the dynamic inelastic behavior of cylindrical shell panels which were fully clamped along two longitudinal edges. The panels were subjected to an impulsive velocity which was distributed uniformly within a rectangular zone on the inner surface. The initial kinetic energy of the dynamic loading was considerably larger than the maximum strain energy which could be absorbed by the panels in a wholly elastic manner. In order to assess the influence of material strain-rate sensitivity on the behavior of these panels, tests were conducted on hot rolled mild steel specimens, which are highly strain rate sensitive, and repeated on aluminum 6061 T6 specimens which are believed to be almost strain rate insensitive.

EXPERIMENTAL ARRANGEMENT

The experimental arrangement which was used in the current investigation into the dynamic plastic behavior of various cylindrical shell panels is illustrated in Figure 1 and is similar to that employed in references [13, 19, 82, 89, 90, 93]. It might be remarked that a ballistic pendulum was utilized in references [5, 22, 23, 55-57] in order to measure the impulsive velocity which was imparted by sheet explosive to initially straight beams and flat plates. However, this particular procedure was not employed here because it would only provide the component of a total impulse acting on a curve surface which is tangential to the path of the pendulum.

A standard rolling machine was used to make the typical cylindrical shell panel shown in Figure 2 from initially flat rectangular plates of either hot rolled mild steel or aluminum 6061. The wide flanges along the longitudinal sides were formed with the aid of a brake press. The mild steel specimens were then annealed while the aluminum ones were heat treated to the T6 condition. In order to obtain a fully clamped condition along the two longitudinal edges of the cylindrical shell panels the gripping surfaces of the clamps, which are shown in Figure 3, were serrated and case hardened and the specimens

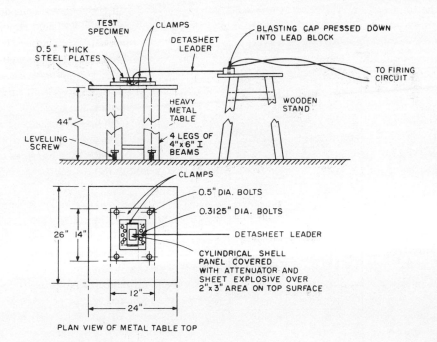

Fig. 1. Experimental Arrangement

were clamped with ten 5/16 in. diameter high strength steel bolts. The wall thickness of the cylindrical shell panels was measured at a number of locations and found to vary ± 0.0003 in. from the average values which are quoted in Table 1 together with various other shell dimensions.

A standard Instron machine was used to conduct tensile tests at strain rates of approximately 8×10^{-4} in/in/sec. on mild steel and aluminum straight parallel side specimens manufactured to ASTM standards. The lower yield stress for mild steel and 0.2% proof stress for aluminum 6061-T6 were determined from specimens which were cut parallel and perpendicular to the direction of rolling. Typical tensile stress-strain curves are presented in Figure 4, while the major experimental values are listed in Table 2.

An attenuator of 0.25 in. thick low density (0.027 gram/cm^3) polyeurethene foam was placed over the shell surface to be loaded and covered with the required amount of

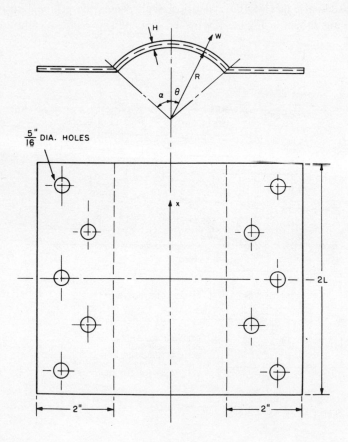

Fig. 2. Cylindrical Shell Panel

DuPont Detasheet EL506D sheet explosive. DuPont 4684 cement was used between the Detasheet, foam and specimen. The sheet explosive was detonated by means of a regular number 6 blasting cap which was shielded from the specimen by a sturdy metal block and connected to the sheet explosive with a Detasheet leader 20 in. long $\times$ 0.125 in. wide $\times$ 0.010 in. thick. In all tests the Detasheet layer was distributed uniformly within a 2 in. (circumferential) by 3 in. (longitudinal) area on the foam attenuator at the center of the inner surface of a cylindrical panel.

The detonation wave front in DuPont Detasheet EL 506D sheet explosive travels at approximately 23,000 ft/sec. according to the manufacturer. Thus the detonation wave front traverses the sheet explosive on the surface of the cylindrical shell panels in approximately 5.5 μsec. since the leader is connected to the center of the loaded zone. It is evident from some photographs taken with a high speed camera during the actual response of a beam loaded dynamically with Detasheet [93, 94] that the rise time

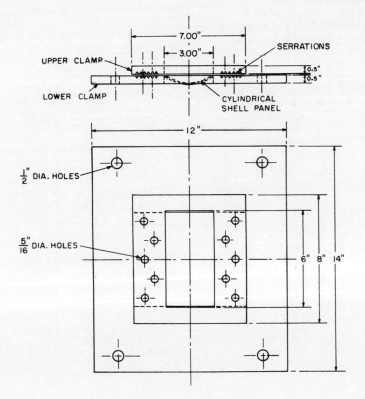

Fig. 3. Clamps

required to reach the maximum pressure is very short compared with the duration of the beam response. Moreover, theoretical predictions indicate that the permanent deflections of some rigid, perfectly plastic structures loaded with rectangular pressure pulses [15, 38, 55, etc.] are dependent on the magnitude of the total impulse and essentially independent of the actual magnitudes of pressures and pulse durations when the maximum pressure considerably exceeds the static collapse pressure of the structure. It is believed, therefore, that the dynamic loading imparted by the Detasheet explosive to the specimens may be considered as impulsive.

In order to calibrate the sheet explosive a series of tests was undertaken on 3 in. diameter by 0.125 in. thick mild steel discs which were loaded dynamically with various

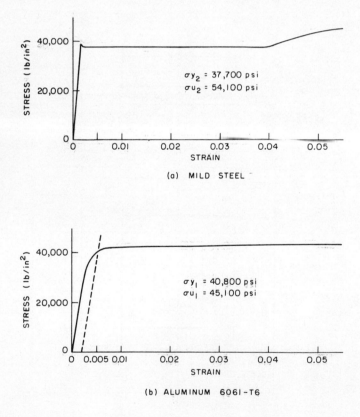

Fig. 4. Typical Static Stress-Strain Curves for (a) Mild Steel and (b) Aluminum 6061 T6

TABLE 1

Data for (a) Aluminum 6061 T6 and (b) Hot Rolled Mild Steel Cylindrical Steel Panels

Specimen Number	H in.	2α deg.	2R in.	2L in.	We grams	I lb. sec.	V_o in/sec.	λ	$\frac{\lambda H}{R}$	E_r	W_o in.	$\frac{W_o}{H}$
1	0.1244	90.0	4.12	5.98	1.34	0.576	2609	11.33	0.68	3.84	0.0212	0.1704
2	0.1248	91.2	4.07	5.95	2.11	0.907	4086	26.95	1.65	9.47	0.0633	0.5072
3	0.1248	90.2	4.10	5.98	2.64	1.135	5119	42.93	2.61	14.82	0.0995	0.7973
4	0.1249	90.1	4.11	5.99	2.44	1.049	4730	36.76	2.23	12.60	0.0930	0.7446
5	0.0906	92.7	4.06	5.95	1.25*	0.516	3229	32.29	1.44	5.96	0.0219	0.2417
6	0.0910	92.1	4.06	5.96	1.40*	0.577	3600	39.78	1.78	7.45	0.0441	0.4846
7	0.0909	92.5	4.06	5.96	1.59*	0.656	4094	51.55	2.31	9.59	0.0659	0.7250
8	0.0909	90.0	4.18	5.97	1.33*	0.549	3441	38.60	1.68	6.72	0.0322	0.3542
9	0.0908	92.0	4.06	5.97	1.13*	0.466	2913	26.15	1.17	4.87	0.0115	0.1267
10	0.0818	90.4	4.15	5.98	1.35	0.581	4052	67.38	2.66	9.96	0.0637	0.7787
11	0.0815	90.4	4.15	5.98	1.81	0.778	5453	122.93	4.83	18.04	0.1410	1.7301
12	0.0815	90.6	4.14	5.97	1.57	0.675	4728	91.97	3.62	13.59	0.0962	1.1804
13	0.0816	91.2	4.11	5.98	1.20	0.516	3605	52.57	2.09	7.90	0.0557	0.6826
14	0.0816	90.6	4.14	5.98	1.67	0.718	5022	103.55	4.08	15.31	0.1068	1.3088

*I_s = 18.42 x 10^4 dyne sec/gram

TABLE 1 (a)

Specimen Number	H in.	2α deg.	$2R$ in.	$2L$ in.	W_e grams	I lb. sec.	V_o in/sec.	λ	$\frac{\lambda H}{R}$	E_r	W_o in.	$\frac{W_o}{H}$
1	0.1206	90.6	4.10	5.99	6.27	2.696	4355	107.84	6.34	110.48	0.1703	1.4121
2	0.1206	90.4	4.11	5.98	4.76	2.047	3308	62.51	3.67	63.80	0.0960	0.7960
3	0.1202	91.2	4.07	5.98	3.37	1.449	2346	31.03	1.83	32.17	0.0470	0.3910
4	0.1209	91.2	4.07	5.98	5.02	2.159	3473	67.26	4.00	70.54	0.1020	0.8437
5	0.1205	90.4	4.11	5.98	5.77	2.481	4013	92.16	5.40	93.90	0.1481	1.2290
6	0.1073	92.4	4.03	5.97	2.29	0.985	1789	22.37	1.19	18.90	0.0259	0.2414
7	0.1076	92.1	4.05	5.98	4.55	1.957	3548	88.32	4.69	74.02	0.1021	0.9489
8	0.1080	92.0	4.06	5.99	6.29	2.705	4888	167.23	8.90	140.03	0.2096	1.9407
9	0.1078	91.7	4.05	5.98	3.82*	1.576	2852	56.86	3.03	48.04	0.0664	0.6160
10	0.1076	90.0	4.15	5.98	5.24*	2.161	3910	112.65	5.84	90.01	0.1182	1.0985
11	0.1081	89.2	4.29	5.98	5.80*	2.392	4361	148.32	7.47	107.94	0.1535	1.4200
12	0.0764	90.6	4.14	5.98	1.84	0.791	2046	63.32	2.34	26.14	0.0273	0.3573
13	0.0760	91.2	4.11	5.96	3.15	1.354	3518	186.37	6.89	77.65	0.1135	1.4934
14	0.0755	90.4	4.15	5.98	2.65	1.139	2984	138.55	5.04	55.55	0.0844	1.1179
15	0.0759	90.4	4.15	5.98	2.17	0.933	2431	90.94	3.33	36.85	0.0480	0.6324

TABLE 1 (b)

$*I_s = 18.42 \times 10^4$ dyne sec/gram

TABLE 2

Mechanical Properties from Tensile Tests on (a) Aluminum 6061 T6 (E = 10.5 x 10^6 p.s.i.) and (b) Hot Rolled Mild Steel (E = 30 x 10^6 p.s.i.)

Nominal Thickness in	σ_{y1} p.s.i.	σ_{y2} p.s.i.	σ_o p.s.i.	σ_{u1} p.s.i.	σ_{u2} p.s.i.	e_1 %	e_2 %
0.125	40800	41900	41350	45100	45200	17.0	17.0
0.091	40500	40900	40700	45000	45200	17.5	17.0
0.080	38000	40700	39350	43300	45100	16.5	16.8

TABLE 2 (a)

Nominal Thickness in	σ_{y1} p.s.i.	σ_{y2} p.s.i.	σ_o p.s.i.	σ_{u1} p.s.i.	σ_{u2} p.s.i.	e_1 %	e_2 %
0.120	36100	37700	36900	51600	54100	35.0	29.0
0.108	36300	37000	36650	51000	52900	31.0	30.0
0.076	33900	36600	35250	49500	52000	28.0	30.0

TABLE 2 (b)

masses of Detasheet explosive covering a 0.25 in. thick foam attenuator in the same manner as the shells. The clamping device for the cylindrical panels was removed from the test table shown in Figure 1 and replaced by an 0.5 in. thick flat steel plate with a 3.50 in. diameter hole cut in the center. A steel rule was erected perpendicular to the plate surface at the edge of the hole. Two narrow strips of paper were sufficient to support the discs which were accelerated downwards while the specimens accelerated upwards simply sat on a piece of cardboard having a central hole with an inner diameter slightly smaller than the outer diameter of a timepiece. A Fastax (Wollensak WF-2) framing camera (approximately 12,000 pictures per second during a test) equipped with a standard Fastax 1000 cycles per second glow tube was used to record the motion of the timepieces on 100 ft. rolls of Eastman Negative Type 7224 film or Kodak Reversal Type 7278 film. Thus the exact time it takes for a timepiece to pass any two marks on the steel rule may be estimated directly from a projection of the film on a screen. The average velocity of a specimen was then calculated in each 0.5 in. interval during the first 6 in. or so of flight. However it was often difficult to calculate the velocity of a specimen during the first 3 in. of flight because the products of detonation usually obscured the film. It may be shown that air drag and gravity have a negligible influence on the specimens over the distances travelled during the tests. Moreover, it appears from the high speed photographs that either insufficient time has elapsed for shock waves to be reflected from the chamber walls and interfere with the motion of the timepieces or their influence is negligible.

TABLE 3

Data for Calibration Tests

Test Number	Specimen Geometry	M grams	W_e grams	U (up) D (down)	V_o cm/sec	I_s 10^4 dyne sec/g
1	DISC	105.48	1.50	U	2470	17.38
2	DISC	107.02	2.68	U	5066	20.25
3	DISC	106.27	2.64	U	4834	19.44
4	DISC	106.23	2.70	U	5080	20.00
5	DISC	105.90	2.69	U	4465	17.60
6	DISC	107.02	2.65	D	4637	18.73
7	DISC	105.87	2.63	D	4834	19.44
8	DISC	105.87	2.67	D	5080	20.10
9	DISC	106.58	2.72	D	4668	18.30
10	DISC	105.70	2.51	D	4834	20.35
11	Half Cylinder	413.51	3.38	D	1577	19.27
12	Spherical Cap	113.85	3.13	U	5080	18.48
13	"	102.42	2.59	D	4922	19.49

The average of all the velocities in each of the intervals examined for a particular specimen was regarded as the impulsive velocity V_o (cm/sec) imparted to a mass M (grams) by a certain amount of explosive W_e (grams). Thus, the specific impulse I_s (dyne sec/gram) of the Detasheet explosive is defined as

$$I_s = \frac{MV_o}{W_e} \quad \text{dyne sec/gram}$$

The values of the specific impulse which were calculated from thirteen different calibration tests are listed in Table 3. In order to provide a check on the calibration procedure test number 1-5, and 12 were conducted on timepieces with a vertically upwards flight while tests 6-10, 11 and 13 were accelerated downwards. It appears from tests 11 to 13 which were conducted on a half cylindrical shell (0.375 in. wall X 6 in. O.D. X 3 in. long) and spherical shell caps (0.115 in. wall X 10 in. O.D. X 3 in. base) that the specific impulse (I_s) is not influenced by the surface curvatures of a timepiece. It was observed from a test conducted on a circular plate without an explosive layer that the Detasheet leader itself has a negligible influence on the specific impulse. This observation agrees with the results of a similar test using a ballistic pendulum which was reported in reference [22].

An overall average value of I_s = 19.2 X 10^4 dyne sec/gm or 0.430 lb sec/gm was assumed for the specific impulse of the Detasheet explosive used in the tests on the cylindrical shell panels. This value was calculated from the results of tests 1 to 10 reported in Table 3. These tests, five of which were accelerated vertically upwards and five downwards, were all conducted on circular specimens with layers of explosive ranging from 0.010 in. to 0.020 in. thick. It is evident from Figure 5 that this value for I_s gives fair agreement with the test results over the whole range of W_e which were examined.

It should be noted that a different batch of sheet explosive was used for the 0.091 in. thick aluminum 6061 T6 and 0.108 in. thick mild steel cylindrical panels marked with an asterisk in Tables 1a and 1b. The calibration procedure was repeated and gave an overall average value of $I_s = 18.42 \times 10^4$ dyne sec/gm or 0.4125 lb sec/gm for this explosive.

Thus once the explosive was carefully weighed (W_e) prior to each test it was assumed that an impulsive velocity having a magnitude $V_o = \dfrac{I_s W_e}{M}$ was imparted to the specimen throughout the area of mass M covered with Detasheet.

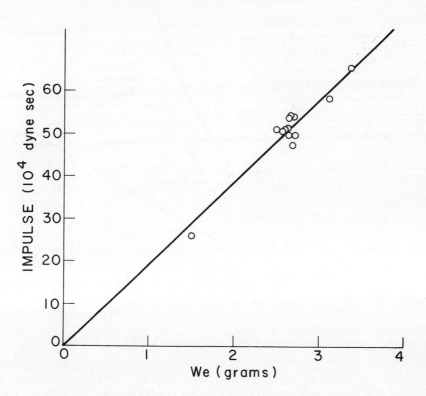

Fig. 5. Total Impulse Imparted to the Calibration Specimens Versus the Corresponding Weight of Sheet Explosive

The permanent transverse deflections of the cylindrical shell panels were measured by placing a special device, which was fitted with 0.0001 in. dial gauges, over the fully clamped shells prior to and after testing.

Average densities of 7.26×10^{-4} lb sec^2/in^4 for mild steel and 2.51×10^{-4} lb sec^2/in^4 for aluminum were obtained by carefully weighing several specimens and measuring their volume using a water displacement method.

DISCUSSION

The maximum permanent transverse deflections of various cylindrical shell panels which were loaded with an impulsive velocity distributed uniformly within a rectangular

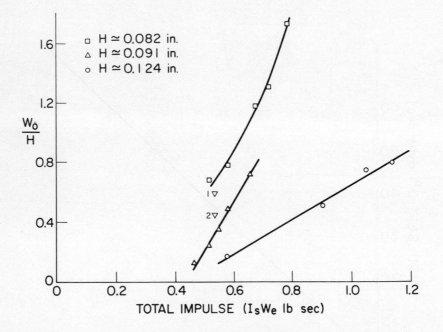

Fig. 6. W_0/H versus Total Impulse (I_sW_e) for Aluminum 6061 T6 Cylindrical Shell Panels with $2\alpha \simeq 90°$

$\nabla1$ Estimated using finite-difference method of [64] for an elastic linear strain hardening material. Data is same as that for specimen number 8 except $V_0 = 3353$ in/sec. 5 meshes in circumferential sense and 10 meshes in axial sense of one-quarter panel.

$\nabla2$ Same as $\nabla1$ except influence of strain rate sensitivity retained with $D = 6500$ sec^{-1} and $p = 4$.

zone located at the center of the inner surface area are presented, together with other details in Tables 1(a) and 1(b). It is evident, from the values of E_r, which is an energy ratio defined in the appendix, that the initial kinetic energy considerably exceeds the

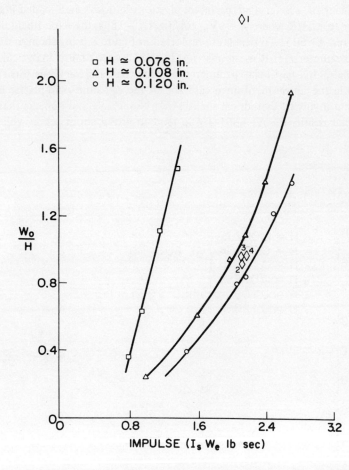

Fig. 7. W_0/H versus Total Impulse (I_sW_e) for Hot Rolled Mild Steel Cylindrical Shell Panels with $2\alpha \stackrel{\triangle}{=} 90°$

◊1 Estimated using finite-difference method of [64] for an elastic perfectly plastic material. Data is same as that for specimen number 10 except V_0 = 3811 in./sec. 5 meshes in circumferential sense and 10 meshes in axial sense of one quarter panel.

◊2 Same as ◊1 except influence of strain rate sensitivity retained with D = 40.4 sec^{-1} and p = 5.

◊3 Same as ◊2 except 10 meshes in both circumferential and axial directions.

◊4 Same as ◊2 except all data is same as specimen number 10.

maximum amount of strain energy which could be absorbed by the cylindrical panels in a wholly elastic manner. It is quite likely, therefore, that material elasticity does not have a significant effect on the overall structural response.

Figures 6 and 7 show the variation of the maximum permanent deflection to shell thickness ratio (w_o/H) with the corresponding total impulse imparted by the sheet explosive to the cylindrical panels. It can be shown that the maximum permanent transverse displacement of a long, rigid, perfectly plastic cylindrical shell loaded impulsively is related linearly to $\lambda H/R$ where $\lambda = \rho V_o^2 R^2 / (\sigma_o H^2)$. Thus, the experimental results for the mild steel and Al 6061 T6 panels are replotted in Figure 8 using the non-dimensionalized impulse parameter $\lambda H/R$ as an abscissa. In the absence of any analytical studies for cylindrical panels, this particular parameter is somewhat arbitrary but was selected because it arises in the above mentioned case and in the dynamic rigid-plastic behavior of another problem involving cylindrical shells [46]. Nevertheless, it appears that almost all the experimental results for Al 6061 T6 do tend to group about a line while those for

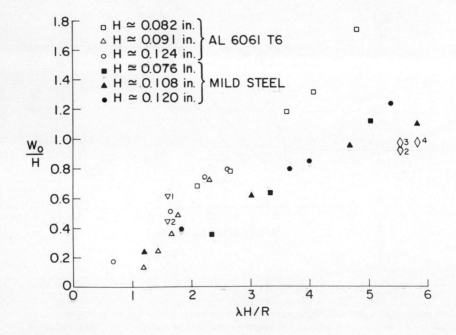

Fig. 8. W_o/H versus the Non-dimensional Impulse Parameter $\lambda H/R$ for all Aluminum 6061 T6 and Hot Rolled Mild Steel Panels ($2 \alpha \simeq 90°$)

▽ and ◇ finite-difference results are described in legends of Figures 6 and 7.

hot rolled mild steel fall close to another line. Clearly, more than the two parameters W_o/H and $\lambda H/R$ are necessary for a complete description of the complex dynamic behavior of a cylindrical panel. It is evident that material strain hardening, strain rate sensitivity, elastic effects and variations in the angle 2α which are indicated in Table 1 would influence the dynamic response in addition to other geometrical and mechanical properties. However, the characteristics of the static stress-strain relations presented in Figure 4 indicate that strain hardening should not be important, at least for small transverse deflections. Furthermore, it has already been observed that elastic effects could be neglected without undue error. It is possible, therefore, that material strain rate sensitivity is the principal reason why the permanent deflections of the hot rolled mild steel panels are smaller than the corresponding A1 6061 T6 ones.

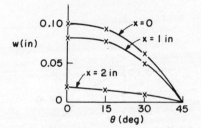
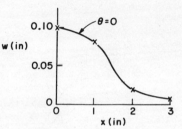

(a) SPECIMEN NUMBER 3

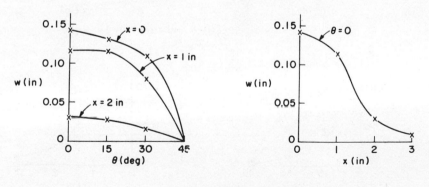

(b) SPECIMEN NUMBER II

Fig. 9. The Permanent Profiles of Two Aluminum 6061 T6 Panels

It is interesting to observe that straight lines could be drawn through the two sets of experimental results presented in Figure 8. It is likely, therefore, that the influence of finite-deflections could be disregarded, at least within the range of deflections considered. This tentative conclusion appears to be consistent with a physical viewpoint since the circumferential membrane forces in a panel would be included in an infinitesimal analysis while the two free edges would prevent the development of large longitudinal membrane forces. Thus, unlike axially restrained beams [5, 21, etc.], plates [13-19, 23, 59, etc.] and cylindrical shells [38], significant additional components of the generalized forces would not be introduced during finite deflections which remained small. Demir and Drucker [96] reported an experimental investigation which showed that geometry changes did not influence the static behavior of an axisymmetrically ring loaded cylindrical shell with free ends.

The spatial distribution of the permanent transverse deflections of some cylindrical shell panels are presented in Figures 9 and 10.

The numerical calculations presented in Figures 6-8 and 11 were kindly supervised by E. A. Witmer, who obtained them using the Petros 3 finite-difference procedure [64].

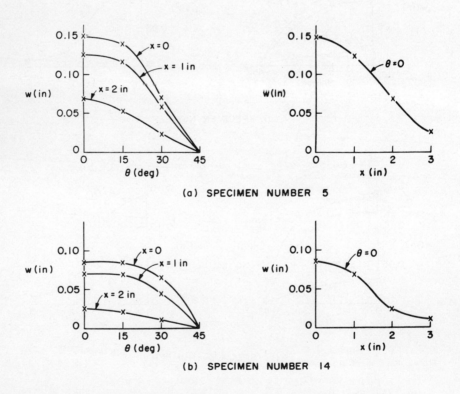

Fig. 10. The Permanent Profiles of Two Hot Rolled Mild Steel Panels

These results illustrate clearly the importance of material strain rate sensitivity on the magnitude of the permanent deflections. It is believed that the uncertainty in the actual values of the Cowper-Symonds [69] strain rate sensitivity coefficients (D, p) even for the uniaxial tests conducted at a constant strain rate [2, 44], and the assumption that this simplified relation is valid for the structural problem at hand is partly responsible for the difference between the experimental values and numerical predictions. As remarked previously, no systematic experimental studies have been conducted into the strain rate sensitive behavior of materials under the influence of multi-dimensional stresses. Moreover, most studies have been conducted at almost constant strain rates and no investigations appear to have been undertaken in order to examine the influence of

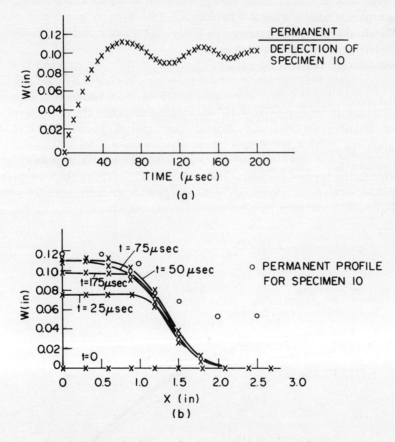

Fig. 11. Finite-Difference Predictions for Hot Rolled Mild Steel Specimen Number 10. Data is same as ◊4 in Figure 7.

(a) Variation of central transverse deflection with time.
(b) Variation of transverse deflection profile along longitudinal axis $\theta = 0°$ with time.

various strains and strain rate histories. It might be remarked that as the finite-difference mesh size tends towards zero, the corresponding magnitude of W_0/H tends to increase. The maximum circumferential strains on the inner and outer surfaces at the centre of an aluminum 6061 T6 panel for case 2 shown in Figure 6 are 0.030 at t = 98.75 μsec and 0.022 at t = 480 μsec, respectively. The corresponding values for the mild steel panel case 4 shown in Figure 7 are 0.046 at t = 114 μsec and 0.049 at t = 74.5 μsec. The maximum axial strains at the centre of the panels for both cases range from -0.009 to 0.009.

It appears from Figures 6 and 7 that the experimental technique described in the last section gives reasonably consistent and reliable results. However, the major drawback with the experimental procedure is the requirement that each new batch of sheet explosive should be calibrated to give the corresponding specific impulse (I_s). Unfortunately, this is an expensive and time-consuming process. Moreover, tests 1 to 10 in Table 3 which were used to calculate the average value of the specific impulse (I_s) for the Detasheet explosive have a standard deviation of 1.04 dyne-sec/gram, even though all tests in which the timepiece rotated, probably due to eccentric detonation of the explosive sheet, etc., have been omitted.

It should be noted that specific impulses from 16.7×10^4 dyne-sec/gram [13] to 21.7 $\times 10^4$ dyne-sec/gram [93] (assuming $\rho = 1.5$ gram/c.c., as quoted by the manufacturer) have been reported for Detasheet covering a 0.125 in. thick neoprene attenuator. The manufacturers quote a value of 20×10^4 dyne-sec/gram, while tests conducted on two batches of Detasheet at Picatinny Arsenal gave average values of 18.74×10^4 dyne-sec/gram and 19.48×10^4 dyne-sec/gram when covering a low density polyethylene attenuator [95]. It was observed in reference [23] that the specific impulse of Detasheet was sensitive to the type of attonuator material since an 0.125 in. thick neoprene layer allowed a sheet of explosive to impart more energy to a specimen than the same sheet covering an 0.50 in. foam layer.

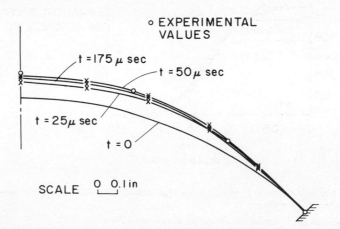

Fig. 11 (c) Variation of transverse deflection profile in circumferential sense at x = 0.

The authors of references [13] and [93] found that the total impulse of Detasheet is related linearly to the corresponding weight or thickness of explosive up to at least 0.060 in. thick layers. However, Wierzbicki and Florence [19] recently conducted some tests using Detasheet covering a rather thick (0.5 in.) neoprene layer and observed that the impulse was not related linearly to the explosive thickness. A six percent reduction in impulse from that predicted by a linear relationship was observed for a 0.040 in. thick layer of Detasheet.

CONCLUSIONS

An experimental investigation was undertaken in order to examine the behavior of various cylindrical shell panels which were loaded with an impulsive velocity distributed uniformly within a rectangular zone located at the center of the inner surface. The panels were fully clamped along the two longitudinal edges and free on the other two. The initial kinetic energy of the dynamic loads was sufficiently large to cause inelastic behavior and to produce maximum permanent transverse deflections of up to nearly twice the corresponding panel thickness. Tests were conducted on mild steel and aluminum 6061 T6 panels which had various thicknesses and an included angle of 90° approximately.

ACKNOWLEDGEMENTS

The work reported herein was supported by the Structural Mechanics Branch of O.N.R. under Contract Number N00014-67-A-0204-0032. The authors are indebted to Professor A. H. Kell for his encouragement and to Captain S. C. Reed for his assistance in obtaining additional support from the Naval Ship Systems Command. The authors also wish to take this opportunity to express their appreciation to Dr. J. W. Leech, F. Merlis and O. E. Wallin of the Aeroelastics Laboratory for their kind cooperation and to Professor E. A. Witmer for permission to use the blast chamber in which the tests were performed and again to Professor Witmer and Susan French for performing the numerical finite-difference calculations.

APPENDIX

Energy Ratio (E_r)

It was mentioned in the introduction that rigid-plastic analyses often give reasonable predictions of the dynamic behavior of structures loaded impulsively when the corresponding energy ratio is large. This energy ratio is defined as the initial kinetic energy divided by the maximum amount of strain energy which can be absorbed by a structure in a wholly elastic manner. In order to assess this particular feature in the current tests an approximate relation for the energy ratio of a cylindrical shell panel is established in this appendix.

The elastic strain energy in a cylindrical shell panel of volume V is

$$S = \int_V \left\{ \frac{1}{2E} \left(\sigma_x^2 + \sigma_\theta^2 - 2\nu\sigma_x\sigma_\theta \right) + \frac{1}{2G} \left(\tau_{x\theta}^2 + \tau_{\theta z}^2 + \tau_{zx}^2 \right) \right\} dV$$

Now, in order to simplify the following work, it is assumed that any plastic flow is controlled by the Maximum Strain Energy Theory of Beltrami. If plastic flow commences throughout the entire volume of a panel then

$$S = \frac{\sigma_0^2}{2E} \int_V dV$$

$$\text{or,} \quad S = \frac{2\alpha\sigma_0^2 HRL}{E}$$

which is considered to be the maximum amount of strain energy a cylindrical panel could absorb in a wholly elastic manner. S is undoubtedly an overestimate of the maximum energy which could be absorbed by a panel in a wholly elastic manner because local plastic deformations would occur at much smaller values of strain energy.

The initial kinetic energy imparted to a cylindrical panel by an impulsive velocity (V_0) distributed uniformly within a zone on the middle plane of circumferential width "a" and axial length "b" is

$$K = \frac{\rho V_0^2 \, Hab}{2}$$

Thus an energy ratio may now be defined as

$$E_r = \frac{K}{S}$$

or,

$$E_r = \frac{\rho V_o^2 abE}{4\alpha R \sigma_o^2 L} .$$

E_r is a measure of the amount by which the initial kinetic energy exceeds the maximum strain energy which can be absorbed by a cylindrical panel in a wholly elastic manner. If E_r is large (3 or larger according to reference [3]) for a given cylindrical panel then it is believed that material elasticity should not influence significantly the overall response.

It has been remarked previously that the sheet explosive used in all the tests on the cylindrical panels measured 2 in. in the circumferential direction and 3 in. axially. Thus, it is assumed that the surface area in the middle plane of a panel which is affected by the initial impulsive velocity is $\dfrac{2}{1 - (\dfrac{H}{2R} + \dfrac{T}{R})} \times 3$ in^2 where T(in) is the thickness of the foam attenuator. The area of the middle plane was used in all the calculations for V_o and E_r presented in Tables 1(a) and 1(b).

It should be noted here that $R \neq 1.5/\sin\alpha$ in Tables 1(a) and 1(b) because the presence of a small radius of curvature at the clamped edge of a panel causes the mid-plane of the shell to deviate slightly from the mean radius R near the supports.

REFERENCES

1. E. H. Lee and P. S. Symonds, Large plastic deformations of beams under transverse impact. *J. App. Mech.*, **19**, 308 (1952).
2. P. S. Symonds *Survey of Methods of analysis for plastic deformation of structures under dynamic loadings.* Brown Univ., Div. of Eng. Report BU/NSRDC/1-67, (1967).
3. S. R. Bodner and P. S. Symonds, Experimental and theoretical investigation of the plastic deformation of cantilever beams subjected to impulsive loading. *J. App. Mech.*, **29**, 719 (1962).
4. P. S. Symonds, Plastic shear deformations in dynamic load problems. *Engineering Plasticity*, C.U.P., Ed. J. Heyman and F. A. Leckie, 647 (1968).
5. P. S. Symonds and N. Jones, Impulsive loading of fully clamped beams with finite plastic deflections and strain-rate sensitivity. *Int. J. Mech. Sci.*, **14**, 49 (1972).
6. H. G. Hopkins and W. Prager, On the dynamics of plastic circular plates. (ZAMP) *J. of App. Math. & Phys.*, **5**, 317 (1954).

7. A. J. Wang and H. G. Hopkins, On the plastic deformation of built-in circular plates under impulsive load. *J. Mech. & Phys. Solids*, **3**, 22 (1954).
8. A. J. Wang, The permanent defection of a plastic plate under blast loading. *J. App. Mech.*, **22**, 375 (1955).
9. P. Perzyna, Dynamic load carrying capacity of a circular plate. *Arch. Mech. Stos.*, **10**, 635 (1958).
10. G. S. Shapiro, On a rigid-plastic annular plate under impulsive load. *J. App. Math & Mech.* (Prik. Mat. i Mek.), **23**, 234 (1959).
11. A. L. Florence, Clamped circular rigid-plastic plates under central blast loading. *Int. J. Solids & Struc.*, **2**, 319 (1966).
12. M. F. Conroy, Rigid-plastic analysis of a simply supported circular plate due to dynamic circular loading. *J. of the Franklin Inst.*, **228**, 121 (1969).
13. A. L. Florence, Circular plate under a uniformly distributed impulse. *Int. J. Solids & Struc.* **2**, 37 (1966).
14. N. Jones, Impulsive loading of a simply supported circular rigid-plastic plate. *J. App. Mech.*, **35**, 59 (1968).
15. N. Jones, Finite deflections of a simply supported rigid-plastic annular plate loaded dynamically. *Int. J. Solids & Struc.*, **4**, 593 (1968).
16. N. Jones, Finite deflections of a rigid-viscoplastic strain-hardening annular plate loaded impulsively. *J. App. Mech.*, **35**, 349 (1968).
17. T. Wierzbicki and J. M. Kelly, Finite deflections of a circular viscoplastic plate subject to projectile impact. *Int. J. Solids & Struc.*, **4**, 1081 (1968).
18. T. Wierzbicki, Large deflections of a strain-rate sensitive plate loaded impulsively. *Arch. Mech. Stos.*, **21**, 67 (1969).
19. T. Wierzbicki and A. L. Florence, A theoretical and experimental investigation of impulsively loaded clamped circular viscoplastic plates. *Int. J. Solids & Struc.*, **6**, 553 (1970).
20. A. D. Cox and L. W. Morland, Dynamic plastic deformations of simply-supported square plates. *J. Mech. Phys. Solids*, **7**, 229 (1959).
21. N. Jones, A theoretical study of the dynamic plastic behavior of beams and plates with finite-deflections. *Int. J. Solids & Struc.*, **7**, 1007 (1971).
22. N. Jones, R. N. Griffin, and R. E. Van Duzer, An experimental study into the dynamic plastic behavior of wide beams and rectangular plates. *Int. J. of Mech. Sci.*, **13**, 721 (1971).
23. N. Jones, T. O. Uran and S. A. Tekin, The dynamic plastic behavior of fully clamped rectangular plates. *Int. J. Solids & Struc.*, **6**, 1499 (1970).
24. P. G. Hodge, The influence of blast characteristics on the final deformation of circular cylindrical shells. *J. App. Mech.*, **23**, 617 (1956).
25. P. G. Hodge, Impact pressure loading of rigid-plastic cylindrical shells. *J. of the Mech. & Phys. of Solids*, **3**, 176 (1955).
26. P. G. Hodge, Ultimate dynamic load of a circular cylindrical shell. *Proc. 2nd. Midwestern Conf. on Solid Mech.*, LaFayette, Ind., 150 (1956).
27. P. G. Hodge, Effect of end conditions on dynamic loading of plastic shells. *J. of the Mech. & Phys. of Solids*, **7**, 258 (1959).
28. G. Eason and R. T. Shield, Dynamic loading of rigid-plastic cylindrical shells. *J. of the Mechs. & Phys. of Solids*, **4**, 53 (1956).
29. P. G. Hodge and B. Paul, Approximate yield conditions in dynamic plasticity. *Proc., 3rd. Midwestern Conf. on Solid Mech.*, **29**, (1957).
30. P. A. Kuzin and G. S. Shapiro, On dynamic behavior of plastic structures. Ed. H. Görtler, *Proc., 11th Int. Congress of App. Mech.*, Munich, 1964, Springer Verlag, New York, 629 (1966).

31. Y. V. Nemirovsky and V. N. Mazalov, Dynamic behavior of cylindrical shells strengthened with ring ribs — Part I Infinitely Long Shell. *Int. J. Solids & Struc.*, **5**, 817 (1969).
32. C. K. Youngdahl, Correlation parameters for eliminating the effect of pulse shape on dynamic plastic deformation. *J. App. Mech.*, **37**, 744 (1970).
33. A. Pabjanek, Dynamic loading of rigid-viscoplastic cylindrical shells. *Arch. Mech. Stos.*, **21**, 199 (1969).
34. T. Wierzbicki, *Viscoplastic flow of rotationally symmetric shells with particular application to dynamic loadings*. PRACE IPPT. IBTP Reports, Warsaw (1968).
35. T. Duffey and R. Krieg, The effects of strain-hardening and strain-rate sensitivity on the transient response of elastic plastic rings and Cylinders. *Int. J. Mech. Sci.*, **11**, 825 (1969).
36. N. Perrone, Impulsively loaded strain hardened rate sensitive rings and tubes. *Int. J. Solids and Struc.*, **6**, 1119 (1970).
37. N. Perrone, On a simplified method for solving impulsively loaded structures of rate-sensitive materials. *J. App. Mech.*, **32**, 489 (1965).
38. N. Jones, The influence of large deflections on the behavior of rigid-plastic cylindrical shells loaded impulsively. *J. App. Mech.*, **37**, 416 (1970).
39. R. Sankaranarayanan, On the dynamics of plastic spherical shells. *J. App. Mech.*, **30**, 87 (1963).
40. R. Sankaranarayanan, On the impact pressure loading of a plastic spherical cap. *J. App. Mech.*, **33**, 704 (1966).
41. W. E. Baker, The elastic-plastic response of thin spherical shells to internal blast loading. *J. App. Mech.*, **27**, 139 (1960).
42. T. Wierzbicki, Impulsive loading of a spherical container with rigid-plastic and strain-rate sensitive material. *Arch. Mech. Stos.*, **15**, 775 (1963).
43. T. A. Duffey, Significance of strain hardening and strain rate effects on the transient response of elastic-plastic spherical shells. *Int. J. Mech. Sci.*, **12**, 811 (1970).
44. P. S. Symonds, Viscoplastic behavior in response of structures to dynamic loading, *Behavior of Materials under Dynamic Loading*, ed. by N. J. Huffington, pub. by ASME, 106 (1965).
45. N. Jones, The influence of strain hardening and strain rate sensitivity on the permanent deformation of impulsively loaded rigid-plastic beams. *Int. J. Mech. Sci.*, **9**, 777 (1967).
46. N. Jones, An approximate rigid-plastic analysis of shell intersections loaded dynamically. To appear *Trans. A.S.M.E.*
47. J. B. Martin, Impulsive loading theorems for rigid-plastic continua. *Proc. Am. Soc. Civ. Engr.*, **90**, 27 (1964).
48. W. J. Morales and G. E. Nevill, Lower bounds on deformations of dynamically loaded rigid-plastic continua. *A.I.A.A. Journal*, **8**, 2043 (1970).
49. D. C. Drucker, Limit analysis of cylindrical shells under axisymmetric loading. *Proc. 1st. Midwest Conf. Solid Mech.*, 158 (1953).
50. P. G. Hodge, Plastic analysis of circular conical shells. *J. App. Mech.*, **27**, 696 (1960).
51. J. B. Martin, Time and displacement bound theorems for viscous and rigid-visco-plastic continua subjected to impulsive loading. *Proc. 3rd, Southeastern Conf., Dev. in Theoret. & App. Mech.*, **3**, 1 (1968).
52. D. N. Robinson, A displacement bound principle for elastic-plastic structures subjected to blast loading. *J. Mech. & Phys. of Solids*, **18**, 65 (1970).
53. S. Kaliszky, Approximate solutions for impulsively loaded inelastic structures and continua. *Int. J. Non-Linear Mech.*, **5**, 143 (1970).
54. J. B. Martin and P. S. Symonds, Mode approximations for impulsively loaded rigid-plastic structures. *Proc. Am. Soc. Civ. Engrs.*, **92**, 43 (1966).

55. T. Nonaka, Some interaction effects in a problem of plastic beam dynamics, Parts 1-3, *J. of App. Mechs.*, **34**, 623 (1967).
56. S. R. Bodner, Deformation of rate-sensitive structures under impulsive loading. *Engineering Plasticity*. Camb. Univ. Press, Ed. J. Heyman and F. A. Leckie, 77 (1968).
57. G. L. Johnson and J. B. Martin, The permanent deformation of a portal frame subjected to a transverse impulse. *Int. J. Solids & Struc.*, **5**, 1171 (1969).
58. N. Perrone, Response of rate-sensitive frames to impulsive load. Proc. A.S.C.E., *J. Eng. Mech. Div.*, **97**, No. EM1, 49 (1971).
59. T. Wierzbicki, Bounds on large dynamic deformations of structures. Proc. A.S.C.E., *J. Eng. Mech. Div.*, **96**, 267 (1970).
60. E. A. Witmer, H. A. Balmer, J. W. Leech and T. H. H. Pian, Large dynamic deformations of beams, circular rings, circular plates and shells. *A.I.A.A. J.*, **1**, 1848 (1963).
61. J. W. Leech, E. A. Witmer, and T. H. H. Pian, Numerical calculation technique for large elastic-plastic transient deformations of thin shells. *A.I.A.A. J.*, **6**, 2352 (1968).
62. L. Morino, J. W. Leech and E. A. Witmer, An improved numerical calculation technique for large elastic-plastic transient deformations of thin shells, Part 1 — Background and theoretical formulation. *J. App. Mech.*, **38**, 423 (1971).
63. L. Morino, J. W. Leech and E. A. Witmer, An improved numerical calculation technique for large elastic-plastic transient deformations of thin shells, Part 2 — Evaluation and applications. *J. App. Mech.*, **38**, 429 (1971).
64. S. Atluri, E. A. Witmer, J. W. Leech and L. Morino, Petros 3: *A finite-difference method and program for the calculation of large elastic-plastic dynamically-induced deformations of multi-layer variable-thickness shells*. Aero & Astro. Res. Lab. Rep. ASRL TR 152-2, (1971).
65. R. D. Krieg and T. A. Duffey, Univalve 11, *A code to calculate the large deflection dynamic response of beams, rings, plates and cylinders*. Sandia Rep. SC-RR-68-303, (1968).
66. W. B. Murfin, Elastic-plastic collapse of structures subjected to a blast pulse. *Shock & Vib. Bull.*, **37**, 177 (1968).
67. C. Lindbergh and D. E. Boyd, Finite inelastic deformations of clamped shell membranes subjected to impulsive loadings. *A.I.A.A. J.*, **7**, 228 (1969).
68. N. J. Huffington, *Large deflection elastoplastic response of shell structures*. Ballistic Res. Lab. Rep. 1515 (1970).
69. G. R. Cowper and P. S. Symonds, *Strain-hardening and strain-rate effects in the impact loading of cantilever beams*. Tech. Rep., O.N.R., Contract Nonr 562 (10), Nr 064-406 (1957).
70. P. Perzyna, Fundamental problems in viscoplasticity. *Advances in App. Mechs.*, **9**, 243, Academic Press, N.Y., (1966).
71. B. Rawlings, The dynamic behavior of mild steel in pure flexure. *Proc. Roy. Soc.*, A., **275**, 528 (1963).
72. R. J. Aspden and J. D. Campbell, The effect of loading rate on the elasto-plastic flexure of steel beams. *Proc. Roy. Soc.*, A., **290**, 266 (1966).
73. J. Klepaczko, The strain-rate behavior of iron in pure shear. *Int. J. Solids & Struc.*, **5**, 533 (1969).
74. J. D. Campbell and A. R. Dowling, The behavior of materials subjected to dynamic incremental shear loading. *J. Mech. Phys. Solids*, **18**, 43 (1970).
75. J. Duffy, J. D. Campbell and R. H. Hawley, On the use of a torsional split hopkinson bar to study rate effects in 1100-0 aluminum. *J. App. Mech.*, **38**, 83 (1971).
76. G. Gerard and R. Papirno, Dynamic biaxial stress-strain characteristics of aluminum and mild steel. *Trans. A.S. for Metals*, **49**, 132 (1957).

77. U. S. Lindholm and L. M. Yeakley, A dynamic biaxial testing machine. *Exp. Mech.*, **7**, 1 (1967).
78. J. N. Goodier, Dynamic plastic buckling. *Proc. Int. Conf. on Dynamic Stability of Structures,* Northwestern Univ., Illinois, Oct. 1965, Ed., G. Herrmann, pub. Pergamon Press, Oxford (1967).
79. N. Perrone, *A new approach to impact attenuation.* NSF Report GK-2802 (1970).
80. N. Perrone, Crashworthiness and biomechanics of vehicle impact. *Colloquium on Dynamic Response of Biomechanical Systems,* pub. by A.S.M.E., New York, pp 1-22, (1970).
81. B. Rawlings, J. B. Burgmann and T. M. Ford, The effects of loading rate on the response of pin-jointed triangulated steel frames. *Proc. Inst. Civ. Engrs.,* **46**, 327 (1970).
82. G. R. Abrahamson and J. N. Goodier, Dynamic plastic flow buckling of a cylindrical shell from uniform radial impulse. *Proc. 4th U.S. Nat. Congress on App. Mech.,* 939 (1962).
83. A. L. Florence and J. N. Goodier, Dynamic plastic buckling of cylindrical shells in sustained axial compressive flow. *J. App. Mech.,* **35**, 80 (1968).
84. D. A. Rodriquez, Influence of strain hardening induced incremental circumferential compression on the inelastic response of impulsively loaded cylindrical shells. *J. App. Mech.,* **35**, 173 (1968).
85. D. L. Anderson and H. E. Lindberg, Dynamic pulse buckling of cylindrical shells under transient lateral pressures. *A.I.A.A. Journal,* **6**, 589 (1968).
86. A. L. Florence and H. Vaughan, Dynamic plastic flow buckling of short cylindrical shells due to impulsive loading. *Int. J. Solids & Struc.,* **4**, 741 (1968).
87. A. L. Florence, Buckling of viscoplastic cylindrical shells due to impulsive loading. *A.I.A.A. Journal,* **6**, 532 (1968).
88. H. Vaughan, The response of a plastic cylindrical shell to axial impact. *Zeits für Ange. Math, und Phys.,* **20**, 321 (1969).
89. H. Vaughan and A. L. Florence, Plastic flow buckling of cylindrical shells due to impulsive loading. *J. App. Mech.,* **37**, 171 (1970).
90. E. A. Witmer, E. N. Clark and H. A. Balmer, Experimental and theoretical studies of explosive-induced large dynamic and permanent deformations of simple structures. *Exp. Mech.,* **7**, 56 (1967).
91. T. H. H. Pian, Dynamic response of thin shell structures. *Proc. 2nd. Symp. Naval Structural Mech., O.N.R.,* Ed. E. H. Lee and P. S. Symonds, Pergamon Press, Oxford (1960).
92. H. E. Postlethwaite and B. Mills, Use of collapsible structural elements as impact isolators, with special reference to automative applications. *J. of Strain Analysis,* **5**, 58 (1970).
93. A. L. Florence and R. D. Firth, Rigid plastic beams under uniformly distributed impulses. *J. App. Mech.,* **33**, 481 (1965).
94. J. S. Humphreys, Experiments on dynamic plastic deformation of shallow circular arches. *A.I.A.A. Journal,* **4**, 926 (1966).
95. E. N. Clark, F. H. Schmitt, D. G. Ellington, R. Engle and S. Nicolaides, *Plastic deformation of structures,* Vol. I − Study of the plastic deformation of beams. Air Force Flight Dynamics Lab. Tech. Doc. Rpt. 64-64, (1965).
96. H. H. Demir and D. C. Drucker, An experimental study of cylindrical shells under ring loading. *Progress in App. Mech.,* The Prager Anniversary Volume, Macmillan, New York, 205 (1963).

PEAK LOAD-IMPULSE CHARACTERIZATION OF CRITICAL PULSE LOADS IN STRUCTURAL DYNAMICS

G. R. ABRAHAMSON and H. E. LINDBERG

Stanford Research Institute, Menlo Park, California

Abstract—Critical pulse loads for structures are characterized in terms of peak load and impulse. Critical load curves are presented for structural elements and the effect of pulse shape is examined for pulse loads with a step rise. It is shown how critical load curves for structural elements can be readily combined to obtain critical load curves for complex structures. The utility of the peak load-impulse characterization scheme in planning experiments and in presenting and interpreting experimental and theoretical results is demonstrated.

The peak load-impulse characterization of critical pulse loads is shown to contain the same information as the well-known dynamic load factor. Whereas the dynamic load factor gives the relation of dynamic loads to an equivalent static load, the peak load-impulse characterization defines equivalent dynamic loads.

A number of references are cited that indicate familiarity with the peak load-impulse characterization and predate the work on which the present writing is based.

INTRODUCTION

Pulse loads occur in a wide variety of situations, for example in the driving of a nail, in the slamming of a ship's hull in heavy seas, and in the loading of structures by blast waves from explosions. In all of these situations the load is a pulse with a rapid rise and subsequent decay.

It is important to distinguish pulse loads from other dynamic loads, such as oscillatory loads, because pulse loads often evoke different response modes, involve different methods of analysis, and are best characterized by different parameters. These differences arise from basic differences in the nature of the dynamic problems. For example, for oscillatory loads it is often the steady-state response to a repetitive load that is of primary interest; for pulse loads the maximum transient response is of primary concern.

The best characterization scheme for critical loads is that which provides the simplest presentation of theoretical and experimental results. Such a scheme can greatly simplify the problem of specifying critical loads and show where further work will be the most fruitful. The peak load-impulse characterization scheme for pulse loads described here evolved from the problem of determining critical pulse loads for structures loaded by blast waves in air. This difficult problem and its underwater counterpart have been studied for decades. As shown below, the peak load-impulse characterization scheme greatly facilitates specification of critical loads for structures subjected to air blast loading and is generally useful for structures subjected to pulse loads.

In presenting the characterization scheme, some general features are described first. Next, a detailed analysis is given for the rigid-plastic system of one degree of freedom to illustrate the calculation of critical load curves in terms of peak load and impulse. This is followed by the presentation of critical load curves for uniformly loaded rigid-plastic beams and plates and for dynamic buckling of cylindrical shells under uniform lateral loads. Finally, the peak load-impulse characterization of critical pulse loads is compared with the dynamic load factor characterization, and some aspects of the history of the peak load-impulse scheme are presented.

CHARACTERIZATION SCHEME

Most engineers first encounter pulse loads in impulsive loading problems, where the load duration is so short that the load may be regarded as producing an instantaneous initial velocity. Another familiar problem is that of a step load, which consists of a sudden rise to a constant magnitude. For both of these loads, the first example studied is usually the linear spring-mass system. The response of the linear spring-mass system to these ideal loads is then generalized to arbitrary pulse loads using superposition methods, such as the Duhamel integral. Multidegree-of-freedom systems are treated by superposition of load elements (Duhamel integral) and orthogonal response modes. These procedures are generally stressed in conventional mechanics courses. For response beyond the linear range, the ideal impulse and step load are also a useful basis for generalization, but superposition can no longer be used and a different approach must be taken. The approach presented here provides a means of handling critical pulse loads for complex structures loaded beyond the linear range and at the same time simplifies the description of critical pulse loads for elementary linear systems. At the end of this section we discuss complex structures loaded beyond the linear range. However, to introduce the main features of the characterization scheme, we consider first the response of a linear spring-mass system; this also facilitates comparison later to the dynamic load factor characterization.

For an ideal impulse acting on a linear spring-mass system, the load is completely described by the magnitude of the impulse, and for a step load it is completely described by the magnitude of the load. Let I_o be the ideal impulse that produces a specified critical displacement and P_o be the magnitude of the step load that produces the same displacement. Then for an arbitrary ideal impulse I, the maximum displacement will be less than critical for $I/I_o < 1$, critical for $I/I_o = 1$, and greater than critical for $I/I_o > 1$. Similarly, for an arbitrary step load of magnitude P, the maximum displacement will be less than critical for $P/P_o < 1$, critical for $P/P_o = 1$, and greater than critical for $P/P_o > 1$.

For loads of intermediate duration (between the short-duration and long-duration extremes) of impulse I and magnitude P, it is reasonable to expect that the maximum displacement will depend on both the impulse ratio I/I_o and the load-magnitude ratio P/P_o. This is shown in Fig. 1 for the linear spring-mass system under pulse loads with a step rise and exponential decay. In this figure, the critical load curve is presented in terms of I/I_o and P/P_o. The curve is the locus of combinations of I/I_o and P/P_o that produce the same maximum displacement; combinations that lie below the curve produce displacements that are less, and combinations that lie above produce displacements that are greater. The simple hyperbolic shape of the critical load curve of Fig. 1 is typical for a wide variety of structures and loads; as we shall see later, this is one of the main features of the present load characterization scheme.

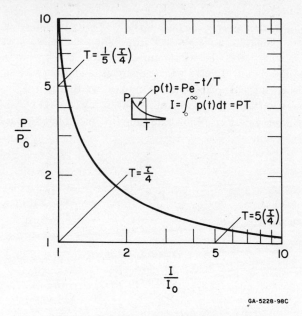

GA-5228-98C

Fig. 1. Critical pulse loads for a linear spring-mass system.

Along a line of unit slope in Fig. 1 passing through a reference point $(I_1/I_o \; P_1/P_o)$ we have

$$\frac{\log \dfrac{P}{P_o} - \log \dfrac{P_1}{P_o}}{\log \dfrac{I}{I_o} - \log \dfrac{I_1}{I_o}} = 1 \tag{1}$$

which reduces to

$$\frac{I}{P} = \frac{I_1}{P_1} \tag{2}$$

Thus, along lines of unit slope the ratio I/P is constant. In particular, along the line passing through $I_1/I_o = P_1/P_o = 1$ in Fig. 1 we have

$$\frac{I}{P} = \frac{I_o}{P_o} \tag{3}$$

The ratio I_o/P_o defines a characteristic time for the system under consideration. Pulse loads with $I/P \ll I_o/P_o$ are impulsive (duration short compared to response time), and pulses with $I/P \gg I_o/P_o$ are quasi-static (duration long compared to response time). This is discussed further below.

For a linear spring-mass system the ratio I_o/P_o is readily found. Under an ideal impulse I the initial kinetic energy of a mass m is $I^2/2m$. When the mass comes to rest the stored energy is $k\hat{w}^2/2$, where $\hat{w}$ is the maximum deflection and $k = 4\pi^2 m/\tau^2$ is the spring constant, τ being the period. Equating these energies, and using zero subscripts to denote the critical displacement and corresponding impulse, we find that $\hat{w}_o = I_o\tau/2\pi m$. Under a step load P the maximum displacement is given by $P = 2k\hat{w}$, and for the critical displacement we have $\hat{w}_o = P_o\tau^2/8\pi m$. Equating the maximum displacements gives, for a linear spring-mass system,

$$\frac{I_o}{P_o} = \frac{\tau}{4} \tag{4}$$

by writing (3) as

$$\frac{I}{P} = \frac{I_1/I_o}{P_1 P_o} \quad \frac{I_o}{P_o} = \frac{I_1/I_o}{P_1/P_o} \quad \frac{\tau}{4} \tag{5}$$

values of I/P for any values of I_1/I_o and P_1/P_o are readily found. As indicated in Fig. 1, for an exponential pulse I/P = T, T being the decay constant. Hence from (4) and (3)

$$T = \frac{\tau}{4} \tag{6}$$

for the decay constant along the line of unit slope passing through $I/I_o = P/P_o = 1$, as indicated in Fig. 1. Using I/P = T in (5), decay constants along other lines of unit slope are obtained. We note in passing that loads with $T = \tau/4$ are the most efficient for producing a given maximum displacement in that extremes in both peak load and impulse are avoided.

Along the critical load curve of Fig. 1 the maximum displacement is constant. Hence, for loads which are off the curve but nearby, the difference in maximum displacement from the critical value depends on the normal distance to the critical load curve. In the vicinity of the vertical asymptote, changes in load magnitude for a given impulse cause relatively small departures from the critical load curve, but changes in impulse for a given load magnitude cause relatively large departures. Hence, for pulses with say $T \leqslant (1/5)$ $(\tau/4)$ the response is insensitive to load magnitude and depends mainly on impulse. In the vicinity of the horizontal asymptote, say for pulses with $T \geqslant 5 (\tau/4)$, the response is insensitive to impulse and depends mainly on load amplitude. In the intermediate knee region, the response depends on both load magnitude and impulse. These three load regions are henceforth called the short-duration region (vertical part of curve), the mid-duration region (knee of curve), and the long-duration region (horizontal part of curve). **1687179**

Critical load curves similar to that of Fig. 1 can be determined for pulse loads that can be described by two parameters, and in particular for pulse loads that differ only in magnitude and time scale, such as the exponential pulses discussed above. Besides exponential pulses, two others of common interest involving two parameters are triangular pulses (step rise, linear decay) and rectangular pulses (step rise, constant magnitude, step decay). In Fig. 2 critical load curves for exponential, triangular, and rectangular pulses acting on a linear spring-mass system are shown. The curves differ most in the mid-duration region and come together in the short-duration and long-duration regions. Along the line of unit slope in Fig. 2, the critical values of P/P_0 and I/I_0 for exponential and triangular pulses differ by about 20%, and for exponential and rectangular pulses they differ by about 40%. This comparison of effects of pulse shape on critical load curves can be very significant in practical work where measurement and control of pulse shape can be very difficult.

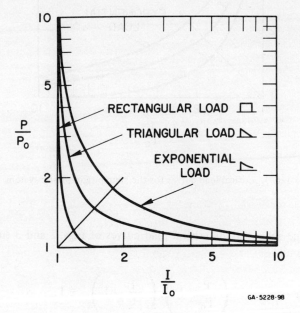

GA-5228-98

Fig. 2. Effect of pulse shape on critical pulse loads for a linear spring-mass system.

Our discussion so far has been limited to the linear spring-mass system. However, as mentioned earlier, the shape of the critical load curves of Figs. 1 and 2 is typical for a wide variety of structures. To illustrate this, in Fig. 3 are shown critical load curves for a rigid-plastic system of one degree of freedom.* Here P_0 is the minimum load required to

*This system is analyzed in the next section.

initiate motion and I_o is the ideal impulse required to produce a specified critical displacement. The important point to notice in Fig. 3 is that the critical load curves are similar in shape to those of Fig. 2. As shown later, similar curves are found for rigid-plastic beams and plates and for elastic and plastic buckling of cylindrical shells under lateral pulse loads.

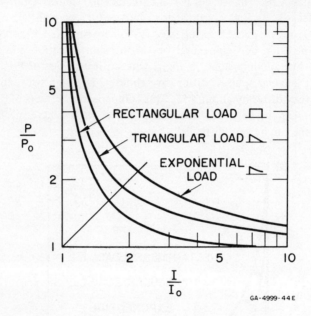

Fig. 3. Critical load curves for the simple rigid-plastic system.

The hyperbolic shape of the critical load curves of Figs. 2 and 3 suggests the simple approximate formula

$$\left(\frac{P}{P_o} - 1 \right) \left(\frac{I}{I_o} - 1 \right) = 1$$

This formula is plotted in Fig. 4 together with the critical load curves for exponential pulses from Figs. 2 and 3. Thus, for simple structures, an approximate critical load curve in terms of P and I can be obtained if P_o and I_o are known, and these are generally relatively easy to estimate.

We now come to a very important feature of the present characterization scheme. Suppose we wish to construct a critical load curve for a complex structure subjected to pulse loads of a given spatial distribution. To do so, we attempt to break the structure into noninteracting elements (beams, plates, shells) and construct critical load curves for

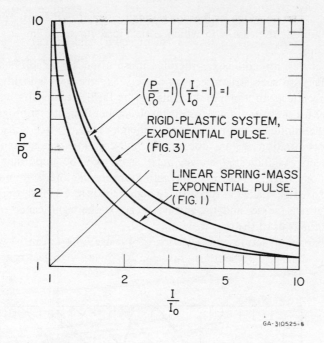

Fig. 4. The rectangular hyperbola as an approximate critical load curve.

the elements. From our past experience we know that the critical load curves for the elements will have the shape shown in Figs. 2 and 3. Hence, the critical load curve for a complex structure will appear as shown in Fig. 5. Here the critical load curve for the

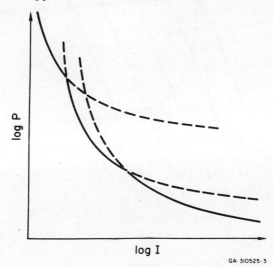

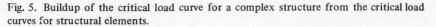

Fig. 5. Buildup of the critical load curve for a complex structure from the critical load curves for structural elements.

structure is taken as the envelope of the critical load curves for the structural elements. The important feature is the ease with which critical load curves for linear and nonlinear elements can be combined.

Critical load curves for complex structures must often be established or verified experimentally. Because such structures may be very expensive, it is highly desirable to keep the number of tests to a minimum. This can be facilitated by first estimating the critical load curves for the structural elements involved; an efficient test program can then be formulated. The minimum test program will generally require three series of tests, one series in the short-duration region, one in the mid-duration region, and one in the long-duration region. As indicated above, pulse shape will be important only in the mid-duration region. Each series of tests will probably require a minimum of two structures, the first to be loaded at successively higher loads to establish the approximate position of the critical curve, and the second to assess the significance of cumulative damage effects in the tests of the first structure.

In tests of complex structures the question of damage criteria can be very difficult. However, if the damage gradient is sufficiently great, damage criteria may be

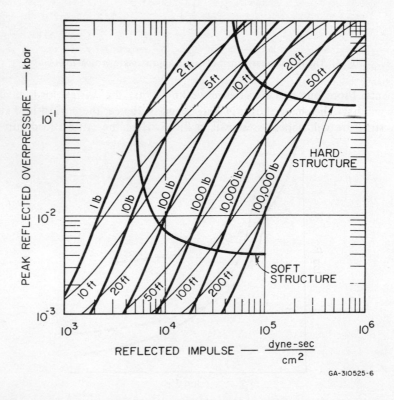

GA-310525-6

Fig. 6. Comparison of critical load curves for hypothetical soft and hard structures with loads developed from spherical explosive charges in air.

unimportant. In the peak load-impulse (P-I) plane, the damage gradient is generally steepest at the long-duration end of the critical load curve and is smallest at the short-duration end. As a rule of thumb, for long-duration blast loads an increase of 25% above the threshold damage load produces greatly enhanced damage, and for short-duration blast loads an increase of 50 to 100% above the threshold impulse is required for a similar enhancement of damage.

The utility of critical load curves in the P-I plane is illustrated in Fig. 6, which shows critical load curves for a relatively soft and relatively hard structure, together with the blast loads developed by spherical explosive charges in air. When plotted in this way, the relationship of possible loads and the critical load curve for a structure is readily seen.

RIGID-PLASTIC SYSTEM OF ONE DEGREE OF FREEDOM

As an example of the determination of critical pulse loads in terms of peak load and impulse, we analyze here the rigid-plastic system of one degree of freedom. This system is chosen because it is the simplest to analyze; however, such a system is not without practical significance, for it has been used in air explosion problems to represent tank and missile structures and in underwater explosion problems to represent the hull plating of surface ships and the cylindrical pressure hull of submarines. With this system (which may represent a beam, plate, shell) the force resisting deformation is constant; thus it is of interest primarily for problems in which the deformation greatly exceeds the elastic range. In the following, the analysis is carried out in detail for rectangular pulse loads (step rise, constant, step decay), and solutions are stated for triangular pulse loads (step rise, linear decay) and exponential pulse loads (step rise, exponential decay).

The equation of motion is simply

$$\ddot{w} = p - P_o \qquad\qquad p > p_o \qquad\qquad (6)$$

where $\ddot{w}$ is the acceleration, and p and P_o are the load and the resisting force, respectively, per unit mass. For rectangular pulse loads

$$p(t) = \begin{cases} P & 0 \leqslant t \leqslant T \\ 0 & t > T \end{cases} \qquad\qquad (7)$$

t being time and T pulse duration. Putting (7) into (6) gives

$$\ddot{w} = \begin{cases} P - P_o & 0 \leqslant t \leqslant T & P > P_o \\ -P_o & t > T & \dot{w} > o \end{cases} \qquad\qquad (8)$$

Integrating the first of these equations with zero initial displacement and velocity gives, at $t = T$

$$\dot{w} = (P - P_o)\,T \qquad\qquad w = \frac{1}{2}\,(P - P_o)\,T^2 \tag{9}$$

Integrating the second equation of (8) with (9) as initial conditions gives, for $t \geqslant T$,

$$\dot{w} = -P_o\,(t-T) + (P-P_o)\,T \tag{10}$$

$$w = -P_o\,\frac{1}{2}\,(t-T)^2 + (P-P_o)\,T\,(t-T) + (P-P_o)\,\frac{1}{2}\,T^2 \quad \dot{w} > 0$$

Putting $\dot{w} = 0$ and $t = \hat{t}$ in the first of the above equations we find that the motion stops at

$$\frac{\hat{t}}{T} = \frac{P}{P_o} \qquad \text{or} \qquad \hat{t} = \frac{I}{P_o} \tag{11}$$

Substituting $t = \hat{t}$ in the second equation of (10) gives for the maximum displacement

$$\hat{w} = \frac{I^2}{2P_o}\left(1 - \frac{P_o}{P}\right) \tag{12}$$

Letting $p \to \infty$ in (12) we find for the maximum displacement under an ideal (zero duration) impulse I_o

$$\hat{w}_o = \frac{I_o{}^2}{2P_o} \tag{13}$$

The characteristic response time $\hat{t}_o$ of the system (under an ideal impulse) is found from (11) and (13) and is

$$\hat{t}_o = \frac{I_o}{P_o} = \left(\frac{2\hat{w}_o}{P_o}\right)^{1/2} \tag{14}$$

Thus, the duration of the motion under an ideal impulse is equal to the duration of the load along the line $I/P = I_o/P_o$. Finally, identifying $\hat{w}$ of (12) with $\hat{w}_o$ of (13) we obtain, with $I = I/I_o$ and $P = P/P_o$,

$$I^2 = \frac{P}{P-1} \tag{15}$$

This defines the values of P and I that give the same maximum deflection of a rigid-plastic system of one degree of freedom under rectangular pulse loads.

For triangular pulse loads (step rise, linear decay), the critical load curve is given by

$$I^2 = \begin{cases} \dfrac{3}{16} \dfrac{P^4}{(P-1)^3} & 1 < P \leqslant 2 \\[4mm] \dfrac{3P}{3P-4} & 2 \leqslant P \end{cases} \tag{16}$$

For exponential pulse loads (step rise, exponential decay) the critical load curve is given by

$$I^2 = \frac{P^2}{2(P-1)\hat{\tau} - \hat{\tau}^2} \tag{17}$$

where

$$\hat{\tau} = P(1 - e^{-\hat{\tau}}) \tag{18}$$

$\hat{\tau} = \hat{t}/T$ being the time at which the motion stops. The critical load formulas (15), (16), and (17) were used to plot the curves of Fig. 3.

ADDITIONAL CRITICAL LOAD CURVES

In the foregoing discussion we considered critical load curves for the linear spring-mass system and the simple rigid-plastic system. Critical load curves in terms of peak load and impulse have also been worked out and reported for uniformly loaded rigid-plastic beams and plates and for dynamic buckling of cylindrical shells under uniform lateral pressure. These are discussed here.

Rigid-Plastic Beams and Plates. Rigid-plastic beams and plates deflect by the action of plastic hinges. The analysis begins with the assumption of deflection mechanisms that afterward are shown to be valid in that the bending moments do not exceed the yield moment.

For rigid-plastic beams under static loads, the deflection mechanism is shown in Fig. 7(a); for clamped beams the support hinges are plastic hinges, for pinned beams they are natural hinges. The deflection mechanisms for uniform impulsive (zero duration) loads are shown in Fig. 7(b). The beam deflects initially by plastic or natural hinges at the supports and by moving plastic hinges that propagate away from the supports toward the center of the beam. The moving hinges eventually reach the center, and further deflection occurs by hinges at the center and at the supports. For pulse loads with a step rise and

GA-310525-4

Fig. 7. Deflection mechanisms for rigid-plastic beams under uniformly distributed pulse loads.

finite duration, the initial position of the moving hinges depends on the peak load P (per unit length) and the static collapse load P_o (per unit length). For $P \geqslant 3P_o$, the initial position of the moving hinges, measured from the supports, is

$$\frac{X_h\,(0)}{L} = \sqrt{3\,\frac{P_o}{P}} \qquad\qquad P \geqslant 3P_o \qquad\qquad (19)$$

where L is the half-length of the beam. For $P = 3P_o$, (19) gives $X_h\,(0) = L$, thus for $P \leqslant 3P_o$ the static deflection mechanism is valid.

For clamped beams the static collapse load is

$$P_o = \frac{4M_o}{L^2} \qquad\qquad (20)$$

where M_o is the yield moment and, for a beam of rectangular cross section, is given by

$$M_o = \sigma_o\,\frac{bh^2}{4} \qquad\qquad (21)$$

b being the width, h the depth, and σ_o the yield stress. For pinned beams, it is readily seen that the static collapse load is half of that for clamped beams. It turns out that, when given in terms of the static collapse load, the critical load curves are identical for clamped and pinned beams.

The analysis of rigid-plastic beams under pulse loads involves rigid-body dynamics of bodies of changing lengths. Several such analyses have been worked out by Florence and are summarized in Reference [1]. Some of his results are presented in the following paragraphs.

The central deflection of a clamped or pinned rigid-plastic beam under an ideal impulse I_o per unit length

$$\delta_o = \frac{2I_o^2}{3mP_o} \qquad\qquad (22)$$

where m is the beam mass per unit length. The duration of the motion is

$$\hat{t}_o = \frac{I_o}{P_o} \tag{23}$$

For a rectangular pulse load of magnitude P and impulse I (each per unit length) the central deflection is

$$\delta = \begin{cases} \dfrac{1}{m} \dfrac{I^2}{P_o} \dfrac{3}{4} \left(1 - \dfrac{P_o}{P}\right) & P_o < P \leqslant 3P_o \\[3mm] \dfrac{1}{m} \dfrac{I^2}{P_o} \dfrac{2}{3} \left(1 - \dfrac{3}{4}\dfrac{P_o}{P}\right) & 3P_o \leqslant P \end{cases} \tag{24}$$

Putting $\delta = \delta_o$ gives

$$I^2 = \begin{cases} \dfrac{8P}{9\,(P-1)} & 1 < P \leqslant 3 \\[4mm] \dfrac{P}{P - \dfrac{3}{4}} & 3 \leqslant P \end{cases} \tag{25}$$

where, as before, $I = I/I_o$ and $P = P/P_o$.
For a triangular pulse load the central deflection is

$$\delta = \begin{cases} \dfrac{4}{m}\dfrac{I^2}{P}\left(1 - \dfrac{P_o}{P}\right)^3 & P_o \leqslant P \leqslant 2P_o \\[3mm] \dfrac{1}{m}\dfrac{I^2}{P_o}\left(\dfrac{3}{4} - \dfrac{P_o}{P}\right) & 2P_o \leqslant P \leqslant 3P_o \\[3mm] \dfrac{1}{m}\dfrac{I^2}{P}\left[-1 + \dfrac{3}{4}\dfrac{P}{P_o} - \dfrac{4}{3}\left(1 - 3\dfrac{P_o}{P}\right)^3\right] & 3P_o \leqslant P \leqslant 6P_o \\[3mm] \dfrac{2}{3m}\dfrac{I^2}{P}\left(\dfrac{P}{P_o} - 1\right) & 6P_o \leqslant P \end{cases} \tag{26}$$

Equating the central deflection of (26) to that of (22), we obtain for the critical loads

$$
I^2 = \begin{cases}
\dfrac{P^4}{6\,(P-1)^3} & 1 < P \leqslant 2 \\[3ex]
\dfrac{8P}{9P-12} & 2 \leqslant P \leqslant 3 \\[3ex]
\dfrac{8\,P^4}{9P^4 - 12P^3 - 16\,(P-3)^3} & 3 \leqslant P \leqslant 6 \\[3ex]
\dfrac{P}{P-1} & 6 \leqslant P
\end{cases}
\tag{27}
$$

Finally, for exponential pulse loads the central deflection is

$$
\delta = \begin{cases}
\dfrac{3}{4m}\,\dfrac{I^2}{P}\,\hat{\tau}\,[\,2 - \dfrac{P_o}{P}\,(2 + \hat{\tau}\,)\,] & P_o < P \leqslant 3P_o \\[3ex]
\dfrac{1}{2m}\,\dfrac{I^2}{P}\,[\,3\,\hat{\tau}\,(1 - \dfrac{P_o}{P}) - \bar{\tau}\,(1 - \dfrac{3P_o}{P}) \\[3ex]
\qquad\qquad - \dfrac{3}{2}\,\dfrac{P_o}{P}\,(\hat{\tau}^2 - \bar{\tau}^2)\,] & 3P_o \leqslant P
\end{cases}
\tag{28}
$$

where

$$
\hat{\tau} = \frac{P}{P_o}\,(1 - e^{-\hat{\tau}})
$$

$$
\bar{\tau} = \frac{1}{3}\,\frac{P}{P_o}\,(1 - e^{-\bar{\tau}})
\tag{29}
$$

$\hat{\tau} = \hat{t}/T$, $\hat{t}$ being the duration of the motion and $T = I/P$, and $\bar{\tau} = \bar{t}/T$, $\bar{t}$ being the time at which the moving hinges reach the center of the beam. Equating central deflections as before, we find for the critical loads

$$I^2 = \begin{cases} \dfrac{8\,P^2}{9\,\hat{\tau}\,(2P - 2 - \hat{\tau})} & 1 < P \leqslant 3 \\[4mm] \dfrac{8\,P^2}{18\,(P-1)\,\hat{\tau} - 6\,(P-3)\,\hat{\tau} - 9\,(\hat{\tau}^2 - \bar{\tau}^2)} & 3 \leqslant P \end{cases} \qquad (30)$$

Equations (25), (27), and (30) give the P-I combinations that produce the final deflection δ_0 of (22). These combinations are plotted in Fig. 6. The critical load curves of Fig. 8 are remarkably similar to those of Fig. 3 for the simple rigid-plastic system.

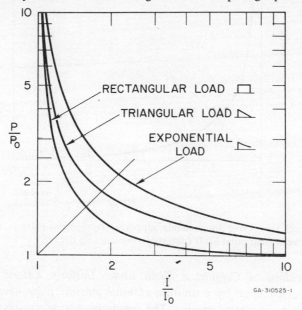

Fig. 8. Critical load curves for pinned and clamped rigid-plastic beams under uniformly distributed pulse loads.

The analysis of rigid-plastic plates is much more complex than that for rigid-plastic beams. So far, critical load curves have been obtained only for rectangular pulses. For simply supported plates, the critical loads for rectangular pulses are [1]

$$I^2 = \begin{cases} \dfrac{3P}{4\,(P-1)} & 1 \leqslant P \leqslant 2 \\[4mm] \dfrac{3P}{3P-2} & 2 \leqslant P \end{cases} \qquad (31)$$

where, as before, P_o is the static collapse load and I_o is the ideal impulse to produce a critical displacement. The critical load curve is plotted in Fig. 9. Again, the similarity to the critical load curve of Fig. 3 for the simple rigid-plastic system is apparent.

The critical load formula for clamped plates is much more complex than that for simply supported plates but, when plotted, the critical curve is nearly identical with that of Fig. 9 [2].

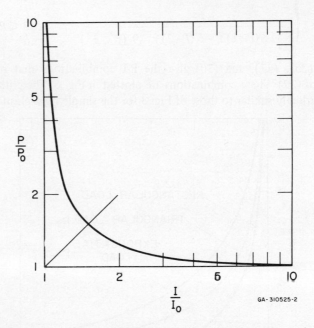

Fig. 9. Critical load curve for a simply supported rigid-plastic plate under uniformly distributed rectangular pulse loads (step rise, constant, step decay).

Dynamic Buckling of Cylindrical Shells under Uniform Lateral Pulse Loads. A cylindrical shell acted upon by a uniform external pressure pulse develops compressive hoop stresses that resist inward motion. The compressive hoop stresses cause elements of the shell that have an initial inward departure from circularity to be thrust ahead of the average motion and elements that have an initial outward departure to be thrust behind. It turns out that certain mode numbers of the initial departure from circularity grow faster than others and this determines the buckling pattern.

Anderson and Lindberg [3] made an extensive analysis of this type of dynamic buckling, called pulse buckling, and reported the results in terms of critical P-I curves. Figure 10 shows the critical load curve for pulse buckling of an aluminum cylinder simply supported at both ends. The lower (elastic) branch of the curve indicates the P-I combinations for which the initial nonuniformities grow by a factor of 1000 during the inward elastic motion. The upper (tangent modulus) branch indicates the combinations for which the nonuniformities grow by a factor of 1000 during inward motion beyond the elastic range. In the dashed part of the curve, the tensile strain rate due to the buckling motion exceeds the compressive strain rate of inward displacement, causing strain reversal and severely complicating the analysis.

The numbers indicated along the curves of Fig. 10 are the most-magnified modes of the initial nonuniformities in the circular shape. The mode numbers are much greater for the tangent modulus branch than for the elastic branch.

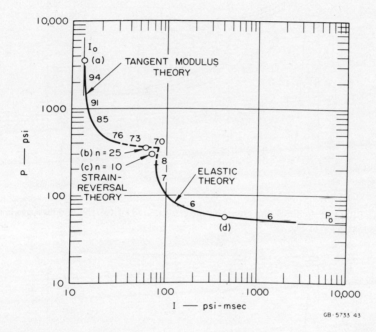

Fig. 10. Critical load curve for exponential pulses for buckling of a 6-inch-diameter aluminum (6061-T6) cylinder with a radius-to-thickness ratio of 100 and a length-to-diameter ratio of 1.

Figure 11 indicates the variation in the critical load curves with the amplification factor chosen for the buckling criterion, and Fig. 12 shows the effects of pulse shape. The figures show that the critical load curve is not very sensitive to the amplification factor or pulse shape.

Figure 13 shows the critical load curves for cylinders of different radius-to-thickness ratios a/h. For $a/h = 24$, the critical load curve consists only of the tangent modulus branch. For larger values of a/h, the critical load curve has an elastic branch which increases with a/h. Figure 14 shows the effect of length to diameter ratio L/D. The tangent modulus branch is unaffected because, over the L/D range indicated, the mode numbers are too high for the ends to matter.

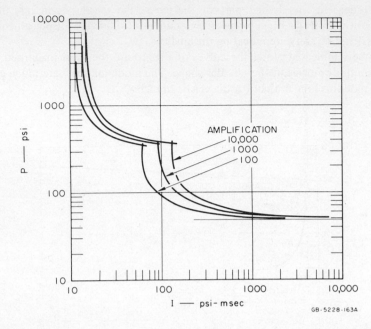

Fig. 11. Effect of amplification factor on the critical load curve. The middle curve is the same as that given in Fig. 10.

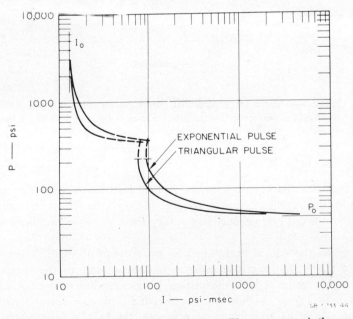

Fig. 12. Effect of pulse shape on critical load curves. The upper curve is the same as that in Fig. 10.

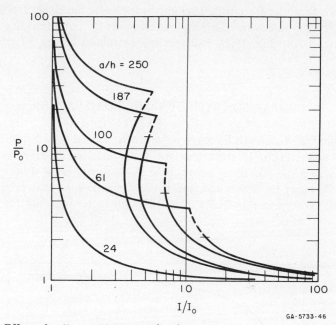

Fig. 13. Effect of radius-to-thickness ratio a/h on the critical load curve for exponential pulses. The curve for a/h = 100 is the same as that in Fig. 10. All the curves are for a length-to-diameter ratio of 1.

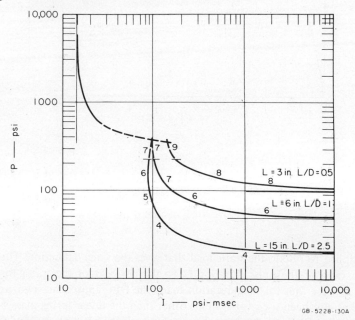

Fig. 14. Effect of length-to-diameter ratio L/D on critical load curves for exponential pulses. The middle curve is the same as that in Fig. 10.

Approximate formulas for critical load curves for pulse buckling of cylindrical shells are given in [3]. The ease with which the critical loads for pulse buckling of cylindrical shells are presented in Figs. 10-14 indicates the usefulness of the P-I characterization scheme.

COMPARISON WITH DYNAMIC LOAD FACTOR

The dynamic load factor (DLF) characterization of critical pulse loads is probably the most widely known. Figure 15 shows the DLF for a linear spring-mass system for pulse loads having a step rise and linear decay. Here the ordinate is the dynamic load factor DLF = P_s/P, P_s being the static load and P the dynamic load that produce the same maximum deflection. Since P_s = DLF P, the DLF is the factor by which the dynamic load

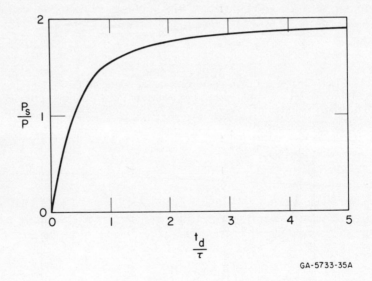

GA-5733-35A

Fig. 15. Dynamic load factor for a linear oscillator under pulse loads with a step rise and linear decay.

must be multiplied to obtain the static load that gives the same deflection. The abcissa in Fig. 15 is the ratio of the pulse duration t_d to the period τ of the spring-mass system. From Fig. 15, we see that for long-duration loads the DLF approaches two, as expected.

The DLF has also been used for systems other than the linear spring-mass system. For example, extensive DLF curves have been obtained for rigid-plastic and rigid-strain hardening systems of one degree of freedom [4].

The DLF curve of Fig. 15 gives the same information as the P-I curve (middle curve of Fig. 2). To construct the P-I curve from the DLF curve, we note that a step load of magnitude P_o produces the same deflection as a static load $P_s = 2P_o$. Thus, the ordinate for the P-I curve is

$$\frac{P}{P_o} = \frac{P_s}{P_o} \quad \frac{P}{P_s} = \frac{2}{DLF}$$

To find the corresponding abcissa, we recall from (4) above that $I_o/P_o = \tau/4$ and note that, for a triangular pulse, $I = Pt_d/2$. Then, putting $\theta = t_d/\tau$, we find that $I/I_o = 4\theta/DLF$.

Although Figs. 2 and 15 contain the same information, they emphasize different aspects. Figure 15 displays the relation of dynamic loads to an equivalent static load, as a function of pulse duration. Figure 2, on the other hand, displays equivalent dynamic loads in terms of the peak load and impulse, the load parameters that are important at the long-duration and short duration extremes. The importance of impulse is suppressed in Fig. 15 (impulsive loads are plotted at the origin), whereas in Fig. 2 it is readily apparent.

HISTORICAL ASPECTS

The peak load-impulse characterization scheme described here was developed at Stanford Research Institute (SRI) over the years from 1959 to 1966 [5, 6]. Since then, it has become apparent that other investigators have had similar ideas. Here the previous work that is known is described.

In 1951, workers at the U.S. Army Ballistic Research Laboratories used iso-damage plots in terms of incident peak pressure and impulse to display results of blast tests on aircraft structures [7]. Mortion [8] in 1954 addressed the problem of blast damage to aircraft structures using reflected peak pressure and impulse. He calculated critical load curves for the simple rigid-plastic system under pulse loads and plotted them in terms of P/P_o and I/I_o.

Kornhauser [9] in 1954 reported a scheme for describing the "inertia sensitivity" of "inertia devices and shock-resistant structures." In this paper and a subsequent book [10], Kornhauser used as parameters for critical loads the acceleration and the velocity change experienced by the structure. These parameters are equivalent to force and impulse in the peak load-impulse scheme. Kornhauser gives theoretical critical load curves for the linear spring-mass system under various pulse loads and in his book has a figure similar to Fig. 5 showing how critical load curves for different structural elements can be combined. In his book, he also gives critical acceleration and velocity change for various animals.

Coombs and Thornhill [11] in 1960 described the response of structural elements to pulse loads using peak load and impulse as the load parameters. They identified the peak load as the important parameter for long-duration loads and impulse as the important factor for short-duration loads.

In 1966, workers at the Lovelace Foundation [12] in describing blast effects on animals recognized the significance of peak load and impulse, and indicated that impulse may be the controlling parameter for short-duration loads.

From this brief review, it is apparent that the concept of describing critical pulse loads in terms of peak load and impulse has been developed independently by many workers. The concept is indeed useful and deserves wider application.

ACKNOWLEDGMENTS

The initial work at Stanford Research Institute on the peak load-impulse characterization of critical pulse loads was carried out in the Engineering Mechanics Group of the Poulter Laboratory under contracts with the Air Force Weapons Laboratory [5]. The technical monitors on this work were M. C. Atkins and D. L. Lamberson. Their guidance and support were important factors in the initial development. Extensions and refinements were undertaken for the Air Force Space and Missile Systems Organization under subcontract with the Lockheed Missiles and Space Company [6]. Subsequent applications have been undertaken for the Defense Atomic Support Agency.

Two persons who contributed to the early formulation of the ideas and whose names do not appear in the references cited are G. E. Duvall and the late J. N. Goodier. Professor Goodier continued as a vital force in the Engineering Mechanics Group throughout the development of the concepts presented here. We dedicate this paper to his memory.

REFERENCES

1. G. R. Abrahamson, A. L. Florence and H. E. Lindberg, *Radiation Damage Study (RADS) Volume XIII — Dynamic response of beams, plates, and shells to pulse loads,* SRI Project FGU-5733, AVMSD-0339-66-RR, September 1966, Contract AF04 (694)-824, BSD TR 66-472.
2. A. L. Florence, Clamped circular rigid-plastic plates under central blast loading, *Int. J. Solids Struct.* 2, 319 (1966).
3. D. L. Anderson and H. E. Lindberg, Dynamic pulse buckling of cylindrical shells under transient lateral pressures, *AIAA J.* 6, 589 (1968).
4. J. M. Biggs, *Introduction to Structural Dynamics,* McGraw-Hill, N.Y. 1964.
5. G. R. Abrahamson and H. E. Lindberg, Estimated bounds on suddenly applied surface loads required to destroy reentry vehicles (U), *J. of Missile Def. Res.,* Limited distribution supplement to Vol. 2, No. 1, JMDR-64-2S, Summer 1964 (SRD).
6. H. E. Lindberg, D. L. Anderson, R. D. Firth and L. V. Parker, *Response of reentry vehicle-type shells to blast loads,* Stanford Research Institute Final Report, Project 5228, Lockheed Missiles and Space Company, P.O. 24-14517 under AF04(694)-655, LMSC-B130200, Vol. IV-C, Sept. 30, 1965.
7. BRL Memo. Report 575, Dependence of external blast damage to A-25 aircraft on peak pressure and impulse (U), September 1951.
8. H. S. Morton, *Scaling the effects of air blast on typical targets,* The John Hopkins Univ., Applied Physics Laboratory, Contract NOw 62-0604-c, Technical Memorandum TG-733, January 1954.

9. M. Kornhauser, Prediction and evaluation of Sensitivity to transient accelerations, *J. Appl. Mech.*, **371** (1954).
10. M. Kornhauser, *Structural Effects of Impact*, Spartan Books, Baltimore, Md., Cleaver-Hume Press (1964).
11. A. Coombs and C. K. Thornhill, *Symp. über wissenschaftliche Grundlagen des Schutzbaues am Ernst-Mach-Institut*, Freidburg i.Br., September 1960.
12. Donald R. Richmond, Edward G. Damon, E. Royce Fletcher, I. Gerald Bowen and Clayton S. White, *The relationship between selected blast-wave parameters and the response of mammals exposed to air blast*, Technical Progress Report DASA 1860 Contract No. DA-49-146-XZ-372, November 1966.

CHLADNI'S PATTERNS FOR RANDOM
VIBRATION OF A PLATE

STEPHEN H. CRANDALL and LARRY E. WITTIG

Acoustics and Vibration Laboratory
Department of Mechanical Engineering
Massachusetts Institute of Technology

Abstract—When a rectangular plate is subjected to a point load whose time history is a stationary wide band random process the distribution of average energy density in the plate is substantially uniform over the plate with the exception of a remarkable pattern of strips in which there is an excess of energy. These patterns are demonstrated by sprinkling grains of salt on the plate in much the same manner as the nodal lines of natural modes are revealed in the classical experiment of Chladni. The existence of strips of excess energy is predicted by analysis of a simplified model and verified by local measurements.

In the classical experiment of E.F.F. Chladni (1756-1827) the nodal lines of a vibrating plate are revealed by sprinkling grains of sand on the vibrating surface. The sand grains dance on those parts of a horizontal surface that have accelerations greater than that of gravity and gradually migrate to the regions of minimum motion. A typical Chladni pattern for pure tone excitation is shown in Fig. 1. In this case the plate is aluminum 38 inches long, 30 inches wide and 1/16 inch thick. The plate is clamped in a steel frame and has a groove milled through two thirds its thickness, around the periphery just inside the frame, in an attempt to stimulate simply supported boundary conditions. If the plate had been truly simply supported its fundamental resonant frequency would have been 10.3 Hz and if it had been perfectly clamped the fundamental resonance would have been 20.5 Hz. The measured fundamental frequency is 13 Hz which indicates an intermediate type of support. The plate was driven by a 2 gm voice coil glued to the underside of the plate. The voice coil moved in the gap of a permanent magnet and was excited by an amplifier driven by a signal generator. The frequency of excitation was about 5000 Hz for the pattern of Fig. 1. To make the pattern stand out clearly in the photo the plate was painted black and grains of table salt were used in place of sand.

When a stationary wide band excitation is substituted for the pure tone the remarkable pattern shown in Fig. 2 is obtained. Four straight strips arranged in a tic-tac-toe pattern emerge substantially free of salt out of a background of uniformly distributed salt grains. The single point of excitation lies underneath one of the four intersections. The voltage into the voice coil had uniform spectral density in the range 20-20,000 Hz. The spectrum of force applied to the place was not measured directly but because of the lightness of the voice coil it can be assumed to be uniform at least for the lower frequencies. We have evidence that the force spectrum is nominally flat to over 5000 Hz and we know that an appreciable roll-off has occurred at 6000 Hz [1]. Since the salt grains tend to migrate toward regions of minimum motion it can be inferred that the four strips in Fig. 2 represent local regions of greater than average energy density. In order to explain this phenomenon we proceed in two steps. We first examine the analagous problem for a taut string where a reasonably complete analysis can be given. We then turn to the plate where, although the ideas are similar, it is necessary to make additional approximations to obtain quantitative results.

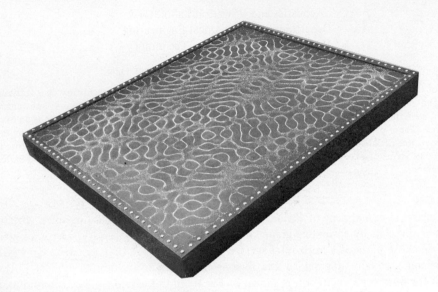

Fig. 1. Chladni pattern for pure tone excitation for a thin plate.

Fig. 2. Chladni pattern for point excitation of a thin plate with stationary wide-band random force.

RANDOM VIBRATION OF A TAUT SPRING

Consider a uniform string of length L and mass per unit length ρA stretched by a tensile force T. If the string is subjected to viscous damping force c_1 per unit length per unit velocity and is excited by a transverse force $f(x,t)$ per unit length the transverse displacement $w(x,t)$ satisfies the following partial differential equation

$$-T \frac{\partial^2 w}{\partial x^2} + c_1 \frac{\partial w}{\partial t} + \rho A \frac{\partial^2 w}{\partial t^2} = f, 0 < x < L. \tag{1}$$

We seek a solution in the classical manner as a sum of products

$$w(x,t) = \sum_{j=1}^{\infty} w_j(t) \; \psi_j(x). \tag{2}$$

For the ψ_j we take the following modes

$$\psi_j(x) = \sqrt{2} \; \sin \frac{j \pi x}{L} \tag{3}$$

which satisfy the free vibration relations

$$-T \frac{d^2 \psi_j}{dx^2} = \omega_j^2 \rho A \psi_j \tag{4}$$

where the ω_j are the natural frequencies

$$\omega_j = \frac{j \pi}{L} \sqrt{\frac{T}{\rho A}} \tag{5}$$

and which also satisfy the orthonormality conditions

$$\int_0^L \rho A \, \psi_j \, \psi_k \, dx = m \, \delta_{jk} \tag{6}$$

where $m = \rho AL$ is the total mass of the string. We substitute Eq. (2) into Eq. (1), multiply through by ψ_j and integrate over x, using Eqs. (4) and (6) to obtain the uncoupled equations for the $w_j(t)$:

$$m \left[\frac{d^2 w_j}{dt^2} + \beta_j \frac{dw_j}{dt} + \omega_j^2 w_j \right] = f_j(t) \tag{7}$$

where the modal exciting force is

$$f_j(t) = \int_0^L f(x,t) \, \psi_j(x)dx \tag{8}$$

and the modal bandwidth β_j is given by

$$\beta_j = \frac{c_1}{\rho A} .$$ (9)

The formal solution to Eq. (7) is given by the convolution integral

$$w_j(t) = \int_{-\infty}^{\infty} h_j(\theta) f_j(t-\theta) d\theta$$ (10)

where the impulse response function is

$$h_j(t) = \begin{cases} 0 & , t < 0 \\ \dfrac{e^{-\frac{1}{2}\beta_j t}}{m p_j} \sin p_j t & , t > 0 \end{cases}$$ (11)

with $p_j{}^2 = \omega_j{}^2 - \beta_j{}^2/4$. The displacement response of the string is thus

$$w(x,t) = \sum_{j=1}^{\infty} \psi_j(x) \int_{-\infty}^{\infty} h_j(\theta) f_j(t-\theta) d\theta.$$ (12)

The corresponding transverse velocity $v = \partial w / \partial t$ is

$$v(x,t) = \sum_{j=1}^{\infty} \psi_j(x) \int_{-\infty}^{\infty} \dot{h}_j(\theta) f_j(t-\theta) d\theta.$$ (13)

We now consider stationary random excitation of the string by a point force located at $x = a$. The space-time correlation of $f(x,t)$ is taken as

$$E[f(x_1,t_1) f(x_2,t_2)] = \delta(x_1 - a) \delta(x_2 - a) R_f(t_2 - t_1)$$ (14)

where $R_f(\tau)$ is the autocorrelation function of the force process. The cross correlation of the modal forces is then evaluated as follows:

$$R_{jk}(\tau) = E[f_j(t) f_k(t+\tau)] = \int_0^L \int_0^L \psi_j(x_1) \psi_k(x_2) E[f(x_1,t) f(x_2,t+\tau)] \, dx_1 dx_2$$

$$= \psi_j(a) \psi_k(a) R_f(\tau).$$ (15)

As a measure of the local time average energy density we consider the mean square velocity at the location x

$$E[v^2] = \sum_{j=1}^{\infty} \sum_{k=1}^{\infty} \psi_j(x)\,\psi_k(x) \int_{-\infty}^{\infty} \int_{-\infty}^{\infty} \dot{h}_j(\theta_1)\,\dot{h}_k(\theta_2)\,R_{jk}(\theta_1-\theta_2)\,d\theta_1\,d\theta_2$$

$$= \sum_{j=1}^{\infty} \sum_{k=1}^{\infty} \psi_j(x)\,\psi_k(x)\,\psi_j(a)\,\psi_k(a)\,I_{jk} \tag{16}$$

where I_{jk} denotes the double integral

$$I_{jk} = \int_{-\infty}^{\infty} \int_{-\infty}^{\infty} \dot{h}_j(\theta_1)\,\dot{h}_k(\theta_2)\,R_f(\theta_1-\theta_2)\,d\theta_1\,d\theta_2 \tag{17}$$

An alternative method of evaluating I_{jk} is available if use is made of the Fourier transform relations for the spectral density $S_f(\omega)$ of the excitation

$$S_f(\omega) = \frac{1}{2\pi} \int_{-\infty}^{\infty} \int_{-\infty}^{\infty} R_f(\tau)\,e^{-i\omega\tau}\,d\tau$$

$$R_f(\tau) = \int_{-\infty}^{\infty} S_f(\omega)\,e^{i\omega\tau}\,d\omega \tag{18}$$

and for the frequency response $H_j(\omega)$ of the j-th mode

$$h_j(t) = \frac{1}{2\pi} \int_{-\infty}^{\infty} H_j(\omega)\,e^{i\omega t}\,d\omega$$

$$H_j(\omega) = \int_{-\infty}^{\infty} h_j(t)\,e^{-i\omega t}\,dt$$

$$= \frac{1}{m(\omega_j^2 + i\omega\beta_j - \omega^2)} \tag{19}$$

Substituting in Eq. (17) for $R_f(\omega)$ from Eq. (18) and using Eq. (19) we find

$$I_{jk} = \int_{-\infty}^{\infty} S_f(\omega)\,\omega^2 H_j(\omega)\,H_k(-\omega)\,d\omega. \tag{20}$$

To model the experimental situation we take $S_f(\omega)$ to be a band limited white noise spectrum

$$S_f(\omega) = \begin{cases} S_o, -\omega_c < \omega < \omega_c \\ \\ 0, \omega_c < |\omega|. \end{cases} \tag{21}$$

We take the cut-off frequency ω_c to lie somewhere between the natural frequencies ω_N and ω_{N+1}. Furthermore we assume that the model overlap ratio

$$r = \frac{\text{modal bandwidth}}{\text{modal spacing}} = \frac{\beta_j}{\omega_j - \omega_{j-1}} = \frac{c_1 L}{\pi \sqrt{\rho AT}} \tag{22}$$

is sufficiently small to permit the following approximations in the sum (16):

$$I_{jk} = 0 \text{ for } j \neq k$$

$$I_{jj} = 0 \text{ for } j > N \tag{23}$$

To evaluate the remaining values of I_{jj} we use Eq. (20) with $S_f(\omega) = S_o$ for $-\infty < \omega < \infty$. The result is

$$I_{jj} = \frac{\pi S_o}{m^2 \beta_j} = \frac{\pi S_o}{m c_1 L}, \; j = 1, 2, \ldots, N \tag{24}$$

which when inserted in Eq. (16) yields

$$E[v^2] = \frac{\pi S_o}{m c_1 L} G(x/L, a/L, N) \tag{25}$$

where $G(x/L, a/L, N)$ stands for the sum

$$G(x/L, a/L, N) = \sum_{j=1}^{N} \psi_j^2(x) \psi_j^2(a)$$

$$= \sum_{j=1}^{N} 4 \sin^2 \frac{j\pi x}{L} \sin^2 \frac{j\pi x}{L}. \tag{26}$$

This sum can be evaluated in closed form [2]. We find

$$G(x/L, a/L, N) = g\left(\frac{x}{L}, N\right) + g\left(\frac{a}{L}, N\right) - \frac{1}{2}g\left(\frac{x-a}{L}, N\right) - \frac{1}{2}g\left(\frac{x-L+a}{L}, N\right) \tag{27}$$

where

$$g\left(\xi, N\right) = N + \frac{1}{2} - \frac{\sin\left(2N+1\right)\pi\,\xi}{2\sin\pi\,\xi} \qquad (28)$$

is a nonnegative, even, periodic function of ξ with unit period. The shape of $g(\xi,N)$ is sketched in Fig. 3. It vanishes at $\xi = 0$ and at $\xi = 1$ and oscillates in the neighborhood of $g = N + \frac{1}{2}$ in the interval between these points. The mean value of $g(\xi,N)$ is N[3].

The shape of $G(x/L, a/L, N)$ can be inferred from that of $g(\xi,N)$. If N is large and a is not too close to $a = 0$ or $a = L$ the function G oscillates in the neighborhood of $G = N$ for most values of x in the interval $0 < x < L$. The exceptional points are those at $x = 0$ and $x = L$ and where $G = 0$ and those at $x = a$ and $x = L - a$ where $G \approx 3N/2$. When $a = L/2$ the two exceptional points at $x = a$ and $x = L - a$ coalesce into a single point where $G \approx 2N$. The mean value of G in every case is $g(a/L,N)$. The spatial distribution of the mean square velocity according to Eq. (25) is shown for three values of N in Fig. 4. Note the local regions of fifty percent excess energy density at $x = a$ and $x = L - a$. Note that the mean square force input increases linearly with the cutoff frequency ω_c and that the average mean square velocity response increases linearly with the mode number N of the largest natural frequency less than ω_c. In Fig. 4 the three response distributions have been normalized to have the same average mean square velocity.

The presence of a local region of higher energy in the neighborhood of the excitation point $x = a$ is not unexpected but the existence of a similar region at the image point $x = L - a$ may well come as a surprise. The symmetry of the response distribution can however be deduced directly from Eq. (26) without performing the indicated summation.

RANDOM VIBRATION OF A THIN PLATE

Consider a simply supported uniform rectangular plate with dimensions L_x by L_y and mass per unit area ρt. If the plate is subjected to viscous damping force c_2 per unit area per unit velocity and is excited by a transverse force $f(x, y, t)$ per unit area the transverse displacement $w(x, y, t)$ satisfies the following partial differential equation

$$D\,\Delta^4\,w + c_2\,\frac{\partial w}{\partial t} + \rho t\,\frac{\partial^2 w}{\partial t^2} = f\left(x, y, t\right), \qquad \left\{ \begin{array}{l} 0 < x < L_x \\[2mm] 0 < y < L_y \end{array} \right. \qquad (29)$$

where D is the flexural rigidity. The procedure we follow to obtain the response of the plate is parallel to that just given for the string. The only differences are associated with the doubly infinite array of modes for the two-dimensional plate as compared with the singly infinite set of modes for the one-dimensional string.

We seek a solution of the form

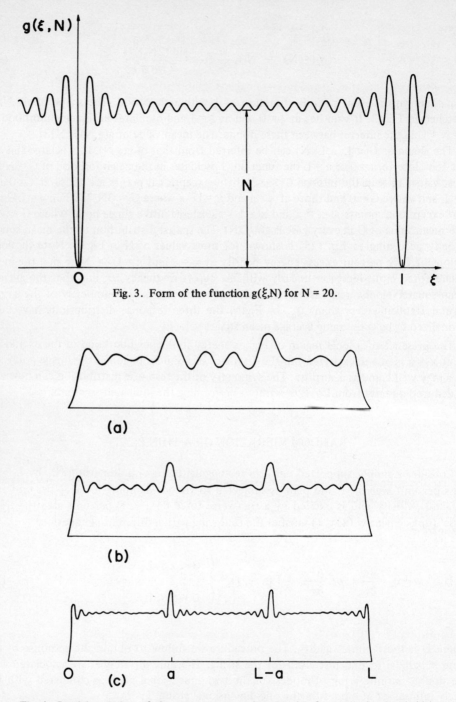

Fig. 3. Form of the function g(ξ,N) for N = 20.

(a)

(b)

(c)

Fig. 4. Spatial variation of the mean square string velocity for excitation bandwidths corresponding to (a) N = 10, (b) N = 20 and (c) N = 40.

$$w(x,y,t) = \sum_{j=1}^{\infty} \sum_{p=1}^{\infty} w_{jp}(t)\, \psi_{jp}(x,y) \tag{30}$$

where the ψ_{jp} are the natural modes of the simply supported plate

$$\psi_{jp} = 2 \sin \frac{j\pi x}{L_x} \sin \frac{p\pi y}{L_y} \tag{31}$$

which satisfy the orthonormality condition

$$\int_0^{L_x} \int_0^{L_y} \rho t\, \psi_{jp}\, \psi_{kq}\, dxdy = m\, \delta_{jp}\, \delta_{kq} \tag{32}$$

where now $m = \rho t L_x L_y$ is the total mass of the plate and which also satisfy the free vibration equation

$$D\Delta^4 \psi_{jp} = \rho t\, \omega_{jp}^2\, \psi_{jp} \tag{33}$$

where the ω_{jp} are the natural frequencies

$$\omega_{jp} = \left[\left(\frac{j\pi}{L_x} \right)^2 + \left(\frac{p\pi}{L_y} \right)^2 \right] \sqrt{\frac{D}{\rho t}} \tag{34}$$

Unlike the string these frequencies are not uniformly spaced. The average modal density n (i.e., average number of natural frequencies per radian per second) is however independent of frequency [4]. We may thus speak of an *average* spacing between resonances

$$\overline{\Delta\omega} = \frac{1}{n} = \frac{4}{L_x L_y} \sqrt{\frac{D}{\rho t}} . \tag{35}$$

Substituting Eq. (30) into Eq. (29) we obtain the uncoupled equations

$$m \left[\frac{d^2 w_{jp}}{dt^2} + \beta_{jp} \frac{dw_{jp}}{dt} + \omega_{jp}^2\, w_{jp} \right] = f_{jp}(t) \tag{36}$$

where the modal exciting force is

$$f_{jp}(t) = \int_0^{L_x} \int_0^{L_y} f(x,y,t) \, \psi_{jp} \, dxdy \tag{37}$$

and the modal bandwidth is

$$\beta_{jp} = \frac{c_2}{\rho t} \tag{38}$$

Since Eq. (36) is identical in form to Eq. (7) for the string we can proceed by analogy to the formal solution for the transverse velocity $v = \partial w / \partial t$ of the plate

$$v(x,y,t) = \sum_{j=1}^{\infty} \sum_{p=1}^{\infty} \psi_{jp}(x,y) \int_{-\infty}^{\infty} \dot{h}_{jp}(\theta) \, f_{jp}(t-\theta) \, d\theta. \tag{39}$$

We now consider stationary random excitation of the plate by a point force located at $x = a$, $y = b$. To model this we take the space-time correlation of the excitation to be

$$E[f(x_1,y_1,t_1) \, f(x_2,y_2,t_2)] = \delta(x_1-a)\,\delta(x_2-a)\,\delta(y_1-b)\,\delta(y_2-b)\,R_f(t_2-t_1) \tag{40}$$

which implies that the cross correlation of the modal forces is

$$E[f_{jp}(t) \, f_{kq}(t+\tau)] = \psi_{jp}(a,b) \, \psi_{kq}(a,b) \, R_f(\tau) \tag{41}$$

By analogy with Eq. (16) the mean square velocity at the point (x, y) due to the excitation at the point (a, b) is

$$E[v^2] = \sum_{j=1}^{\infty} \sum_{p=1}^{\infty} \sum_{k=1}^{\infty} \sum_{q=1}^{\infty} \psi_{jp}(x,y) \, \psi_{kq}(x,y) \, \psi_{jp}(a,b) \, \psi_{kq}(a,b) \, I_{jpkq} \tag{42}$$

where I_{jpkq} may be evaluated in terms of the input autocorrelation function and the modal impulse response functions

$$I_{jpkq} = \int_{-\infty}^{\infty} \int_{-\infty}^{\infty} \dot{h}_{jp}(\theta_1) \, \dot{h}_{kq}(\theta_2) \, R_f(\theta_1-\theta_2) \, d\theta_1 \, d\theta_2 \tag{43}$$

or in terms of the input spectral density and the modal frequency response functions

$$I_{jpkq} = \int_{-\infty}^{\infty} S_f(\omega)\, \omega^2 H_{jp}(\omega)\, H_{kq}(-\omega)\, s\omega. \tag{44}$$

As before, we take the input spectrum to be band-limited white noise as given by Eq. (21). We also assume that the average modal overlap ratio

$$\bar{r} = \frac{\text{modal bandwith}}{\text{average modal spacing}} = \frac{\beta_{jp}}{\Delta\omega} = \frac{c_2 L_x L_y}{4\pi\sqrt{\rho t D}} \tag{45}$$

is sufficiently small to permit approximations analogous to Eq. (23); i.e., we only include in the summation (42) those terms I_{jpkq} for which $j = k$ and $p = q$ and for which the associated natural frequency ω_{jp} is less than the cutoff frequency. For all such terms

$$I_{jpjp} = \frac{\pi S_o}{m^2 \beta_{jp}} = \frac{\pi S_o}{mc_2 L_x L_y} \tag{46}$$

Evaluation of the plate summation (42) is however fundamentally more difficult than that for the string summation (16) because of the irregular limits for j and p which are required to include only those modes whose natural frequencies lie below the cutoff frequency. We write

$$E[v^2] = \frac{\pi S_o}{mc_2 L_x L_y} \sum_{j,p}^{\omega_{jp} < \omega_c} \psi_{jp}^2(x,y)\, \psi_{jp}^2(a,b). \tag{47}$$

APPROXIMATIONS FOR THE MODAL SUM

The indices j and p to be included in the summation (47) must satisfy

$$\omega_{jp} = \left[\left[\frac{j\pi}{L_x}\right]^2 + \left[\frac{p\pi}{L_y}\right]^2 \right] \sqrt{\frac{D}{\rho t}} < \omega_c \tag{48}$$

This is most conveniently visualized in terms of the wave number variables

$$k_x = \frac{j\pi}{L_x} \ , \ k_y = \frac{\rho\pi}{L_y} \ , \ k_c{}^2 = \omega_c \ \sqrt{\frac{\rho t}{D}} \tag{49}$$

for which the requirement (48) becomes

$$k_x{}^2 + k_y{}^2 < k_c{}^2 \tag{50}$$

Figure 5 (a) is a k-space display of the plate modes. Each point represents a mode; e.g., the jp mode is indicated by the point where $k_x = j \pi/L_x$ and $k_y = p \pi/L_y$. The modes which satisfy Eq. (50) lie within the quarter circle of radius k_c.

We have studied two approximate procedures for evaluating Eq. (47). In the first we replace the quarter circle region in Fig. 5(a) by the square region of equal area in Fig. 5(b). This means that we alter the composition of the population of modes included in the summation. While maintaining approximately the same total number of modes we take in additional modes in the neighborhood of $k_x = k_y$ at the expense of leaving out modes in the neighborhood of $k_x = k_c$ and $k_y = k_c$. The summation over the square region separates into two independent summations having the same form as that for the string. Using the notation (26) we find

$$E\,[v^2] = \frac{\pi S_o}{mc_2\,L_x L_y} \ G\,(x/L_x, a/L_x, M) \ \bullet \ G\,(y/L_y, b/L_y, N) \tag{51}$$

where the summation limits M and N are the greatest integers satisfying

$$\left(\frac{M\pi}{L_x}\right)^2 < \frac{\pi}{4} \ k_c{}^2 , \ \left(\frac{N\pi}{L_y}\right)^2 < \frac{\pi}{4} \ k_c{}^2 \ . \tag{52}$$

The distribution of mean square velocity predicted by (51) is readily visualized as the product of two independent orthogonal distributions, each having the form of Fig. 4. Throughout most of the plate the mean square velocity fluctuates in the neighborhood of

$$E[v^2] = \frac{\pi S_o}{mc_2\,L_x L_y} \ MN = \frac{\pi S_o}{mc_2\,L_x L_y} \ \frac{\omega_c}{\overline{\Delta\omega}} \tag{53}$$

where $\omega_c/\Delta\omega$ is approximately the total number of modes with natural frequencies below the cutoff frequency ω_c. Standing out from this background of substantially uniform energy density are four strips of higher energy density which form a tic-tac-toe pattern with intersections at (a,b) and its image points $(L_x$-a,b), and $(a,L_y$-b) and L_x-a,L_y-b). In each strip the energy density increases to about $3/2$ the average level and at the intersections the energy peaks to about $9/4$ the average level.

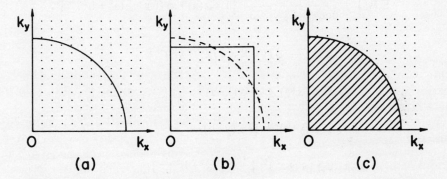

Fig. 5. Array of plate modes in wave number space: (a) resonant modes lie within quarter circle of radius k_c; (b) in separable approximation, sum is taken over modes in square of equal area; (c) in continuous approximation, sum is replaced by integral over quarter circle.

In the second approximation we replace the sum over the isolated points in the semicircle of Fig. 5(a) by a continuous integration over the same semicircle as indicated in Fig. 5 (c); i.e., we approximate the sum in Eq. (47) by the integral

$$\int_0^{k_c} \frac{L_x}{\pi} \, dk_x \int_0^{\sqrt{k_c{}^2 - k_x{}^2}} \psi_{jp}{}^2 \,(x,y)\, \psi_{jp}{}^2 \,(a,b)\, \frac{L_y}{\pi} \, dk_y$$

$$= \frac{L_x L_y}{2} \int_0^{k_c} k\,dk \int_0^{\frac{\pi}{2}} 16 \sin^2\,(kx\cos\theta)\, \sin^2\,(ky\sin\theta)\, \sin^2\,(ka\cos\theta)\, \sin^2\,(kb\sin\theta)\, d\theta \quad (54)$$

In passing to this approximation we lose the symmetry inherent in Eq. (47) with respect to the center of the place. To provide a good approximation we must restrict Eq. (54) to the quadrant where $0 < x < L_x/2$ and $0 < y < L_y/2$ (it is also assumed that the axes have been established so that $0 < x < L_x/2$ and $0 < b < L_y/2$). Since Eq. (54) is even with respect to x and y we can reconstruct the solution for the entire plate by applying the periodicity requirements of Eq. (47) to the approximate solution obtained from Eq. (54) in the region $-L_x/2 < x < L_x/2$, $-L_y/2 < y < L_y/2$. The integration in Eq. (54) can be carried out in closed form [1]. The result for the approximate mean square velocity is

$$E[v^2] = \frac{\pi S_0}{mc_2 L_x L_y} \frac{\omega_c}{\Delta \omega} \Big\{ 1 + \phi(a) + \phi(b) + [(a^2 + b^2)^{1/2}]$$

$$- \phi(x) - \phi(y) + \phi[(x^2 + y^2)^{1/2}]$$

$$+ \tfrac{1}{2}\phi(x+a) + \tfrac{1}{2}\phi(x-a)$$

$$+ \tfrac{1}{2}\phi(y+b) + \tfrac{1}{2}\phi(y-b)$$

$$+ \phi[(y^2 + a^2)^{1/2}] + \phi[(x^2 + b^2)^{1/2}]$$

$$- \tfrac{1}{2}\phi[(y^2 + (x+a)^2)^{1/2}] - \tfrac{1}{2}\phi[(y^2 + (x-a)^2)^{1/2}]$$

$$- \tfrac{1}{2}\phi[(x^2 + (y+b)^2)^{1/2}] - \tfrac{1}{2}\phi[(x^2 + (y-b)^2)^{1/2}]$$

$$- \tfrac{1}{2}\phi[(b^2 + (x+a)^2)^{1/2}] - \tfrac{1}{2}\phi[(b^2 + (x-a)^2)^{1/2}]$$

$$- \tfrac{1}{2}\phi[(a^2 + (y+b)^2)^{1/2}] - \tfrac{1}{2}\phi[(a^2 + (y-b)^2)^{1/2}]$$

$$+ \tfrac{1}{4}\phi[((x+a)^2 + (y+b)^2)^{1/2}] + \tfrac{1}{4}\phi[((x+a)^2 + (y-b)^2)^{1/2}]$$

$$+ \tfrac{1}{4}\phi[((x-a)^2 + (y+b)^2)^{1/2}] + \tfrac{1}{4}\phi[((x-a)^2 + (y-b)^2)^{1/2}]$$

where

$$\phi(\xi) = \frac{J_1(2k_c\xi)}{k_c\xi} \tag{56}$$

is an even function of ξ which takes its maximum value of unity at $\xi = 0$ and which oscillates about zero with rapidly decaying amplitude as ξ increases. A little study of Eq. (55) reveals [1] that the distribution of mean square velocity predicted its qualitatively similar to that predicted by Eq. (51). The average background energy densities are about the same, the increase in energy density in the four strips making up the tic-tac-toe pattern is about fifty percent and the peak energy density at each intersection is about 9/4 the background level.

COMPARISON OF THEORY AND EXPERIMENTS.

To test the predictions of the approximations just described we examined the particular case where the loading point was at $a = L_x/2$ and $b = L_y/3$. The distribution of local mean square velocity was evaluated from the separable sum approximation (51) and from the continuous integral approximation (55). These were compared with a computer evaluation of the exact summation (47) and with measured values on the plate described in the introduction. The salt-grain pattern obtained for this case is shown in Fig. 6. Here the strips at $x = a$ and $x = L_x - a$ have coalesced into a single strip at $x = L_x/2$ where the energy density is about 100% greater than the average background level. In order to compare the analytical predictions with the experimental measurements the cutoff frequency was taken to be 6000 Hz which corresponds to an approximate total number of modes excited $\omega_c/\Delta\omega = 920$. The average modal overlap factor (45) varied between 0.041 at 60 Hz and 0.28 at 6000 Hz for the experimental plate but was taken to be constant for the analytical model.

Fig. 6. Chladni pattern for wide band excitation applied at the point $x = L_x/2$, $y = L_y/3$.

The predicted distributions of $E[v^2(x,y,t)]$ as a function of y for x fixed at $x = L_x/4$ are shown in Fig. 7. Note that both approximations agree quite well with the exact sum in the neighborhoods of y = 0 and y = b. In the background region the separable sum approximation (51) is noticeably inferior to the continuous integral approximation (55).

A traverse was made across the y dimension of the experimental plate (L_y = 38") with $x = L_x/4$ (L_x = 30"). At one-eighth inch intervals the mean square velocity was measured (a non-contact optical sensor produced a signal proportional to place displacement which was differentiated, squared and averaged [1]). The resulting distribution shown in Fig. 8 agrees quite well with the analytical predictions in spite of the over-simplifications in the analytical model with respect to the excitation spectrum, the actual damping mechanism in the plate [1] and the actual boundary conditions of the plate.

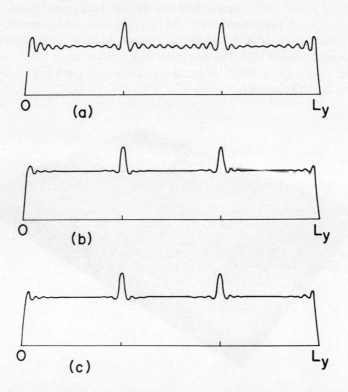

Fig. 7. Predicted distribution of mean square velocity as a function of y for $x = L_x/4$ by (a) separable sum approximation (51), (b) continuous integral approximation (55) and (c) computer evaluation of exact sum (47).

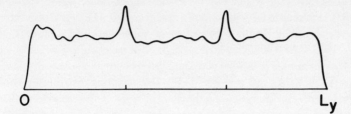

Fig. 8. Experimentally measured distribution of mean square velocity as a function of y for $x = L_x/4$.

ACKNOWLEDGMENT

Support for this research was provided by the National Aeronautics and Space Administration under Contract NAS 8-25317.

REFERENCES

1. Larry E. Wittig, Random Vibration of Point-Driven Strings and Plates, Ph.D. thesis, Dept. of Mechanical Engineering, Massachusetts Institute of Technology, May 7, 1972.
2. I. S. Gradshteyn and I. M. Ryzhik, *Tables of Integrals, Series and Products*, Academic Press, N. Y., 1965, p. 30.
3. See p. 366 of Reference 2.
4. P. W. Smith, Jr. and R. H. Lyon, *Sound and structural vibration*, National Aeronautics and Space Administration Contractor Report CR-160, March, 1965.

ON THE APPLICATION OF THE BOUNDING THEOREMS
OF PLASTICITY TO IMPULSIVELY LOADED STRUCTURES*

J. B. MARTIN

Brown University

*The work reported in this paper was supported by the Office of Naval Research under Task No. NR 064-424.

Abstract—A clearer understanding of the work and complementary work bounding theorems of time independent plasticity has recently been achieved, at least for structures in which the strains remain small. The application of the bounding theorems is now seen to be equivalent to the application of a deformation theory of plasticity which is based on extremum work and complementary work paths in strain and stress space respectively, and in consequence the deformation theory bears a consistent relation to the incremental theory. The basic concepts involved will be reviewed briefly.

The problem of impulsive loading in structural plasticity is one in which the bounding theorems can be readily applied. Displacement bounds can be obtained by applying the deformation theory based on extremum paths rather than the incremental theory. Because the use of deformation theory is equivalent to the use of a non-linear elastic theory, problems in which the displacements are large can be incorporated into the procedure without difficulty. This permits the application of bounding theorems to be extended to the technologically important class of dynamic problems in which the strains are small but the displacements must be taken into account in writing the dynamic equations. Examples are presented.

INTRODUCTION

Over the past seven or eight years the writer has had a continuing interest in the development of methods for bounding displacements which occur in elastic and elastic-plastic structures subjected to impulsive loading. Steady progress has been made in understanding the essential ideas which contribute to these methods, and it now seems that we can place in the hands of the designer a technique which at the cost of a computation of the order of difficulty of a single static problem, will permit him to make conservative estimates of the displacement which a structure will undergo when it is subjected to dynamic loads of a wide range of intensity. We can potentially apply this technique to beam, arch, frame, plate or shell structures, of any configuration, which are subjected to impulsive loading or short duration high intensity pulses which are conservative in nature, and take into account moderately large displacement changes even when the structure exhibits instability when subjected to static loads. Experience to date would indicate that the bound will typically overestimate the actual displacement at a critical point in a realistic structure by a factor of about 0.5 to 1.5, although its accuracy is sensitive to the nature of the loading, the configuration of the structure and the choice of the point at which a displacement bound is required. While from an analytical point of view a bound of this accuracy cannot be considered to be of great value, it could be of substantial assistance to the designer, particularly in the early stages of a design process when the cost of a detailed numerical analysis of dynamic behavior may not be justified, or in situations where dynamic response is an important but not dominant factor in the design.

Previous applications of the bounding technique have been limited to fairly simple illustrative examples, and it seems clear that future efforts must be directed towards the study of realistic structural problems. This would seem to be a suitable opportunity, therefore, to give a simple exposition of the major concepts involved in the bounding procedure as they are currently understood, along with a review of the literature which has led to our present position. Some new developments will be included, involving in particular the application of the bounding method to structures involving instabilities of certain types. A simple example which involves snap buckling will be presented to illustrate this latter case.

Throughout this paper we shall be concerned with problems in which the strains are infinitesimal (or small in the engineering sense). Thus while an element of a structure may undergo moderately large displacements, and the relation between strain and displacement may be non-linear, it is assumed that conventional small strain constitutive relations may be applied to provide the relations between strain or generalized strain and the conjugate stress or generalized stress in the current configuration. It is our intention to present the basic ideas involved in the bounding method in simple terms, and in consequence the arguments required to establish the results rigorously will not be emphasized. While this approach has the advantage of clarity, it does mean that the limitations of the present approach in specific situations are not fully discussed. It follows then that care must be taken in applying the concepts which are described in situations where displacements and strains may be large.

PREVIOUS STUDIES

The initial work in the bounding method [1] involved rigid, perfectly plastic structures subjected to impulsive loading in situations where the small displacement assumptions are applicable. The impulsive loading problem has been particularly easy to deal with since it is a case where the energy stored elastically and dissipated in plastic work in the structure can easily be bounded from above. The bounding method was based on the conventional restrictions which are placed on plastic constitutive relations. It proved comparatively simple to extend the method to elastic and elastic, perfectly plastic structures [2], [3].

The method was next extended [4] to elastic-plastic materials (i.e., including hardening), again for small displacements, by introducing the concept of a path of maximum complementary work in stress or generalized stress space. It was realized at this time that the bounding method was in fact an application of a more general *complementary work bounding theorem* [5], [6] which applied to a wide class of materials and was a generalization of a theorem given for elastic, perfectly plastic materials by Hodge [7].

While it seemed probable that the bounding method could be extended by relaxing the small displacement assumptions (at least to some degree), the complementary work bounding theorem was inappropriate for this generalization. Wierzbicki [8] provided an extension of the method for rigid, perfectly plastic problems, but his approach is rather specialized and not readily capable of extension. On the other hand, it was shown the bounding method could be applied to impulsive loading problems in which displacements are large [9] by using the potential energy theorem, but again this approach could not be generalized without a better understanding of some fundamental concepts in plasticity.

A significant advance was provided by Ponter [10] who showed that the maximum complementary path in stress space implied a minimum work path in strain or generalized strain space, and that the complementary work and work integrals along these paths are potential functions. These results have been extended and clarified by Martin and Ponter [11] by casting the bounding theorems of plasticity into the form of a relation between an incremental theory of plasticity and a deformation theory with particular properties. These results are summarized in Section 3.

These concepts permit us to formulate a potential work bounding theorem whereby the displacement at any point in a structure can be bounded in terms of the energy stored elastically and dissipated in plastic work. This was carried out by Martin and Ponter [12], but limited unnecessarily to structures which are stable under quasistatic loading. It now seems apparent that this restriction is not necessary, and the results are summarized in Section 4.

The final step in establishing the bounding method for dynamically loaded structures requires that we should be able to bound from above the work stored elastically and dissipated in plastic work in the structure at any instant. This is trivial in the case of impulsive loading, and with the exception of [8] the studies above were limited to impulsive loading. However, in independent studies on bounds for elastic structures, Plaut [13] has given a method of bounding the work done by conservative external loads, and hence of bounding the work stored and dissipated in the structure. This process, which is essential to complete the application of the bounding theorem to the dynamic loading problem, is summarized in Section 5.

In Section 6 a brief summary of results available for determining minimum work paths is given, and in Section 7 it is shown how the imposition of the small displacement assumption leads to the earlier results. Finally in Section 8 we give a simple one degree of freedom example intended to elucidate the general ideas.

Some progress has also been made in other aspects of the determination of bounds for dynamically loaded structures. Robinson [14] has considered more general dynamic loads, and Plaut [15] and Infante and Plaut [16] have considered a variety of problems involving dynamic instability in elastic structures. Morales and Neville [17] have provided lower bounds on displacement for rigid, perfectly plastic structures. The application of the bounding method to rigid, viscoplastic materials has been studied [18] and extended by Wierzbicki [19]. Further work in the area of time-dependent materials would appear to be necessary, however; in particular, the analogous potential work bounding theorem given by Ponter [20] has not yet been exploited.

MINIMUM WORK PATHS

Since we are concerned primarily with one- and two-dimensional structures, let us characterize the generalized strains by q_j and the generalized stresses by Q_j. Having in mind conventional elastic-plastic relations, we assume the existence of a time independent, path dependent relation between Q_j and q_j; thus the generalized stress Q_j may be a function of the current value of generalized strain q_j and the strain path, but does not depend on the strain rate. It is also assumed that the constitutive relation satisfies Drucker's postulate in its strongest form. Thus if the strain is changed from q_j^a to q_j^b along any path in strain space (Fig. 1), we require that

$$\int_{q_j^a}^{q_j^b} (Q_j - Q_j^a)\, dq_j \geqslant 0 \tag{1}$$

where Q_j^a is the stress associated with q_j^a at the beginning of the program.

Now consider a strain path which starts at the virgin state $q_j = 0$ and terminates at a given strain q_j. The specific work done along this path (Fig. 2),

$$W(q_j) = \int_0^{q_j} Q_j'\, dq_j' \tag{2}$$

will of course depend on the path, as will the final stress associated with q_j. Of all the possible paths between the origin and q_j, let us choose a path which makes $W(q_j)$ a minimum. We do not require that the minimum work path should be unique; indeed it is generally not. However, the least specific work required to produce a strain q_j in the material, which we shall denote by $W^o(q_j)$, will be uniquely defined. By definition, therefore

$$W(q_j) \geqslant W^o(q_j) \tag{3}$$

where $W(q_j)$ is computed along any path.

We can define the minimum work required to produce any strain, and consequently the function $W^o(q_j)$ can be defined for each point in the strain space. If Q_j is taken to be the stress associated with q_j when the deformation occurs along a minimum work path, we can likewise define a stress Q_j for each point in the strain space;

$$Q_j = Q_j(q_k) \tag{4}$$

This relation between stress and total strain can be regarded as a deformation theory of plasticity; however it is a deformation theory with a particularly important property. The fact that $W^o(q_j)$ is defined as the minimum specific work to produce a strain q_j is a sufficient condition to insure

$$Q_j = \frac{\partial W^o}{\partial q_j}(q_j) \tag{5}$$

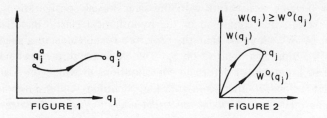

FIGURE 1   FIGURE 2

Further, if the constitutive relation satisfies inequality (1) it can be shown that W^o (q_j) is a convex function. Thus, if q_j^a, q_j^b are any two states of strain,

$$W^o(q_j^b) - W^o(q_j^a) \geqslant (q_j^b - q_j^a) \left.\frac{\partial W^o}{\partial q_j}\right|_{q_j^a} = (q_j^b - q_j^a) Q_j^a \tag{6}$$

Equations (5) and (6) imply that we can treat equation (4) as the constitutive relation for a non-linear elastic material, with $W^o(q_j)$ being the strain energy of this material. The only relation between this hypothetical elastic material and the original inelastic material is contained in inequality (3); the work done in producing a strain q_j in the inelastic material is not less than $W^o(q_j)$, the strain energy of the hypothetical elastic material when the same strain is imposed.

The arguments which lead to these conclusions have been given by Ponter [10] and Martin and Ponter [11], and will not be restated. The principal difficulties in exploiting the hypothetical elastic material lie in the computation of the function $W^o(q_j)$; this requires that we determine the minimum work path for the inelastic material, and this in turn depends on the exact form of the constitutive relations. In order that the continuity of our main argument should not be broken, we will defer a discussion of this point until Section 6.

THE DISPLACEMENT BOUNDING PRINCIPLE

Let us now consider a structure of a given configuration. Suppose that the spatial variables which identify a point on the structure are denoted by s, and that the entire area or length of the structure is represented by S. Further, suppose that on part of S, denoted by S_u, the displacements are specified to be zero. The remainder of S we shall denote by S_P.

Let the structure be composed of an inelastic material satisfying the constitutive relations postulated in Section 3. As the result of dynamic loading on S_P, let the displacement field in the structure at time t be $u(s,t)$, and let the associated strains, derived by whatever equations are appropriate, be $q_j(s,t)$. No restrictions are imposed on the magnitude of the displacements, although the displacement field is assumed to satisfy the appropriate continuity conditions. The specific work done in deforming an element of the structure is taken to be $W(q_j)$, as before. $W(q_j)$ depends on the path of loading and consequently is not a function of q_j alone.

Let us now imagine a structure with the same initial configuration and with zero displacements on S_u, but composed of the hypothetical elastic material whose strain energy is given by W^o. Suppose that the structure is subjected to a single conservative point load $P(s^*)$ acting at a point s^* on S_p. The loads are zero elsewhere on S_p. There may in general be a number of deformed configurations in which the load $P(s^*)$ may be equilibrated; let us select that configuration, denoted by displacements $\hat{u}(s)$ and associated strains $\hat{q}_j(s)$, *for which the potential energy of the structure is an absolute minimum.*

It follows then that, since $u(s,t)$, $q_j(s,t)$ represents a possible configuration of the structure,

$$\int_S W^o (q_j)\, ds - P(s^*) \cdot u(s^*,t) \geqslant \int_S W^o (\hat{q}_j)\, ds - P(s^*) \cdot \hat{u}(s^*) \tag{7}$$

However, by definition,

$$\int_S W(q_j)\, ds \geqslant \int_S W^o(q_j)\, ds \tag{8}$$

Let us also denote the magnitude of P by P, and the components of $u(s^*,t)$ and $\hat{u}(s^*)$ in the direction of P by $w(t)$, $\hat{w}$ respectively. Then, using (8) and rearranging (7)

$$\int_S W(q_j)\, ds - \int_s W^o(\hat{q}_j)\, ds \geqslant P\,[\,w(t) - \hat{w}\,] \tag{9}$$

If we can determine a quantity E(t) such that

$$E(t) \geqslant \int_S W(q_j)\, ds \tag{10}$$

inequality (9) can be written as

$$E(t) - \int_S W^o (\hat{q}_j)\, ds \geqslant P\,[\,w(t) - \hat{w}\,] \tag{11}$$

Thus, by choosing P such that

$$E(t) \leqslant \int_S W^o (\hat{q}_j)\, ds \tag{12}$$

We define a value of $\hat{w}$ such that

$$\hat{w} \geqslant w(t) \tag{13}$$

If possible we would of course choose P such that there is an equality sign in (12).

$w(t)$ is the component of the displacement at s^* in the direction of P at time t resulting from the dynamic load on the inelastic structure; it is bounded from above by ŵ, which is computed from a static analysis of the structure assuming that it is composed of the hypothetical elastic material whose constitutive equation is given by equation (5).

The process by which the appropriate value of P is chosen can easily be demonstrated in diagrammatic form. Suppose that we carry out an analysis of the structure composed of the hypothetical elastic material for a range of values of P, determining in each case the configuration of lowest potential energy. The results may be plotted as a relation between P and ŵ, as shown in Fig. 3, and $\int_S W^o(\hat{q}_j)ds$ and ŵ, as shown in Fig. 4. Supposing that these relations are continuous, we select the appropriate upper bound on w(t) for a given E(t) by entering Fig. 4. If the relations are not continuous, as may happen when there are values of P for which more than one equilibrium configuration exists (Figs. 5 and 6), a value of $\int_S W^o(\hat{q}_j)ds$ which is larger E(t) must be chosen. This is equivalent to completing the full line in Fig. 6 by means of the dotted line: we then again select the appropriate upper bound on w(t) by entering E(t) on Fig. 6.

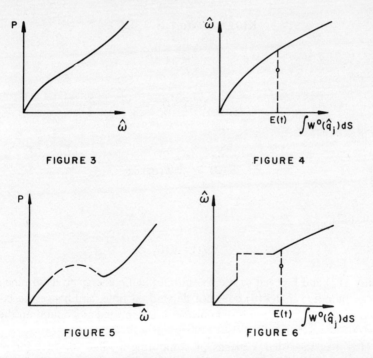

FIGURE 3   FIGURE 4

FIGURE 5   FIGURE 6

DYNAMIC PROBLEMS FOR WHICH E(t) CAN BE COMPUTED

Returning now to the dynamic problem in the inelastic structure, it is necessary that we should be able to bound

$$\int_S W(q_j)\, ds$$

from above, in order that the bounding principle can be applied.

The simplest case in which this can be done is the impulsive loading problem, where the loads are zero on S_P, the displacements are zero on S_u, the displacements on S_P are zero at time $t = 0$ and the initial velocity field is $\dot{u}(s, 0)$. Defining

$$K(0) = \int_s \frac{1}{2} m\dot{u}(s, 0) \cdot \dot{u}(s, 0)\, ds$$

$$K(t) = \int_s \frac{1}{2} m\dot{u}(s, t) \cdot \dot{u}(s, t)\, ds \tag{14}$$

where m is the specific mass, it is evident that since no work is done on the structure by the external loads or reactions on S_u,

$$K(0) = \int_s W(q_j)\, ds + K(t) \tag{15}$$

Further, since

$$K(t) \geqslant 0, \tag{16}$$

it follows that

$$K(0) \geqslant \int_s W(q_j)\, ds \tag{17}$$

We can thus put

$$E(t) = K(0) \tag{18}$$

in inequality (12) and proceed to use the initial kinetic energy to determine a bound on w(t). It may be noted that K(0) does not depend on time, and hence the bound $\hat{w}$ will not depend on time. The bound is then properly interpreted as a bound on the maximum value of w(t).

Plaut [13] has suggested a means of computing a value for E(t) for pulse loaded structures. Suppose that the inelastic structure is subjected to conservative loads $P(s)\, \alpha(t)$ on S_P with $\alpha(t) \geqslant 0$, displacements zero on S_u initial displacements $u(s, 0) = 0$ on S_p and initial velocities $\dot{u}(s, 0)$ on S_p. In this case an energy balance gives

$$\frac{d}{dt}\left[\int_s K(t) + \int_s W(q_j)\, ds\right] = \int \alpha(t)\, p(s) \cdot \dot{u}(s,t)\, ds \tag{19}$$

The right hand side may be written as

$$\alpha (t) \int_s p(s) \cdot \dot{u} (s, t) \, ds \leqslant \alpha (t) \int_s |\frac{p \cdot p}{m}|^{\frac{1}{2}} \cdot |m \dot{u} \cdot \dot{u} |^{\frac{1}{2}} \, ds \qquad (20)$$

By an application of the Schwartz inequality

$$\alpha (t) \int_s |\frac{p \cdot p}{m}|^{\frac{1}{2}} \, |m \dot{u} \cdot \dot{u} |^{\frac{1}{2}} \, ds \leqslant \alpha (t) \left\{ \int_s \frac{p \cdot p}{m} \, ds \right\}^{\frac{1}{2}} \left\{ \int_s m \dot{u} \cdot \dot{u} \, ds \right\}^{\frac{1}{2}}$$

$$= \sqrt{2} \, \alpha (t) \left\{ K (t) \right\}^{\frac{1}{2}} \left\{ \int_s \frac{p \cdot p}{m} \, ds \right\}^{\frac{1}{2}} \qquad (21)$$

$$\leqslant \sqrt{2} \, \alpha (t) \left\{ \int_s \frac{p \cdot p}{m} \, ds \right\}^{\frac{1}{2}} \left\{ K (t) + \int_s W (q_j) \, ds \right\}^{\frac{1}{2}}$$

In these expressions all square roots are taken to be positive. A combination of (19), (20) and (21) gives

$$\frac{d}{dt} \left\{ K (t) + \int_s W (q_j) \, ds \right\} \leqslant \sqrt{2} \, \alpha (t) \left\{ \int_s \frac{p \cdot p}{m} \, ds \right\}^{\frac{1}{2}} \left\{ K (t) \right.$$

$$\left. + \int_s W (q_j) \, ds \right\}^{\frac{1}{2}} \qquad (22)$$

Since

$$K(t) + \int_s W(q_j) \, ds \geqslant 0, \qquad (23)$$

inequality (22) may be integrated to give

$$\int_s W (q_j) \, ds + K (t) \leqslant [\left\{ K (0) \right\}^{\frac{1}{2}} + \frac{1}{\sqrt{2}} \int_o^t \alpha (\tau) \, d\tau \left\{ \int_s \frac{p \cdot p}{m} \, ds \right\}^{\frac{1}{2}} \qquad (24)$$

It may be noted that inequality (20) will be a reasonable approximation only when $p(s) \cdot \dot{u}(s,t)$ is greater than zero. This suggests that inequality (14) can be used with confidence only when the pulse is of short duration, with

$$\alpha(t) = 0 \text{ for } t > T \tag{25}$$

and

$$p(s) \cdot \dot{u}(s,t) > 0 \text{ for } t < T \tag{26}$$

If we then limit our attention to cases in which the maximum displacements of interest occurs for $t > T$, we can by virtue of the observation that $K(t) > 0$ put

$$E(t) = [\{ K(0) \}^{\frac{1}{2}} + \frac{1}{\sqrt{2}} \int_o^T \alpha(\tau) d\tau \{ \int_s \frac{p \cdot p}{m} ds \}^{\frac{1}{2}}] \tag{27}$$

E is again independent of time, and the bounding procedure will give a bound on the maximum value of $w(t)$ for $t > T$.

THE DETERMINATION OF MINIMUM WORK PATHS

The application of the bounding method described above requires that the minimum work path in strain or generalized strain space should be known for the particular material under consideration. At the present time, minimum work paths are known for some particular models of time independent small strain constitutive relations, but a general prescription for wider classes of constitutive relations is not known.

We divide the strain q_j into elastic and plastic components, q_j^e and q_j^P respectively. The work integral W^o along the minimum work path is then written as

$$W^o(q_j) = \int Q_j dq_j^e + \int Q_j dq_j^P = W^e(q_j^e) + W^{oP}(q_j^P) \tag{28}$$

The elastic strain energy W^e is path independent, and is a function of q_j^e alone. W^{oP} is the least plastic work required to impose a plastic strain q_j^P on the material. The problem of determining $W^o(q_j)$ can then be broken into two parts: first we find $W^{oP}(q_j^P)$ for any terminal plastic strain, and then we divide any given total strain q_j into elastic and plastic parts such that $W^o(q_j)$ has its least value. It can be shown (Martin and Ponter [11]) that this second condition is satisfied when

$$\frac{\partial W^e}{\partial q_j^e} = \frac{\partial W^{oP}}{\partial q_j^P} \tag{29}$$

The principal problem is therefore the determination of the plastic strain path which minimizes W^{op}. By applying all our previous arguments to a rigid-plastic material, it can be argued that W^{op} is a convex potential function, and that $\partial W^{op}/\partial q_j^p$ gives the stress or generalized stress at the end of the minimum work path in plastic strain space. The terminal stress, however, need not be unique. Indeed, $\partial W^{op}/\partial q_j^p$ at the origin ($q_j^p = 0$) is discontinuous and contains all stress states which lie within the initial yield surface.

The path in plastic strain space which gives maximum plastic work is known for a variety of specific models. These are: uniaxial relations (i.e., cases where only one component of q_j^p is non-zero), perfect plasticity, isotropic hardening (where the yield) surface expands without change in shape), kinematic hardening, where the yield function $\phi = \phi(Q_j - cq_j^p)$, and Koiter or slip theory type of models where there exist a number of independent linear yield functions each contributing to the total plastic strain. In each of these cases the plastic strain path for minimum plastic work is a radial path in plastic strain space. The associated stress path may not be unique, but in each case can be taken to be a straight line path in stress space starting at a point on the hield surface. In the case of perfectly plastic materials the stress remains constant while plastic strain takes place. These known paths are comparatively simple, but it must be added that we have at present no reason to expect that the radial path in plastic strain space applies in more complex materials.

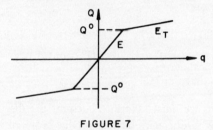

FIGURE 7

As a simple example consider a uniaxial relation between generalized stress Q and total strain q which is bilinear for monotonic loading (Figure 7). This bilinear relation is the relation between stress and strain along the minimum work path (which can be confirmed by inspection), and is treated as the stress-strain relation for the deformation theory.

$$Q = Eq \text{ for } -\frac{Q_o}{E} \leqslant q \leqslant \frac{Q_o}{E}$$

$$Q = E_T \left(q - \frac{Q_o}{E}\right) + Q_o \text{ for } q \geqslant \frac{Q_o}{E} \text{ or } q \leqslant -\frac{Q_o}{E} \tag{30}$$

Putting

$$\frac{1}{E_T} = \frac{1}{E} + \frac{1}{E_p} \tag{31}$$

it can be seen that

$$W^e = \frac{E}{2} (q^e)^2 \tag{32}$$

while for monotonically increasing q^p

$$W^{op} = \quad (q^p)^2 + Q^o \mid q^p \mid \tag{33}$$

For any given q, we can generate equation (30) from equation (31) and (33) by solving simultaneously for q^e and q^p the equations

$$q = q^e + q^p \tag{34}$$

and

$$\frac{dW^e}{dq^e} = \frac{dW^{op}}{dq^p}$$

or

$$E q^e = E_p q^p + Q^o \text{ sgm } (q^p) \tag{35}$$

In solving these equations we must take into account the discontinuity in the derivative of W^{op} when $q^p = 0$; dW^{op}/dq^p may have any value between $+Q^o$ and $-Q^o$ when $q^p = 0$.

Further examples of relations of this type have been given by Martin and Ponter [11], [12].

CONSEQUENCES OF THE SMALL DISPLACEMENT ASSUMPTIONS

Let us consider briefly alternative forms of the bounding method when the displacements of the structure under the dynamic loading are sufficiently small that the dynamic and kinematic relations are linear and the principle of virtual work is applicable.

First we note that the strain point following a minimumwork path in strain space for the incremental relation traces out a stress path in stress space. If the strain path is a minimum work path the stress path is a maximum complementary work path, defined in the sense that

$$\Omega = \int_0^{Q_j} q_j' \, dQ_j' \tag{36}$$

has its greatest value Ω^o compared to all stress paths between the origin and the given terminal stress Q_j. Further, the terminal strain q_j for the maximum complementary work path is

$$q_j = \frac{\partial \Omega^o}{\partial Q_j} \tag{37}$$

We can interpret Ω^o as the complementary energy of the hypothetical elastic material whose strain energy is W^o; equation (37) is the inverse of the constitutive relation given by equation (5). Ω^o is also convex, and for an associated stress and strain Q_j, q_j we have

$$W^o(q_j) + \Omega^o(Q_j) = Q_j q_j \tag{38}$$

Now consider inequality (11). Defining $\hat{Q}_j$ as the stress field associated with $\hat{q}_j$, the strain field in the hypothetical elastic structure subjected to the point load P at point s*, we have

$$P(s^*) \cdot \hat{u}(s^*) = P\hat{w} = \int_S \hat{Q}_j \hat{q}_j \, ds \tag{39}$$

if the principle of virtual work is applicable. Using equation (38) in an integrated form

$$P\hat{w} = \int_S \hat{Q}_j \hat{q}_j \, ds = \int_S W^o(\hat{Q}_j) \, ds + \int_S \Omega^o(\hat{Q}_j) \, ds \tag{40}$$

Substituting equation (40) into inequality (11) and rearranging, we obtain

$$P\hat{w}(t) \leqslant \int_S \Omega^o(\hat{Q}_j) \, ds + E(t) \tag{41}$$

Further, let Q_j^* be a stress field which is statically admissible i.e. it satisfies the equilibrium requirements for the hypothetical elastic structure with a point load $P(s^*)$. The classical elastic complementary energy theorem gives

$$\int_S \Omega^o(\hat{Q}_j) \, ds < \int_S \Omega^o(Q_j^*) \, ds \tag{42}$$

Thus inequality (41) may be written as

$$w(t) \leqslant \frac{\int_S \Omega^o(Q_j^*) \, ds}{P} + E(t) \tag{43}$$

The magnitude of P may be chosen so that the right hand side of P has its least value. This is the form of the bounding method presented in reference [4]. Noting that Ω^o is the actual elastic complementary energy for elastic and elastic, perfectly plastic materials and is zero for rigid, perfectly plastic materials, the formulations of references [1], [2], and [3] are obtained.

AN ILLUSTRATIVE EXAMPLE

Consider the structure shown in Fig. 8. Two massless bars support a point mass m. It is assumed that the bars are hinged at each end, so that they are subjected to axial forces only. It is also assumed that the vars remain straight during deformation, implying that the bars do not buckle. The initial configuration is shown in Fig. 8; u_o may be either positive or negative. It is supposed that the mass m is given an initial velocity $\dot{u}_o$ in the downward direction; clearly there is a possibility that the dynamic response will involve a snap through phenomenon.

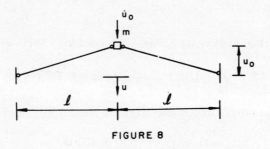

FIGURE 8

The displacement u will be measured from the origin shown in Fig. 8. Let us define

$$\epsilon = \sqrt{\left\{ 1 + \left(\frac{u}{l}\right)^2 \right\}} - 1 \tag{44a}$$

and

$$\epsilon_o = \sqrt{\left\{ 1 + \left(\frac{u_o}{l}\right)^2 \right\}} - 1 \tag{44b}$$

The strain in the bar will then be defined as

$$e = \frac{\epsilon - \epsilon_o}{1 + \epsilon_o} \tag{45}$$

We shall treat an elastic and an elastic, perfectly plastic material. It is assumed that for the elastic case the axial force in the bar N is linearly related to e i.e.

$$N = Ee, \tag{46}$$

while for the elastic perfectly plastic case

$$\dot{e} = \frac{\dot{N}}{E} \text{ for } -Ee_o < N < Ee_o$$

$$\dot{e} = \lambda \text{ for } N = \pm Ee_o \tag{47}$$

where λ is of unspecified magnitude but of the same sign as N (tension taken as positive). Bar forces such that $|N| > Ee_o$ are not admitted. Although the restrictions imposed on this model are severe and not physically justifiable, the model is self-consistent and can adequately illustrate the application of the bounding method.

Let us introduce the dimensionless variables

$$\bar{u} = \frac{u}{l} , \; \bar{u}_o = \frac{u_o}{l} , \; t = \sqrt{\frac{E}{ml}} \; t, \; \bar{N} = \frac{N}{E} \tag{48}$$

The equation of motion for the dynamic problem then takes the form

$$\ddot{\bar{u}} + \frac{2\,\bar{N}\,\bar{u}}{\sqrt{(1 + \bar{u}^2)}} = 0 \tag{49}$$

This equation can be solved comparatively easily for the appropriate initial conditions $\bar{u}(0) = \bar{u}_o, \dot{\bar{u}}(0) = \dot{\bar{u}}_o$.

Consider now the problem of bounding the maximum positive displacement. We apply a dimensionless load $\bar{P}$ (= P/E) to the mass in the downward direction and consider for each value of $\bar{P}$ the configuration of minimum potential energy. In this calculation we replace the elastic, perfectly plastic material by a hypothetical elastic material with a constitutive equation given by

$$e = \bar{N} \text{ for } -e_o \leqslant \bar{N} \leqslant e_o$$

$$e = e_o + \mu \text{ for } \bar{N} = e_o$$

$$e = -e_o - \mu \text{ for } \bar{N} = -e_o \tag{50}$$

where μ is an unspecified non-negative scalar. The strain energy of the hypothetical elastic material is then

$$W^0 = \frac{e^2}{2} \quad \text{for} \ -e_0 \leqslant e \leqslant e_0$$

$$W^0 = \frac{e_0^2}{2} + e_0 + e_0 \{|e| - e_0\} \quad \text{for} \ |e| > e_0 \tag{51}$$

It is evident that when $\bar{u}_0 \geqslant 0$ there are unique equilibrium positions for all positive values of $\bar{P}$, and thus the $\bar{P} \sim \hat{u}$ curve obtained from a direct static analysis provides (within our constraints) the positions of minimum potential energy. For $\bar{u}_0 < 0$, however, snap buckling will occur as $\bar{P}$ is increased, and more than one equilibrium position can be found for small values of $\bar{P}$. It is evident that for $\bar{P}$ vanishingly small the equilibrium position of least potential energy is given by $\hat{u} = -u^0$ i.e. by allowing the structure to snap through before the load is applied. Monotonic increase in $\bar{P}$ will then give other minimum potential energy configurations by a direct static analysis. Figs. 9, 10 and 11 show the $\bar{P} \sim \hat{u}$ relations for the configuration of minimum potential energy for some typical cases. Fig. 9 shows the elastic case, for $\bar{u}_0 = 0$, $+0.2$ and -0.2. The curves for $\bar{u}_0 = \pm 0.2$ coincide, with the dotted line showing the snap through behavior for $\bar{u}_0 = -0.2$ which is neglected. Fig. 10 shows the same information for an elastic, perfectly plastic case in which the snap through occurs with $|N| < e_0$. Fig. 11 shows a case for $\bar{u}_0 = -0.2$ where $|N| = e_0$ during the snap through.

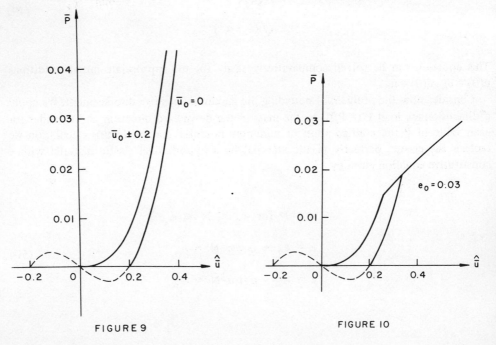

FIGURE 9   FIGURE 10

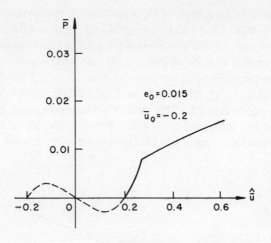

FIGURE II

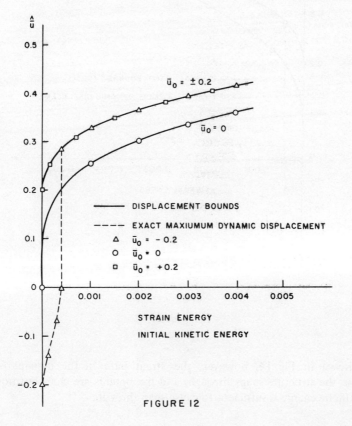

FIGURE 12

The dimensionless strain energy of the hypothetical elastic structure is plotted against $\hat{u}$ for these cases in Figs. 12, 13 and 14 respectively, and is shown by the full lines. The results of dynamic analysis are also plotted: the dimensionless initial kinetic energy is entered on the $\hat{u}$ axis. It may be observed in all cases that when the initial kinetic energy is not sufficient to cause the structure to snap when $\bar{u} = -0.2$ the bound is an extremely conservative estimate. In all other cases in Figs. 12 and 13 the bound and the actual maximum displacement coincide. This occurs primarily because the structure is of one degree of freedom and the actual strain paths between the initial instant and the instant at which the displacement first reaches its maximum value are minimum work paths. In

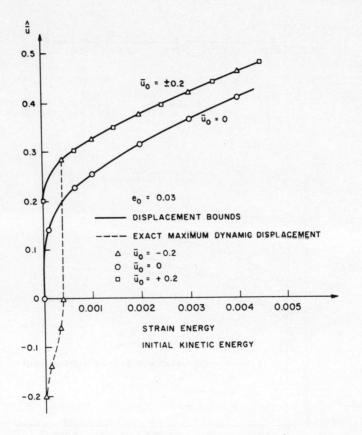

FIGURE 13

the case shown in Fig. 14, however, the strain path in the dynamic case involves unloading as the structure snaps through, and the bounds are close but not exact when the initial kinetic energy is sufficient to cause snap through.

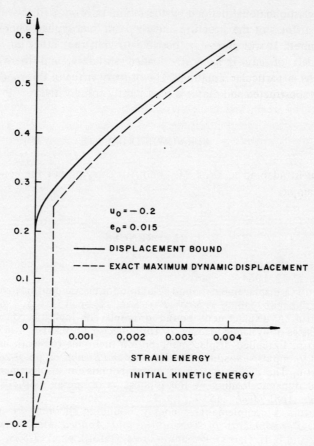

$$\hat{u}$$

$u_0 = -0.2$
$e_0 = 0.015$

——— DISPLACEMENT BOUND

- - - - EXACT MAXIMUM DYNAMIC DISPLACEMENT

STRAIN ENERGY

INITIAL KINETIC ENERGY

FIGURE 14

CONCLUSIONS

This paper offers a review of the major elements which contribute to a method of bounding displacements in impulsively loaded structures. Space does not permit a detailed analysis of each of these elements, but it is hoped that an overall view, despite its superficiality, will provide the reader with a guide to the somewhat meandering development which the method has undergone and with an indication of its potential applications.

While the one degree of freedom illustration is misleading in the sense that it is an example in which the bounds are far more accurate than can be expected for realistic structures, it does underscore the fact that the bounding method does in fact construct a conservative one degree of freedom model of the structure. The model is composed of the hypothetical elastic material defined by the minimum work paths, with a spring stiffness given by the stiffness of the structure under a point load applied where the displacement bound is required. It will always be possible to construct other *ad hoc* one degree of freedom models of any dynamically loaded structure, and there may give better approximations in particular situations. The primary virtue of this model is that it can be automatically constructed and that it is consistently conservative.

ACKNOWLEDGMENTS

The author is indebted to Dr. J. M. Chern for carrying out the computations for the illustrative example.

REFERENCES

1. J. B. Martin, Impulsive loading theorems for rigid-plastic continua, *Proc. ASCE,* **EM 5**, pp. 27-42, 1964.
2. J. B. Martin, a displacement bound principle for elastic continua subjected to certain classes of dynamic loading, *J. Mech. Phy. Sol.,* **12**, pp. 165-175, 1964.
3. J. B. Martin, A displacement bound principle for inelastic continua subjected to certain classes of dynamic loading, *J. Appl. Mech.,* **32**, pp. 1-6, 1965.
4. J. B. Martin, Extended displacement bound theorems for work hardening continua subjected to impulsive loading, *Int. J. Solids and Struct.,* **2**, pp. 9-26, 1966.
5. J. B. Martin, The determination of upper bounds on displacements resulting from static and dynamic loading by the application of energy methods, *Proc. Fifth U.S. Nat. Congr. Appl. Mech., ASME.,* pp. 221-236, 1966.
6. J. B. Martin, A complementary energy bounding theorem for time independent materials, *Developments in Theoretical and Applied Mechanics,* (Edited by D. Frederick and E. H. Harris), Pergamon Press, Oxford, pp. 517-528, 1970.
7. P. G. Hodge, Jr., A deformation bounding theorem for flow law plasticity, *Q. Appl. Math.,* **24**, pp. 171-179, 1966.
8. T. Wierzbicki, Bounds on large dynamic deformations of structures, *Proc. Eng. Mech. Div., ASCE,* **EM 3**, pp. 267-276, 1970.
9. J. B. Martin, Displacement bounds for dynamically loaded elastic structures, *J. Mech. Eng. Sci.,* **10**, pp. 213-218, 1968.
10. A. R. S. Ponter, Convexity conditions and energy theorems for time independent materials, *J. Mech. Phys. Sol.,* **16**, pp. 283-288, 1968.
11. J. B. Martin and A. R. S. Ponter, Some extremal properties and energy theorems for inelastic materials and their relation to the deformation theory of plasticity. To be published in *J. Mech. Phys. Sol.*
12. J. B. Martin and A. R. S. Ponter, Bounds on large deformations of impulsively loaded elastic-plastic structures. Proc. ASCE *(EMI),* pp. 107-120, 1972.
13. R. H. Plaut, On minimizing the response of structures to dynamic loading, *ZAMP.,* **21**, pp. 1004-1010, 1970.
14. D. N. Robinson, A displacement bound principle for elastic-plastic structures subjected to blast loading, *J. Mech. Phys. Sol.* **18**, p. 65, 1970.
15. R. H. Plaut, Displacement bounds for beam-columns with initial curvature subjected to transient loads, *Int. J. Solids and Struct.,* **7**, pp. 1229-1235, 1971.

16. R. H. Plaut and E. F. Infante, Bounds on motions of some lumped and continuous dynamic systems, to appear in *J. Appl. Mech.*
17. W. J. Morales and G. E. Neville, Lower bounds on deformations of dynamically loaded rigid-plastic continua, *J. A.I.A.A.* 8, pp. 2043-2046, 1970.
18. J. B. Martin, Time and displacement bound theorems for viscous and rigid-viscoplastic continua, *Dev. in Theoretical and Applied Mechanics*, Pergamon Press, Oxford, 3, pp. 1-22, 1967.
19. T. Wierzbicki, On the region of admissible deformations in impulsive loading problems, *J. Mech. Phys. Sol.*, 19, pp. 1-10, 1971.
20. A. R. S. Ponter, An energy theorem for time dependent materials, *J. Mech. Phys. Sol.*, 17, pp. 63-71, 1969.

FLUTTER PREVENTION IN DESIGN OF THE SST

M. J. TURNER and J. B. BARTLEY

Commercial Airplane Group, The Boeing Company

Abstract—The general characteristics of flutter problems affecting the structural design of the SST are discussed in relation to configuration constraints resulting from mission performance requirements. Combined analytical and experimental methods that have been used in solving these problems are outlined in detail. Because of structural complications resulting from a long and slender body, thin low aspect ratio wing, and aft location of wing mounted engines a very detailed finite element representation is employed in deriving the analytical model of the complete aircraft structure. Applications of subsonic lifting surface theory and supersonic mach box methods in the treatment of oscillatory airloads are described. Numerical optimization methods to satisfy flutter requirements, based on substructuring of the finite element model, are described and application of these methods in determining efficient distributions of stiffening material and of nonstructural masses are presented. Recent advances in the design and construction of high speed flutter models and utilization of these models, in conjunction with analytical methods, in solution of SST flutter problems are discussed.

INTRODUCTION

Design work on the recently terminated SST Project required detailed analysis of two configurations. The Prototype Point Design (PPD) was a paper airplane designed to meet specified payload/range requirements representative of transatlantic commercial airline operations. The two Prototype (P/T) aircraft, although identical to the PPD in external geometry and structural arrangement, were intended for ground testing, proof of concept flight testing, and demonstration of operational capability. The engineering work to be described in this paper is concerned entirely with wing flutter problems that were encountered in the design of P/T aircraft. This part of the flutter program is believed to be of general interest, since it was found necessary to add a significant amount of weight to the primary structure over that required for strength in order to achieve satisfactory flutter margins. This represents a marked departure from previous experience in design of subsonic transport aircraft. Digital methods of analysis, based on a finite element structural model and lifting surface aerodynamic theory, have been employed as basic design tools to find efficient solutions to the flutter problem and to evaluate relative technical risk of alternate approaches. Wind tunnel tests of scaled dynamic models have been conducted to evaluate the analytical methods.

The results to be presented are obviously derived from a large team effort, involving essential contributions by many individuals. Unfortunately the number is so great that we cannot mention them individually; however we do wish to express our great appreciation for the dedicated efforts of the flutter and loads engineers who were directly associated with the SST Structures Technology Staff and for able assistance provided by members of the following Boeing organizations: Wind Tunnel Model Design Group, Weights Staff,

Computer Services, Flutter Research Group, and Dynamics Laboratory. Valuable consulting services were provided by Dr. Holt Ashley of the Department of Aeronautics and Astronautics, Stanford University, and important contributions as technical monitor in the areas covered in this paper were made by Mr. Robert Rosenbaum of the Department of Transportation. Also we should like to express our appreciation for support provided in flutter model testing by Mr. A. G. Rainey and his associates in the Aeroelasticity Branch of the NASA Langley Research Center and by the staff of the 6' x 6' Supersonic Tunnel at the NASA Ames Research Center.

The development of a large aircraft to cruise efficiently at M = 2.7 with an additional requirement for extensive subsonic operation over populated areas has introduced configuration constraints which tend to make the flutter problem more critical. The configuration of the aircraft is shown in Figure 1. The combination of a relatively thin wing (thickness/chord ≈ .03 from the outboard boundary of strake to wing tip) long, slender body, and wing mounted engines tend to create a multiplicity of complex mode shapes with relatively low natural frequencies. Spanwise separation of individual engines was considered advisable to avoid the possibility of mutual unstart interaction between adjacent engine inlets. Further limitations on spanwise positioning of engines result from

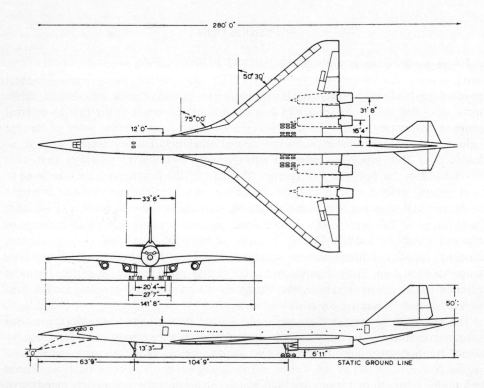

Fig. 1. External Configuration

the requirement to avoid inlet interference from the landing gear in extended position and to minimize hot exhaust impingement on the horizontal tail, Aft location of the engines relative to the wing leading edge was dictated by the requirement to maintain satisfactory engine inlet efficiency throughout the flight envelope. As a result of these constraints it has not been possible to obtain beneficial dynamic coupling effects from engine inertias comparable to gains that have been realized on some subsonic aircraft with pod mounted engines. In fact some degredation of the wing flutter speed is attributed to the outboard engines; efforts to counteract this effect by modification of engine mount flexibility will be discussed in a later section.

Some alleviation of the flutter penalty could have been achieved by a reduction in aspect ratio and/or by an increase in leading edge sweep angle. However, these changes were considered unacceptable because of unfavorable effects on subsonic cruise efficiency and engine noise during takeoff.

THE FLIGHT ENVELOPE

The portion of the flight envelope that is most likely to be affected by intrusion of a critical flutter condition may be identified by reference to Figure 2, showing the variation of design dive speed, V_D, with altitude for the P/T airplane and a computed wing flutter envelope for an early strength-designed version. The V_D envelope for the 747 airplane is also shown for comparative purposes. Clearly the crux of the flutter clearance problem for the subsonic airplane is to demonstrate adequate margin near the corner of the flight envelope, were max q and max M occur simultaneously. For the SST, flutter analyses for high subsonic and low supersonic speeds, must be given equally careful attention and a transonic model program is particularly essential in order to bridge the gap in analytical capability near $M = 1$.

INITIAL PHASE OF THE FLUTTER PROGRAM

Before proceeding to a more comprehensive discussion of later work accomplished during detail design of the Prototype it seems advisable to review earlier stages of the program briefly. The final fixed wing configuration, previously shown in Figure 1, was selected in August, 1968, at the conclusion of an intensive configuration review, including both fixed wing and variable sweep concepts. Flutter calculations were conducted on the basis of a finite element representation of an initial strength-designed structure, and results of that analysis indicated the need for sizeable modifications to achieve satisfactory flutter margins. Development of a flutter optimization process was then initiated, utilizing a modularized representation of the structure to evaluate relative sensitivity of flutter speed to variations of structural stiffness and weight in various local regions.

On the basis of that study it was decided to add material to the main wing box, primarily to increase torsional stiffness between the side of the body and the outboard

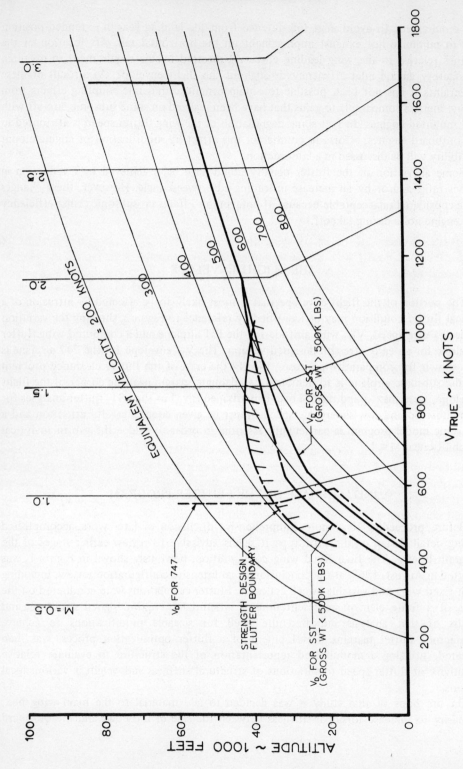

Fig. 2. Flight Envelopes

engine nacelle. Thickness of the wing was increased slightly near the root, and stiffening was added locally in the vicinity of the wheel well. When the configuration was validated in February 1969, it was estimated that approximately 12000 lbs. of added structural weight, beyond strength requirements, were needed to satisfy wing flutter requirements. A full span subsonic flutter model had been tested in December, 1968. Although this model did not incorporate all of the structural changes described above, it did provide increased confidence in the analysis, since measured flutter speeds were in reasonable agreement with flutter calculations for the structural configuration represented by the model.

ANALYTICAL METHODS FOR DETAIL DESIGN STAGE

Structural design efforts on the wing during the initial design stage prior to validation were concerned mainly with the definition of primary structure in the main wing box. As additional detailed information on wing leading edge and trailing edge structures and on the engine support beams became available it was found that these elements had an important effect on flutter characteristics. Therefore it became necessary to incorporate additional detail in the analytical model of the structure. The general level of complexity of the resulting structural model is shown in Figure 3. Wing and fuselage skins are represented by flat quadrilateral membrane elements. Rib and spar segments are represented as beams with shear flexibility included. The finite element model of the half airplane has 3400 structural nodes. The reduced models employed in vibration mode analysis have 146 and 171 nodes for symmetric and antisymmetric cases, respectively. Modal coordinates are employed in the flutter analysis since this makes it possible to limit the complex matrix eigenvalue problem to manageable order. Symmetric and anti-symmetric analyses are performed separately; approximately twenty degrees of freedom (rigid body displacements plus free-free airplane modes) are required in each case. Space limitations make it impractical to present a comprehensive display of modal character-istics. However the second, third, and fourth symmetric modes for a relatively heavy weight condition (GW = 565,000 lbs) are displayed in Figures 4, 5, and 6. All of these modes contribute significantly to the critical flutter condition, and they should serve to demonstrate the plate-like character of wing deformations, elastic interaction between wing and fuselage, and the effect of engine inertias on wing deformation shapes. Also the pronounced wing camber changes give an indication of the need for lifting surface theory in the analysis of incremental oscillatory airloads.

Neutral stability boundaries for the aeroelastic system are determined by numerical solution of a complex matrix eigenvalue problem. The familiar characteristic equation is expressed as follows, for reference in subsequent discussions:

$$\left([M] + \frac{1}{\mu k^2} [A] - X[K] \right) \{q\} = \{0\} \tag{1}$$

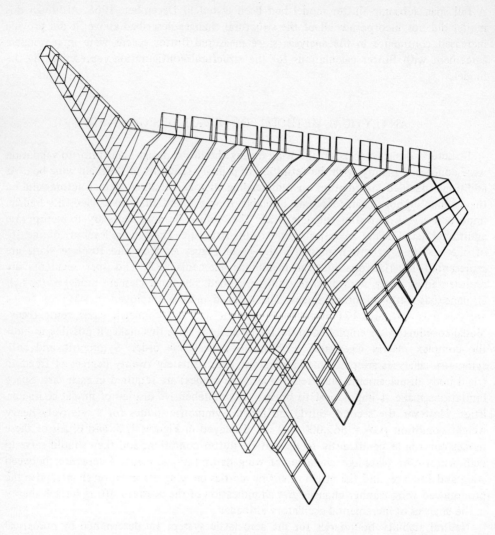

Fig. 3. Analytical Structural Model

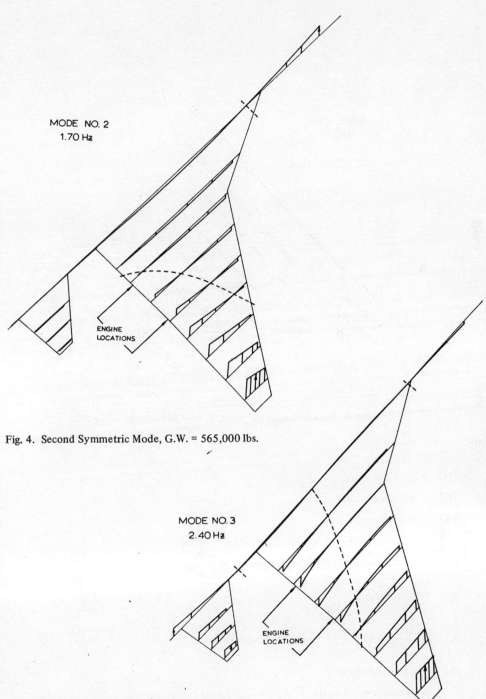

MODE NO. 2
1.70 Hz

ENGINE
LOCATIONS

Fig. 4. Second Symmetric Mode, G.W. = 565,000 lbs.

MODE NO. 3
2.40 Hz

ENGINE
LOCATIONS

Fig. 5. Third Symmetric Mode, G.W. = 565,000 lbs.

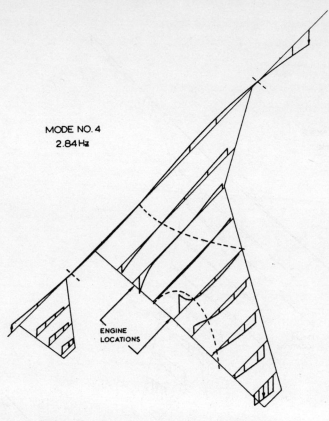

MODE NO. 4
2.84 Hz

ENGINE
LOCATIONS

Fig. 6. Fourth Symmetric Mode, G.W. = 565,000 lbs.

where

$[M]$ = inertia matrix

$[K]$ = stiffness matrix

$[A]$ = airforce matrix

$\left\{ \begin{matrix} q \end{matrix} \right\}$ = column matrix of model coordinates

k = nondimensional reduced frequency = $\dfrac{\omega b_r}{v}$

μ = mass ratio = $\dfrac{m_r}{\pi \rho b_r{}^2 s}$

X = $\dfrac{\omega_r{}^2}{\omega^2}$ $(1 + ig)$ = characteristic parameter

m_r = reference mass

ρ = atmospheric density

b_r	= reference wing semichord length
s	= wing semispan
ω_r	= reference frequency
g	= structural damping parameter
ω	= flutter frequency
v	= airspeed

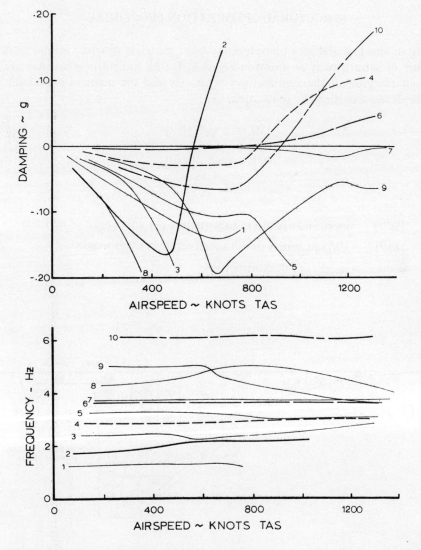

Fig. 7. Solutions of Complex Eigenvalue Problem for m = 0.9 and ρ = 0.001267 slugs/ft^3 (Altitude = 20,000 feet)

Typical results of solving the eigenvalue problem for fixed Mach number and altitude (which determines μ) and for a succession of k-values are shown in Figure 7; only the modes having the 10 lower frequencies are shown. Neutral stability conditions are indicated at successively increasing speeds in modes 2, 6, 4, 10; mode 2 with frequency of approximately 2 cps is clearly the critical one. In order to achieve consistency between airspeed and Mach number it is generally necessary to repeat the eigenvalue analysis for several assumed altitudes.

STRUCTURAL OPTIMIZATION PROCEDURE

To establish a practical basis for optimization the complete structure is subdivided into a number of substructures as shown in Figure 8. Inertia and stiffness matrices are then computed for each of the substructures separately and the matrices representing the entire airplane are obtained by summation, i.e.

$$[M] = \Sigma \, w_i \, [M^{(i)}] \tag{2}$$

$$[K] = \Sigma \, w_i \, [K^{(i)}] \tag{3}$$

where

$[M^{(i)}]$ = inertia matrix of i'th substructure per unit weight of material

$[K^{(i)}]$ = stiffness matrix of i'th substructure per unit weight

w_i = weight of i'th substructure

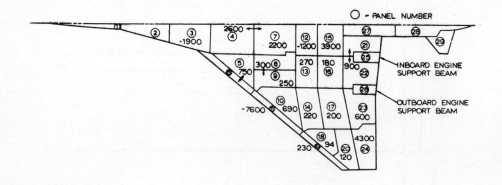

Fig. 8. Definition of Substructures and Weight Gradients (lbs/knot)

In dynamic analysis of complex structures the value of matrix assembly procedures based on this kind of representation can hardly be over-estimated. Apart from its usefulness in optimization studies it also provides increased efficiency in accommodating local design changes. Considerable judgment must be exercised by the analyst in defining substructures so that the elements that are critical for the problem at hand are separately identified. Also it will be evident that the suggested treatment is not limited to the "cookie-cutter" approach that is suggested by the diagram in Figure 8, since skin panels, ribs, spars, groups of stiffeners, etc., may be separately identified as "substructures." Nonstructural items, e.g. balance weights, may be considered by setting $[K^{(i)}] = [0]$.

The numbers identified with the substructures in Figure 8 (apart from the identification indices) are weight gradients, representing the weight that must be added in the various regions (per airplane — both sides included) to obtain a 1 knot increase in critical flutter speed for one flight condition. Obviously these trends cannot be extrapolated indefinitely, since variations in flutter speed with weight are generally nonlinear; however, in the present case the linear approximation for individual substructures has proved to be quite accurate over a range of several hundred pounds. The procedure that has been followed in practice is to analyze effects of graded changes of primary wing structure extending over several adjacent substructures, to simulate modifications that maintain reasonable continuity. Variations in stiffness of individual engine support beams have also been explored; results of these studies will be discussed in the following section.

Variations in primary wing structure, with the original strength-designed engine beams, produced the interesting trends shown in Figure 9. Initial optimization efforts were structured to raise the critical crossover point on the V-g plot. This indicated a need for extensive stiffening of the outer portion of the wing that tended to produce a more explosive type of instability (as indicated by increasing steepness of the V-g plot at the crossover point). Furthermore, the weight required to produce an acceptable flutter margin in this way was excessive. An alternative criterion for optimization was then adopted by seeking ways to progressively reduce the value of g_{max} at the peak of the V-g plot. This led in the direction of stiffening the inboard region of the wing.

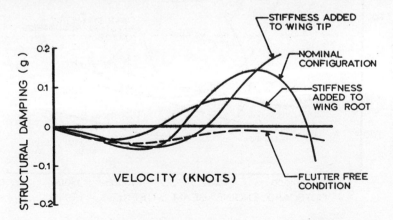

Fig. 9. Optimization Trends for Primary Wing Structure

Although the critical flutter speed was actually *reduced* during the initial stages of this process, this trend eventually reversed as g_{max} became negative and the unstable condition was completely eliminated. A structural weight increment of about 12000 lbs. over strength requirements was needed to accomplish this. However, the damping margin that could be achieved in this way appeared to be somewhat marginal.

ENGINE BEAM STIFFNESS AND WING MASS BALANCE STUDIES

Extensive analyses were conducted to evaluate effects of variations in stiffness of the cantilever beams supporting the aft engine mounts. It was concluded from these studies that the outboard engine was a primary factor in wing flutter behaviour and that the inboard engine was relatively ineffective. Initial studies indicated that a substantial reduction in stiffness of the outboard engine beam would eliminate the critical flutter condition; results are shown in Figure 10. It was also found that airloads induced by flexing of the wing trailing edge through its attachment to the engine beam had a detrimental effect on flutter speed.

Further analysis showed that an antisymmetrical mode became critical with decreasing engine beam stiffness. Although this instability was relatively mild and might have been overcome by relatively modest wing structural changes, this approach was abandoned because of practical design considerations and high risk associated with possible uncertainties in the actual location of the left hand boundary of the region of instability.

Studies of mass balance effects with a boom extending forward of the wing leading edge near the tip had yielded rather discouraging results on configurations with engine beams in the low to intermediate stiffness range. However it was found that a significant

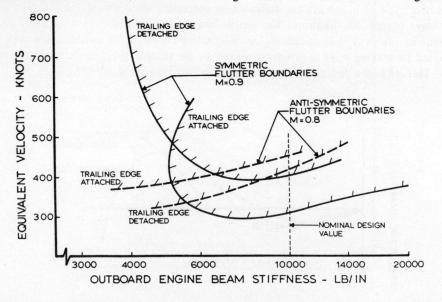

Fig. 10. Effect of Variations in Outboard Engine Beam Stiffness

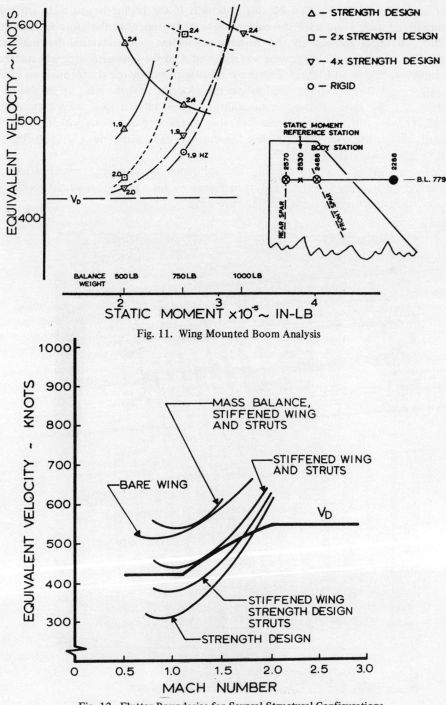

Fig. 11. Wing Mounted Boom Analysis

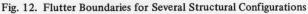

Fig. 12. Flutter Boundaries for Several Structural Configurations

weight saving could be realized by this approach if the engine beams were stiffened substantially. Inboard and outboard engine beam stiffnesses were therefore increased to 24K and 32K lbs/in, respectively. At the same time some of the material that had been added earlier to stiffen the inner wing was removed, and the outboard wing was stiffened in Regions 14, 17, 18 and 20 (cf. Figure 8). Results of symmetric flutter analyses of the modified wing with a boom mounted ballast mass located near the wing tip are shown in Figure 11. At the time of program cancellation the estimated total weight increment required for wing flutter clearance was 7000 lbs., distributed as follows: (1) 3000 lbs. for outboard wing stiffening; (2) 1500 lbs. for engine beam stiffening; (3) 2500 lbs. for ballast weight and boom installations.

Results of flutter analyses showing the influence of successive modifications to increase the wing flutter speed are shown in Figure 12 for a high gross weight condition. Comparative results are also shown for the bare wing.

Fig. 13. Internal Structure of Transonic Wing Model

Fig. 14. Dynamic Calibration of Transonic Wing Model

WIND TUNNEL MODEL TESTING

The considerations that have led to the adoption of a complex finite element structural model in flutter analysis of the SST have imposed similar complications in the construction of scaled flutter models. Previously, in designing low speed flutter models of subsonic aircraft, it had been possible to employ a simplified structural representation consisting of an assemblage of spars with properly scaled flexural and torsional stiffnesses. The required external shape was provided by segmented fairings attached to the spars. To date no comparable simplified representation has been devised for configurations similar to the SST, and it has been necessary to develop modeling techniques which closely approximate the geometry of the actual aircraft structure on a reduced scale.

The internal structure of a transonic flutter model of the SST wing is shown in Figure 13. Skins are of laminated fiberglass, and ribs and spars are of sandwich construction with polyurethane core and fiberglass faces. A manifold system is provided for changing the fuel load simulation. The completed model is shown in Figure 14 undergoing a resonance test in preparation for wind tunnel testing. Typical wind tunnel test results and comparative analytical data are shown in Figure 15 for a configuration with stiffened engine beams and outboard wing stiffening. The analysis of the model wing

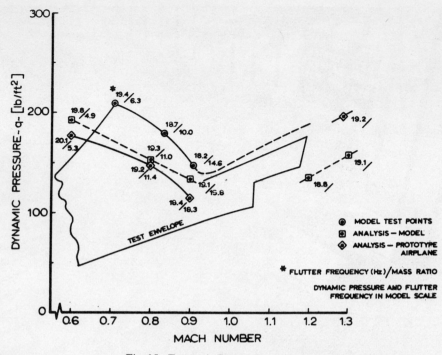

Fig. 15. Transonic Flutter Test Results

was based on finite element methods, using a compilation of measured material properties which has been prepared specifically for use in model design. The scaled values from an airplane wing analysis were obtained from the finite element model described earlier. Measured deflection influence coefficients have also been used in vibration mode calculations and in flutter analysis; these calculations are generally in reasonably good agreement with finite element analyses and resonance tests. Comparative data on deflection influence coefficients and natural frequencies of the transonic wing model are given in the following tables.

TABLE I

Errors in Analytical Prediction of Measured
Deflection Influence Coefficients

		Average Error %	Maximum Error %
Diagonal Elements	ϵ	-2.6	-11.0 (1)
	$|\epsilon|$	4.83	
Nondiagonal Elements	ϵ	-1.67	16.5
	$|\epsilon|$	9.38	

(1) Negative sign implies high predicted stiffness

TABLE II

Comparison of Measured and Calculated Natural Frequencies

Mode	Measured Frequency Hz	Calculated from Measured Infl. Coefficients Hz	Calculated from Theoretical Infl. Coefficients Hz
Base Wing			
1	15.72	15.24	16.82
2	42.25	40.67	40.87
3	50.8	51.7	56.29
4	78.6	72.4	76.22
5	81.4	88.22	89.54
Wing Plus Nacelles			
1	13.10	13.18	14.47
2	20.18	20.18	19.84
3	20.84	20.89	20.91
4	38.9	37.67	38.64
5	53.5	52.74	56.65
6	76.5	69.66	71.63

Flutter tests of a supersonic wing model were conducted at the NASA Ames Research Center in March of this year, and it was found that predicted flutter speeds, based on Mach box aerodynamics, were consistently conservative in the low supersonic range. The degree of conservatism was in fact greater than that shown in the high subsonic range, cf. Figure 15. However there was a rather sharp decrease in measured flutter speed in the immediate vicinity of $M = 1$, extending below the interpolated value from subsonic and supersonic analyses. This behaviour was attributed to tunnel wall effects in the supersonic wind tunnel, and arrangements were being made to verify that conjecture by retesting the supersonic model in the Transonic Dynamics Tunnel at the Langley Research Center. This task has not been completed.

UNFINISHED BUSINESS

At program termination work was under way on a transonic flutter model of the complete airplane for confirmation of the final Prototype design and to provide response data for use in interpretation of the airplane flutter test. This model was to be tested in the NASA Transonic Dynamics Tunnel on a cable mounting system that provides dynamic simulation of free flight conditions. An excitation system was also being provided to obtain subcritical response data for direct comparison with data from flight flutter tests of the Prototype. Similar information was also to be obtained by analysis.

Work was also underway on detail design of the full scale exciter system for the flight flutter test program. This consisted of two servo-driven vanes at the wing tips and a two-axis inertia exciter in the aft body. Instrumentation was being provided for direct measurement of excitation forces, and it was planned to employ a telemeter link and an on-line computer to determine transfer functions from fast Fourier transforms of force input and response. By using transient excitation it was expected that substantial reductions in flight time could be realized, by comparison with earlier programs using a slow sinusoidal sweep.

CONCLUDING REMARKS

The preceding discussion has been concerned with only one phase of the flutter prevention program on the SST, namely the solution of the primary wing flutter program. E.g., no attention has been given to interesting and important work on problems of aeroelastic instability involving stability augmentation and control systems. Obviously it was not possible to present a complete account in a paper of this kind, and the wing flutter problem was thought to be of special interest because of its impact on structural weight. Several of the tools employed in this work (improved dynamic modeling techniques, methods of unsteady airload analysis, finite element structural analysis, and digital methods of dynamic analysis of systems with many degrees of freedom) have been developed over the past twenty years specifically to deal with this kind of problem. Although our analytical capability is vastly improved, response time is still inadequate for design purposes. Hence the present need is to develop an integrated design system based on these improved analytical capabilities. It is hoped that the experiences described here may be of some value to those who are involved in that important task.

Since we have been solely concerned with applications of existing technology, we have not felt that it was necessary to include a historical survey of the evolution of aeroelastic science. However, for the benefit of interested nonspecialists we have included a selected list of basic references which provide systematic discussions of basic principles, extensive bibliographies, and chronological accounts of basic contributions.

REFERENCES

1. R. L. Bisplinghoff, H. Ashley and R. L. Halfman, *Aeroelasticity*. Addison-Wesley, Cambridge, Mass. (1955).
2. R. L. Bisplinghoff and H. Ashley, *Principles of Aeroelasticity*. Wiley, New York (1962).
3. Y. C. Fung, *An Introduction to the Theory of Aeroelasticity*. Wiley, New York (1955).
4. I. E. Garrick, *Aerodynamic Flutter*. AIAA Selected Reprint Series, Vol. V, American Institute of Aeronautics and Astronautics (1969).

5. W. P. Jones (Ed), *AGARD Manual on Aeroelasticity*, NATO Advisory Group for Aeronautical Research and Development (1961). Revised ed. (1968), R. Mazet, editor.
6. H. C. Martin, *Introduction to Matrix Methods of Structural Analysis.* McGraw Hill, New York (1966).
7. J. S. Przemieniecki, *Theory of Matrix Structural Analysis*, McGraw Hill, New York (1968).
8. R. H. Scanlan and R. Rosenbaum, *Introduction to the Study of Aircraft Vibration and Flutter.* Macmillan, New York (1951).

SPACE SHUTTLE — A NEW ARENA FOR THE STRUCTURAL DYNAMICISTS

HARRY L. RUNYAN AND ROBERT C. GOETZ

NASA Langley Research Center, Hampton, Virginia

Abstract — The Space Shuttle represents the focus of NASA development endeavors for the next decade. The purpose of the shuttle is to provide a general launch capability for a variety of payloads ranging in weight to as high as 65,000 pounds. The shuttle is to be reusable, and every effort is being made to provide a relatively low-cost system. With these ground rules, it represents one of the most challenging technological projects by NASA. This paper will concentrate on one aspect of the shuttle development, namely, dynamics and aeroelasticity. The objective of our research during the first year of activity has been directed at searching out problems peculiar to the Space Shuttle, reviewing the state of the art, and developing a program to satisfy the needs. Some of the specific items covered in this paper are vibration modes, thermal protection system dynamics, fluid dynamics, ground winds, flutter, buffet, and noise.

INTRODUCTION

The Space Shuttle represents one of the greatest and most interesting technological challenges the aerospace technologist has faced up to the present time. This arises from two factors, namely, the effort to develop a reusable (100 mission) vehicle as well as a low-cost space launch system. As now envisioned, the Space Shuttle is a two-stage vehicle, with the second stage riding "piggyback" on the side of the booster as shown in Figure 1. (The dynamicists can immediately envision the many coupling problems created by this off-set orbiter.) The sequence of the flight events shown in Figure 2 involves a vertical take-off in a launch vehicle mode, separation of the booster and orbiter where the orbiter continues up to a nominal altitude of from 100 to 270 nautical miles; there, the orbiter will now function as a spacecraft, with ability to remain in orbit for 7 days. The orbiter will accomodate a payload size of 60 ft long by 15 ft in diameter and a maximum payload capability of 65,000 pounds. The reentry is accomplished in a semiballistic mode, coming in at angles of attack varying between $20°$ to $60°$, and then finally, a transition to an airplane mode with a horizontal landing capability as well as a short ferry capability. Thus we have combined into one vehicle, a launch vehicle, a spacecraft, a reentry craft and finally, an airplane.

Some of the major dynamic problems faced by the dynamicist are the external thermal protection system, liquid dynamics, POGO, vibration, ground winds, noise and buffeting, and flutter as well as the transient phenomena of lift-off, separation, docking, and landing. The remainder of the paper will indicate some of the research work now underway in order to provide a technological base for the successful development of the Space Shuttle.

115

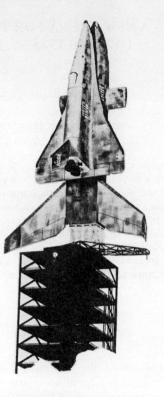

Fig. 1. Space Shuttle launch configuration.

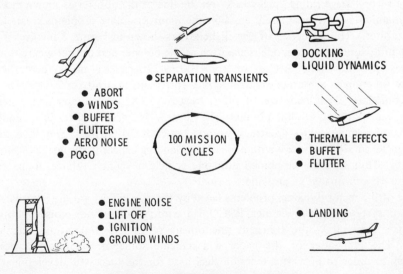

Fig. 2. Space Shuttle dynamic problems.

THERMAL PROTECTION SYSTEM

Of all the systems on the Space Shuttle, the thermal protection system (TPS), is one of the more critical, particularly since it has a large impact on the overall vehicle weight; thus, every effort must be made to reduce the weight as well as the cost. At the present time, there are two main contenders for a TPS system, the metallic heat shield using some such exotic materials as Thoria disbursed-nickel-chrome, and the so-called "hardened compacted fiber" concept. In both cases, the outer skin is separated from the primary load-carrying structure by posts which are of the order of 2-3 inches, as shown in Figure 3. Temperatures ranging up to 2500°-3000° F may be experienced at the stagnation point, however, temperatures below 2000° F exist over most of the under side of the craft. The TPS is required to carry only the external aerodynamic load, which is relatively low, thus the thickness of the exterior skin can approach minimum gage. Therefore, the many dynamic inputs may well be the principal design element in configuring the TPS. It seems axiomatic that when a structure is required to carry only the external aerodynamic loads and not the main vehicle loads, we are likely to be faced with many dynamic problems such as panel flutter, panel response to noise, fatigue, and so forth.

MULTIPLE CLIP SUPPORTS

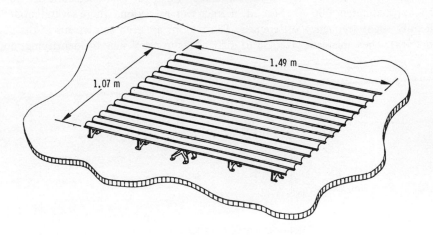

Fig. 3. Temperature protection panel concept.

Fundamental to the analysis of the response of these panels is the knowledge of the vibration modes. As a part of the general effort of LRC, we are working to establish analytical and experimental procedures to predict basic vibration characteristics, such as natural vibration modes and damping of TPS panels. The vibration characteristic of the panel systems in thermal environments will be important as inputs to flutter and fatigue investigations. This work is being carried out by William Walton and Hugh Carden of the LRC. A combined analytical and experimental study of the vibration modes and damping of a typical TPS panel configuration is first being conducted at room temperature. The panel, shown in Figure 3, is a thin skin corrugated metal heat shield mounted on multiple clip supports. This panel is one of several developed at LRC to attempt to determine realistic cost, fabrication procedures, and minimum weight of thin metal heat shields. These panels will also be tested in a heated environment under realistic Space Shuttle aerodynamic loads in the LRC 8-foot high-temperature structures tunnel to prove the feasibility and integrity of such metallic heat-shield panels.

The NASTRAN computer program has been applied to compute vibration modes of the panel illustrated in Figure 3. Two analytical approaches are being used. In the first approach, the panel was analyzed as an orthotropic flat plate with the corrugations approximately accounted for in the calculated orthotropic constants which are inputs to NASTRAN. In the second approach, the panel will be analyzed using detailed geometric representations of the corrugations. The general element capability of NASTRAN will be used to input static influence coefficients of portions of the corrugations treated as substructures of the entire panel.

Typical computer results from the orthotropic analysis are presented in Figure 4. Given in Figure 4(a) is a plot of the modal density of the panel. The figure emphasizes two points. First, the density of modes is very high; for example, there are 80 modes below approximately 600 Hz. Second, there are many regions where several modes occur with only slight frequency differences. The regions are seen as plateaus in the curve of Figure 4(a). These plateaus give rise to difficulties in classifying or identifying modes or

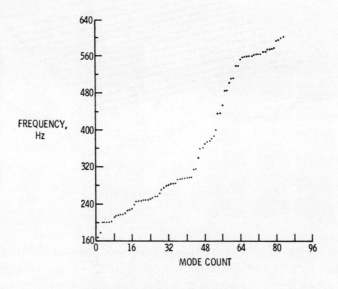

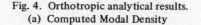

Fig. 4. Orthotropic analytical results.
(a) Computed Modal Density

nodal patterns which occur experimentally. For example, in Figure 4(b) four computed nodal patterns are shown to lie within a 2-Hz bandwidth. Experimentally, these modes would give rise to some complex combination and the nodal pattern may not exhibit any recognizable pattern.

ORTHOTROPIC ANALYSIS

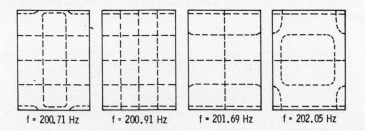

f = 200.71 Hz   f = 200.91 Hz   f = 201.69 Hz   f = 202.05 Hz

Fig. 4. Concluded.
(b) Typical Computed Modal Patterns

In addition to the in-house work described above, we are also pursuing the development of techniques for measuring vibration modes of plates under heated conditions. Under contract to TRW, research is underway to determine the possible use of holography. Previous work has demonstrated that continuous laser holograph may be used to very accurately determine the vibration modes of plates at room temperature; an example of this is shown on the left of Figure 5. Unfortunately, the continuous laser

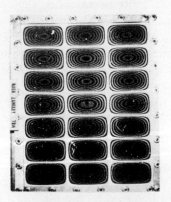

CONTINUOUS LASER   PULSED LASER
ROOM TEMP   1040° F TEMP

Fig. 5. Holographic measurements of panel vibration deflections.

technique is more of a laboratory technique than a practical and useful tool in a development sense. However, they have found that the pulsed laser technique will operate in a thermal convective environment and may be adaptable in a heated and vibratory environment since the pulse is so rapid that the air motions all seem essentially in a static condition, and thus does not distort the passage of the pulse. A hologram of a plate which is vibrating and heated to 1040° F is shown on the right. Although the contour lines are not as distinct as for the cold plate, we still find excellent results and this technique will probably be explored further.

LIQUID/STRUCTURE INTERACTION

Since the vehicle at lift-off is comprised of from 70-80 percent liquid, the efficient management of this dynamic mass has constituted a serious problem for all launch vehicles, but is even more so in view of the weight criticality of the shuttle. Just the stabilization of the liquid mass with respect to coupling with the rigid body modes and the automatic control system has required many pounds of fuel slosh baffles to insure a stable vehicle. For a circular tank with a hemispheric bottom, we feel that the state of the art of the prediction of the frequency of oscillation is adequate. However, for the Space Shuttle, the tanks may not be cylindrical and analytical procedures for determining the natural frequencies and motion must be developed.

In an effort to reduce the weight of baffles required for vehicle stabilization, considerable research has been done by Southwest Research Institute on the use of flexible baffles. Research has indicated that in the ideal case, a rigid baffle may weight as much as 100 times more than a flexible baffle for the same value of damping produced. The major problem revolves around material compatibility, one that remains flexible at cryogenic temperatures, and does not constitute a fire hazard when exposed to liquid oxygen. Southwest Research Institute has studied the problem of material selection under contract to Langley and has found that Kapton and Teflon FEP maintain their material properties at liquid nitrogen temperatures. Both materials also passed the lox compatibility test, that is, the materials did not ignite when immersed in liquid oxygen and hit with a standard impact.

Other times of flight also require extensive study such as abort when the vehicles are still mated and the flight is still in the sensible atmosphere sufficient to cause a high drag and consequent deceleration when all or most of the main engine might be shut down. Not only do we have the possibility of dome impact due to the liquid rapidly streaming up to the top of a partially filled tank, resulting from the negative acceleration, but we have the problem of vehicle stabilization due to the rapid shift in the center of gravity and the turning moment induced as the liquid is turned by the ellipsoidal tank dome. Some procedures are available for computing the axisymmetric case as well as the two-dimensional cases. These flow types are illustrated in Figure 6(a) and some typical force predictions obtained from these procedures are shown in Figure 6 (b). This figure shows the forces exerted by the liquid on the tank during the flow types illustrated in the previous figure. For each of the two flow types, curves representing the total force on the

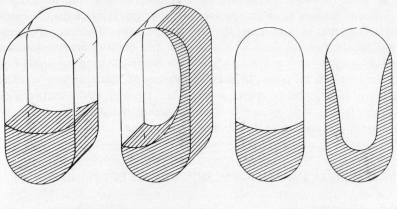

TWO DIMENSIONAL   AXISYMMETRIC

Fig. 6. Liquid/tank interaction.
(a) Flow Types

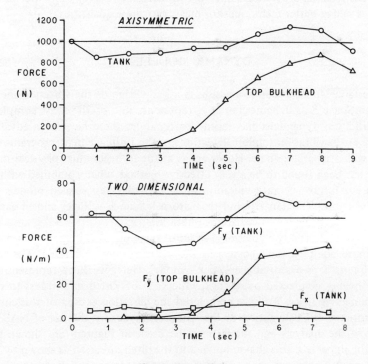

Fig. 6. Concluded.
(b) Force Results

tank and the force on the top bulkhead are a function of time. For the axisymmetric case, the flow was initiated from an equilibrium low gravity interace shape by applying a continuous acceleration of 0.1g along the tank axis. The two-dimensional channel flow problem was begun from the same interface shape, but the 0.1g unsettling acceleration was applied at an angle of 6° to the tank axis, with the result that more liquid flowed up one side of the tank than the other. In assessing a variety of "dome impact" problems, for most conditions these problems appear to be minor; however, if the shuttle is designed for abort during launch, these problems could become critical but, as noted before, new techniques are required for the nonaxisymmetric tank shape.

LIQUID-STRUCTURAL-PROPULSION SYSTEM COUPLING (POGO)

One of the dynamic instabilities that has been plaguing the launch vehicle designer for a number of years has been a phenomenon termed POGO. This involves an interaction of the structures, having a feedback network phased such that the engine can put energy into the system, usually in one of the lower longitudinal vibration modes. The general block diagram of the feedback system seems to be understood; the principal deficiency in predicting POGO is an accurate determination of the transfer function of the various elements. Our research program is aimed at improving our ability to predict these transfer functions, as well as better techniques for measuring these quantities.

DYNAMIC MODELING

Of importance to all dynamic analyses is a knowledge of the vibration modes of the vehicle. The Space Shuttle, undoubtedly, represents one of the most complex systems that the structural dynamicist has faced. Fortunately, there have been advances in the state of the art in vibration mode calculations — the finite-element programs now being developed will play a large part. However, as a parallel technique, physical modeling of the system has been found to be a very effective method, when combined with analytical approaches, to insure accurate vibration modes. In the past, scaled models have been made of the Saturn I, Titan III, and the Saturn V launch vehicles and in each case the physical model has provided essential information.

For the Space Shuttle, we envision a series of three physical models each progressively becoming more complex (and costly) as the vehicle is better defined. The first model, shown in Figure 7, is essentially a "stick" model. Here, we have represented the two vehicles by beams, connected by springs, which can be varied in stiffness to represent a range of connecting links. We have completed the vibration testing of this configuration and have initiated the calculation of the vibration modes by the use of NASTRAN. To illustrate how the analysis and model may be used, in Figure 8 are shown two modal responses, the upper due to a shaker applied in the pitch direction as shown by the arrow. The lower modal response was obtained with the force applied in the longitudinal direction. In our initial calculations of the modal functions, we calculated only the pitch

1/15-SCALE SPACE SHUTTLE DYNAMIC MODEL

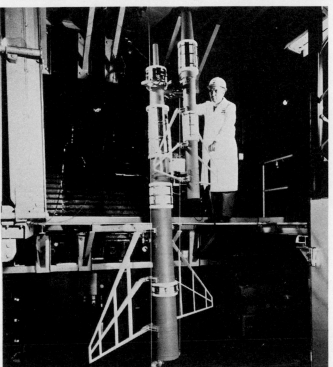

Fig. 7. 1/15-scale Space Shuttle dynamic model.

model, and this particular experimental mode shown in Figure 8 did not correspond to any analytical models. However, examination of the experimental longitudinal modal data did indicate the result shown in the lower part of the figure. Thus, it appears that our first calculated endeavor is not complete, and we shall have to include all possible motions in order to analytically determine all the modes. Essentially, we cannot think in terms of pitch mode or longitudinal mode, but must work in the domain of three-dimensional modes.

MODE COMPARISON WITH PITCH OR LONGITUDINAL INPUT

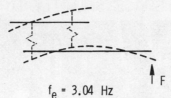

f_e = 3.04 Hz

PITCH INOUT FORCE

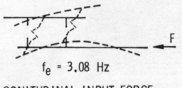

f_e = 3.08 Hz

LOGNITUDINAL INPUT FORCE

NORMALIZED LENGTH

Fig. 8. Typical complex mode shapes.

For the second model (which we plan to base on the results of the Phase B studies), we will include tanks which will contain propellants. The third and final model we recommend to be a near replica model, of a scale sufficient to represent most of the important stiffnesses and inertias.

GROUND-WIND LOADS

A vehicle mounted on its launching stand will be exposed to winds and gusts which are essentially normal to the vehicle longitudinal axis. In the past, most vehicle cross sections have been cylindrical and the phenomena they experience are understood to the extent that wind-tunnel testing techniques are becoming standardized. For the Space Shuttle, however, the presence of large lifting surfaces flexibly connected to noncylindrical fuselages and the side mounting of the orbiter on the booster have raised the possibility of new instabilities associated with ground winds. We have geared our research program to search out and identify these new phenomena and the associated vehicle responses.

Two of these new areas have been identified by Reed [1] and are currently being investigated in order to gain a better understanding of their mechanisms and their relevancy to current Space Shuttle concepts. The first is called "stop-sign" flutter, and it is analogous to wing stall flutter, since its origin is associated with separated flow on the large area of the wings. The resulting motions are torsional oscillations about the vehicle longitudinal axis. The second is a "galloping" instability. This phenomenon derives its name from the galloping of communication lines resulting from ice depositing unsymmetrically on them. The classical shape used to describe "galloping" in communication engineering textbooks is a "D" cross section. The cross sections of some evolving Space Shuttle designs for both the booster and the orbiter range from circular to "D" shapes. Unlike "stop-sign" flutter, which is an unsteady aerodynamic phenomenon, "galloping" instability has its origin associated with static aerodynamics. For "galloping" to exist, the static aerodynamic forces produced by a cross-section contour tend to reinforce lateral motions and become critical when they are oriented so as to produce a self-sustained oscillation normal to the wind direction.

In order to investigate "galloping," one may either study vehicle response normal to the wind direction or the static aerodynamic forces in the horizontal plane normal to the vehicle axis. Both of these approaches have been undertaken with similar results. Typical results will be shown obtained in low-speed wind tunnels where static aerodynamic forces have been measured on a variety of small models of boosters, orbiters, and the mated combinations of both. Figure 9 shows 1/300-scale modles of two such mated configurations. The models are shown mounted perpendicular to the flow on a sting which went through a simulated ground plane and was connected to a balance which measured the two components of horizontal force normal and parallel to the wind. The primary

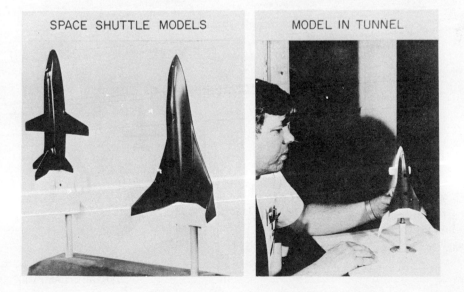

Fig. 9. Space Shuttle ground-wind-load study.

variable of the test was wind azimuth angle. Utilizing these static force measurements, regions of possible "galloping" instability were identified. As expected, for all configurations, those which had a flat lower surface oriented to the wind were the most critical.

Some "galloping" stability characteristics associated with a delta wing booster are shown in Figure 10(a). The results are presented in terms of an aerodynamic damping

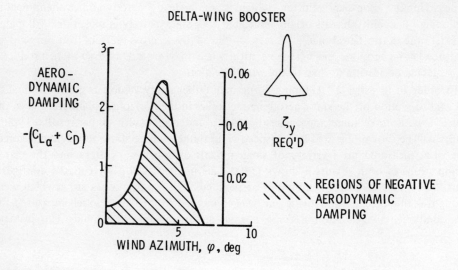

Fig. 10. Galloping stability characteristics of Space Shuttle configuration.
(a) Booster Alone

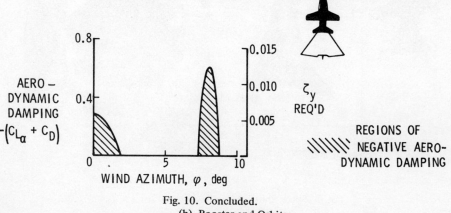

Fig. 10. Concluded.
(b) Booster and Orbiter

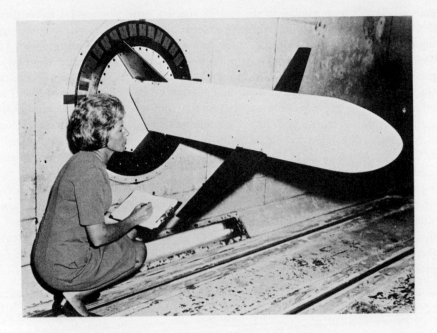

Fig. 11. Space Shuttle "stop-sign" flutter model.

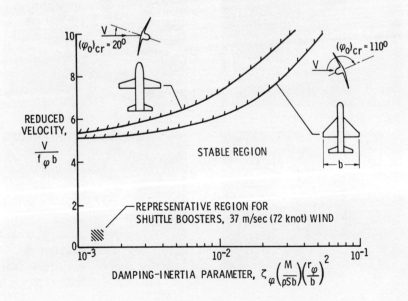

Fig. 12. "Stop-sign" flutter boundaries.

parameter which is proportional to the sum of the lift curve slope and drag coefficients, plotted as a function of wind azimuth angle. An additional ordinate given on the right of the figure shows the amount of structural damping that would be required to prevent "galloping" instability for the most critical design conditions; that is, for an empty fuel weight condition at the designed maximum wind speed of 37 m/sec (72 knots). In Figure 10 (b) are shown results of how the "galloping" instability of this booster would be affected by the addition of a straight-winged orbiter. These results indicate that mounting the orbiter on the booster has a significant stabilizing influence on "galloping" stability. This is illustrated by the amount of negative damping, which is reduced by a factor of about 4 and also by the range of critical wind azimuth angles which are also reduced. It should be noted that the levels of structural damping required to prevent "galloping" appear to be less than the inherent structural damping expected in typical launch vehicle structures. It has been recognized that these results are based on low Reynolds number data, and might not be representative of full-scale results. Recent tests confirm this suspicion that "galloping" stability predictions can be altered by Reynolds number effects. This emphasizes the need for subsequent ground-wind-load studies in high Reynolds number facilities.

BODIES + WINGS + TAILS

BODIES + TAILS

BODIES ONLY

DELTA WINGS

Fig. 13. Aeroelastic model configurations studied.

To determine the susceptibility of Space Shuttle configurations to "stop-sign" flutter an experimental program has been conducted to measure the aerodynamic roll damping derivatives by the forced oscillation technique. The generalized booster model used in this investigation is shown in Figure 11. It was sting mounted to the sidewall of the tunnel with its longitudinal axis (roll axis) perpendicular to the flow direction. It could

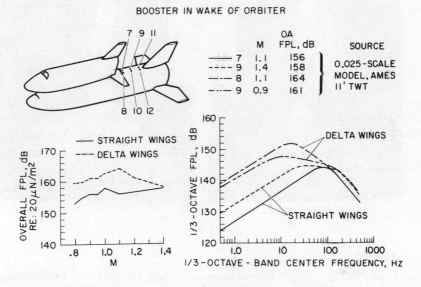

Fig. 14. Estimated pressure fluctuations during ascent.

accomodate a variety of interchangeable lifting surfaces. The model was oscillated in roll at a constant amplitude at various frequencies ranging from 2 to 8 Hz and with wind speeds ranging from 20 to 44 meters per second. Again, the primary variable was wind azimuth angle which was varied from -10° to +190°. Results from this study are summarized in Figure 12. This figure illustrates "stop-sign" flutter boundaries for a generalized straight wing and delta wing booster. The boundaries are presented in terms of reduced wind velocity as a function of the nondimensional damping-inertia parameter. These boundaries are for the most critical azimuth angles for each configuration and are illustrated by the sketches in the figure.

As you can see, the delta wing configuration is slightly more susceptible to "stop-sign" flutter than the straight wing configuration. However, the significant finding illustrated in the figure is that the "stop-sign" flutter boundaries are well removed from the representative region for shuttle boosters, which is shown by the shaded area. Thus, from these preliminary results, one might conclude that "stop-sign" flutter for the Space Shuttle booster configuration in the cantilever position is not a problem. However, it should be remembered that anything which may reduce the vehicle roll frequency would

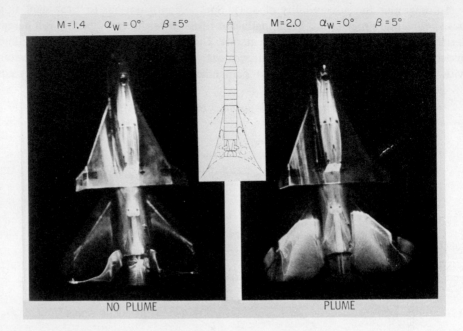

Fig. 15. Plume-induced flow separation.

tend to lower the "stop-sign" flutter boundaries thus bringing this instability into the realm of possibility. One such condition would be the mounting of the orbiter on the side of the booster. Thus, the susceptibility of the flexibly coupled two-body configuration to "stop-sign" flutter remains a question.

From this initial exploratory phase of ground-wind-load studies we are currently progressing to more sophisticated investigations as baseline configurations emerge. More specifically, larger models will be tested in high Reynolds number facilities. These lightweight rigid models of the mated configuration, flexibly coupled, will be mounted on a flexible support system to simulate the fundamental modes and damping of the launch configuration. These studies should answer the question of the combined problems singled out in the first phase of the investigation previously described. The final phase will simulate ground-wind-load effects on a complete aeroelastic model of the prototype vehicle. Included in this phase will be testing of the launch vehicle in the presence of its launch tower apparatus and a study of possible ground-wind-load alleviation methods.

AERODYNAMIC NOISE

In addition to the vehicle response, the aeroelastician needs a complete knowledge of the aerodynamic inputs which load the vehicle. Aeroacoustic loads are caused by a variety of sources, most of which are highly configuration dependent. These include zones of turbulence, regions of separated flow, shock-wave boundary-layer interference, and others. The input loads caused by these sources are usually defined in terms of dynamic fluctuating pressure over areas of concern on the vehicle. These areas can be identified either intuitively by vehicle configuration inspection or by more detailed analysis of the local flow on the vehicle. Suspicious regions of early configurations have been identified by inspection and fluctuating pressures measured on an aeroelastic model at Ames Research Center. Data were obtained on the model shown in Figure 13. Coe et al. [2] measured fluctuating pressures over the orbiter cockpit, on the orbiter heat shield in the vicinity of flow interference from the booster, and also on the booster in the region of the wake flow from the orbiter. Some results of the full-scale pressure fluctuations for both straight wing and delta wing configurations in this wake flow region are presented in Figure 14. The full-scale pressure fluctuations shown on the figure represent the maximum data obtained at any of the measurement locations over the full range of test angles of attack and sideslip. It might be noted that the data show that the pressure fluctuation on the booster in the wake flow of the orbiter were about two times higher for delta wing than for straight wings. The significant finding, however, shows that the maximum overall fluctuating pressure level with delta wing in this region was about 164 dB.

Illustrated on Figure 15 schematically is the phenomenon of plume-induced separated flow as observed on the S-IC stage of the Saturn V vehicle. Quantitative measurements of pressure fluctuations were also made on this vehicle in the past giving confirmation to the potentially serious problem of plume-induced unsteady loads on boosters. Figure 15 also illustrates the results of a florescent oil-flow study which shows a region of severe turbulence caused by plume-induced separated flow on a model of the Space Shuttle launch configuration. Shown is a top view of a delta wing configuration with and without the plume. While the plume simulation was not correct for the wind-tunnel Mach number (M = 2) the intent here is only to show the large area of the wing surface that can be affected at supersonic Mach numbers. Regions such as these will have to be studied in more detail and measurements obtained to determined the impact of this phenomenon.

BUFFET RESPONSE

Input loads such as these can cause the vehicle to experience severe buffeting during launch. Muhlstein [2] has presented and discussed preliminary results of a buffet investigation utilizing the aeroelastic model shown in Figure 13. The 0.025-scale model that was used in this investigation was tested with and without straight and delta wings. The four configurations were tested in the 11-foot by 11-foot Ames transonic wind tunnel and some typical results are shown in Figure 16. These results are presented in terms of the root-mean-square bending-moment response as a function of angle of attack at a Mach number of 0.9. As the data illustrate, the response is relatively low for the bodies-only configuration and is relatively insenstive to angled attack and yaw. However, for the configuration with straight wings, the response is much larger and a strong function of angle of attack. This result of rapid increase in buffet with increasing lift is

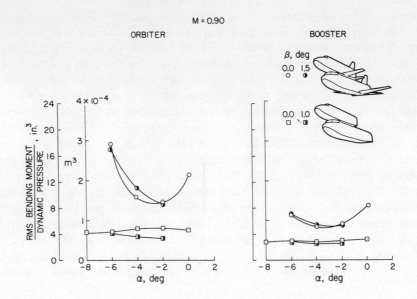

Fig. 16. Buffet response in the pitch plane.

typical of thick straight wings in this Mach number range. The main conclusion from these preliminary results is that coupling the orbiter and booster bodies together in parallel does not in itself seem to present a major buffet problem. However, if the bodies have large lifting surfaces there can be a significant increase in buffet. It cannot be discarded that some of the early launch vehicle failures are believed to be the result of buffet. Therefore, recognizing that buffet is strongly configuration dependent, it is essential to continue studying Space Shuttle configurations as they evolve to insure that potentially serious and detrimental buffet problems are not overlooked.

In addition to buffeting during launch, the problem area of buffet during reentry must also be considered. During reentry, both the booster and later the orbiter will be subjected to hostile thermal environments which will degrade the structure and thus make it less able to resist buffet loads. This problem is receiving attention particularly in the area of wing buffet at high angles of attack. A recent program was underaken by Goetz [2] to determine the susceptibility of a booster wing with large tip fins to buffet. The 1/20-scale semispan model investigated is shown in Figure 17 mounted on the sidewall of the Langley transonic dynamics tunnel. The general testing procedure was with the model at zero angle of attack as shown on the left of the figure; flow was established at a given Mach number and dynamic pressure. The model was then rotated, as shown on the right of the figure, to 90°. Resulting buffet boundaries are shown in Figure 18(a) as functions of angle of attack and Mach number. The boundary implies that this booster wing configuration will buffet during launch over the Mach number range from about 0.8 to 1.0. Also included on the figure and depicted by the circular symbols are the angles of

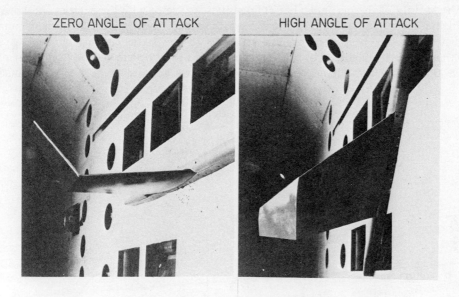

Fig. 17. Booster wing concept with large tip fins.

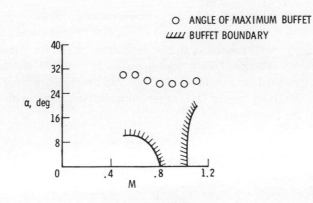

Fig. 18. Wing buffet.
(a) Buffet Boundaries

attack at which the wing experienced maximum relative buffet loads. Over the entire Mach number range these angles were consistently between about 25° and 30°.

The severity of the maximum buffet intensities associated with these angles are shown in Figure 18(b). The results on the left of the figure show the ratio of maximum predicted full-scale bending-moment fluctuations to the static bending moment as a function of Mach number. The highest value of this ratio occurs at a Mach number of about 0.8, where the fluctuating intensities are about 25 percent of the static values

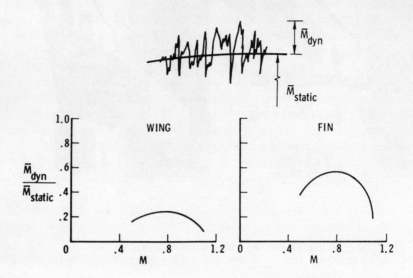

Fig. 18. Concluded.
(b) Predicted full scale maximum buffet intensities

measured at the wing root. However, a similar evaluation of the fin response, shown on the right of the figure, indicates fluctuating intensities of about 55 percent of the static bending moment measured at the fin root. The point to be amplified from these results is that these buffet loads are higher than those normally associated with thinner, more conventional, wing configurations.

LIFTING SURFACE FLUTTER

During the launch phase of flight, there is also the possibility for an instability to occur, such as flutter. The transonic and maximum dynamic pressure regimes of launch flight are expected to be the most critical in terms of flutter and it is in this region that the ability to predict unsteady aerodynamic forces is most uncertain. Consequently, an experimental wind-tunnel program has been undertaken to provide preliminary trend data in the transonic regime for lifting surface components to provide early insight into the flutter susceptibility of these surfaces and to uncover new problem areas associated with the unique configurations evolving for Space Shuttle vehicles.

One such program was conducted by Goetz [2] to determine the effect of large tip fins on the flutter characteristics of a proposed booster wing (fig. 17). Flutter boundaries have been defined for this wing configuration and also for an identical wing (not shown) with the fin removed and replaced with a ballast which simulated the mass and torsional inertia of the fin in an effort to isolate the aerodynamic influence of the fin on the wing flutter behavior. Some flutter results are presented on Figure 19 in terms of the flutter velocity-index parameter $V/b\omega_r\sqrt{\mu}$ over the Mach number range from about 0.5 to 1.3. The velocity-index parameter is proportional to the flutter velocity, V; having been nondimensionalized by the product of a reference semi-chord, b; a reference circular frequency, ω_r; and the mass ratio, μ. The solid curve represents the flutter boundary for the wing with tip fin, and the dashed curve, the boundary for the wing above. A comparison of these boundaries reveals two predominant effects associated with the fin on the wing flutter behavior. First, the Mach number at which the minimu flutter speed is encountered shifts to a lower value. Second, the flutter speed is substantially reduced over most of the Mach number range, up to about 1.15. For the lower range of subsonic Mach numbers, this reduction is about 27 percent; and at transonic speeds, the value of minimum flutter velocity is reduced about 45 percent. Consequently, it is concluded that large tip fins have a detrimental effect on wing flutter behavior of these proposed booster wing designs.

A new potential flutter problem is depicted in Figure 20. When the orbiter is mounted on top of the booster, the wing of one vehicle may be in close proximity to the wing or tail of the other, thus resulting in a "biplane" type configuration. The figure shows the effect of one wing placed directly above the other. The solid curve on the plot of flutter

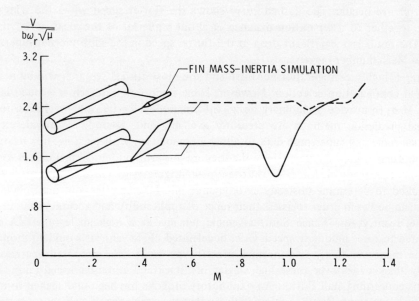

Fig. 19. Effect of wing tip fin on wing flutter characteristics.

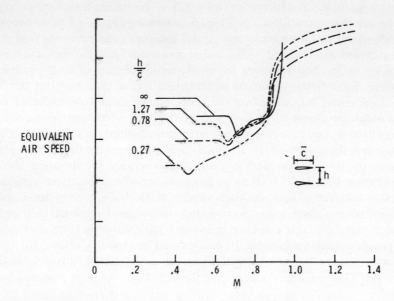

Fig. 20. Transonic flutter characteristics of wings in close proximity.

speed versus Mach number shows the flutter speed when the wings are well separated and
no interference occurs. The dashed curve shows the flutter speed when the wings are
brought together to a separation distance of about a quarter of the mean aerodynamic
chord. The result is a significant drop in the flutter speed in the subsonic and supersonic
region of Mach number range.

Flutter is highly configuration sensitive and the most critical research will be needed
after final configuration selection. However, some important research is needed in the
interim, first, to provide preliminary design information in the transonic regime, since no
analytical prediction methods are presently available and, second, to provide experimental subsonic and supersonic data to allow a comparison with existing, but relatively
new, nonplanar flutter theories. These theories are powerful numerical techniques for the
prediction of unsteady air forces on shuttle-type configurations. These techniques need to
be exercised to determine unsteady aerodynamic interference effects of multiple lifting
surfaces and bodies in order to assess their range of applicability and accuracy.

During reentry, the Space Shuttle vehicle remains at a high angle of attack until
decelerated to some moderate speed. Once decelerated, these vehicles would go through a
transition maneuver to fly in a normal aircraft-type mode. The proposed operation of
relatively thick wings over these high angles of attack and extensive speed range raises
questions concerning stall flutter. An exploratory program has been undertaken to try to
answer such questions. Results to date indicate that the avoidance of stall flutter for thick
straight wings will dictate the design torsional stiffness of the wing. While no such definite

conclusion has been reached concerning delta wings, the reuse requirement for shuttle makes it imperative that stall flutter be avoided. Consequently, delta wings will continue to be studied to assure that they are free from this instability.

PANEL FLUTTER

Because of reuse requirements and fatigue considerations, freedom from panel flutter will be a design requirement for the thermal-protection panels of the shuttle. Gaspers [2] has pointed out the importance of the boundary layer in the design of such panels. The boundary layer has a stabilizing effect which is largest at low supersonic speeds which are the most critical and, therefore, minimum panel weight can be achieved by sizing panels according to the local boundary-layer thickness. Based on theoretical results, it appear that it may be possible to size panels so that flutter is completely suppressed. Although static divergence could still occur, it would be limited in amplitude by nonlinear in-plane forces and would not present a fatigue problem.

A program has also been conducted in an effort to determine the effects of flexible supports and flow angularity on the flutter susceptibility of panels, since both of these conditions are inherent to the shuttle vehicle and flight. Results obtained by Bohon and Shore [2] for typical heat-shield designs are compared with the shuttle trajectory in Figure 21. The comparison is shown in terms of the aerodynamic paramter, $q_l / \sqrt{M_l^2 - 1}$

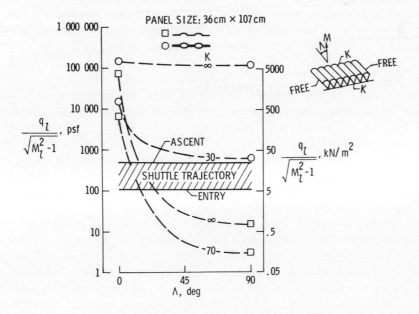

Fig. 21. Flutter comparison of shuttle surface structure.

(where q_l = local dynamic pressure; M_l = local Mach number) as a function of flow angularity. The range of this aerodynamic parameter for the shuttle is shown by the shaded band; the upper boundary is the maximum value on ascent, and the lower boundary is the maximum value during entry of the orbiter.

Flutter calculations are presented for two panel designs (a single corrugated panel and a double corrugated panel) over flow angles from $0°$ to $90°$. For each panel configuration, two support conditions were investigated: (1) the panel rigidly supported along two of its edges ($K = \infty$) and (2) the panel flexibly supported on the same edges on equal springs. Comparison of the flutter claculations with the shuttle trajectory when $\Lambda = 0°$ indicates a large flutter margin for both panels regardless of edge support. However, the single corrugated panel would encounter flutter at small flow angles even for rigid supports. On the other hand, the flutter region of the double corrugated panel is completely out of the range of the shuttle trajectory. Based on present theories, lightweight designs for Space Shuttle application are possible, provided sufficient attention is given to flutter prevention. While a broad technology base has been developed for panel flutter analysis, wind-tunnel verification tests of actual panel configurations will be required.

CONCLUDING REMARKS

The Space Shuttle represents one of the most interesting systems from the standpoint of the dynamicist. In almost every corner of this multifaceted arena, significant advances in the state of the art will be required for the successful development of the shuttle. A vigorous research program is now underway to insure that the designer will have in his hands adequate tools. This paper has given an overview of some parts of the research program.

REFERENCES

1. Anon: *Space Transportation System Technology Symposium.* NASA TM X-52876, Vol. II, 1970.
2. Anon: *NASA Space Shuttle Technology Conference. NASA TM X-2274, Vol. III, 1971.*

ANALYSIS OF THE TRANSIENT RESPONSE OF SHELL STRUCTURES BY NUMERICAL METHODS

T. L. GEERS and L. H. SOBEL

Lockheed Palo Alto Research Laboratory

Abstract—This paper examines some basic considerations underlying dynamic shell response analysis and the impact of these considerations upon the practical aspects of solution by numerical methods. Emphasis is placed on the solution of linear problems. The present states of development of the finite difference and finite element methods are reviewed, and techniques for the treatment of temporal variation are discussed. An examination is made of the frequency parameters characteristic of thin shell theory, applied excitations and spatial mesh geometries, and the significance of these parameters with respect to computational convergence is illustrated. The application of smoothing techniques to improve computational convergence is discussed.

INTRODUCTION

The exponential development of sophisticated numerical methods of solution has enabled structural analysts to utilize increasingly realistic models in studies of shell structures subjected to static and dynamic loads. This paper discusses the most widely used of these methods, the finite difference method and the finite element method. It also restricts its attention to the application of these methods in the transient response analysis of structures involving thin shells, i.e., shells whose characteristic radii of curvature greatly exceed their corresponding thicknesses. Finally, it emphasizes the seemingly obvious point that judgments regarding the accuracy and convergence of a given set of response computations must be based upon the use to which those computations are to be put.

RESPONSE VARIABLES

With regard to the last point above, it is of interest to examine the two most common applications of transient shell response computations. First, they may be used as excitation inputs to small structural systems that are attached to the shell structure. Second, they may be used to predict failure or survival of the shell itself.

In connection with the first use, let us examine briefly the response of a damped, single-degree-of-freedom oscillator excited at its spring-dashpot attachment point. Such an oscillator may be used to represent a particular mode of an attached structural system. The modal response quantity on which the failure or survival of such a system most directly depends is the relative displacement across the spring-dashpot pair. Thus we write the governing equation for the oscillator in the form [1]

$$\ddot{r} + 2\zeta\omega_o\,\dot{r} + \omega_o{}^2 r = -\ddot{x}_o \tag{1}$$

where r is relative displacement, x_o is attachment point displacement, ω_o and ζ are the oscillator's fixed-base undamped natural frequency and critical damping ratio, respectively, ($\zeta \ll 1$ in the vast majority of cases) and a dot denotes single differentiation in time. If we now introduce the Fourier transform [2]

$$\bar{f}(\omega) = \int_{-\infty}^{\infty} f(t)e^{-j\omega t}dt \tag{2}$$

the relative displacement response for quiescent initial conditions is given by

$$r(t) = \frac{1}{2\pi} \int_{-\infty}^{\infty} \frac{\omega^2\,\bar{x}_o(\omega)\,e^{j\omega t}}{\omega_o{}^2 + 2j\zeta\omega_o\omega - \omega^2}\,d\omega \tag{3}$$

Let us now consider three frequency regions in the (positive) frequency domain: (1) the region $0 \leqslant \omega \leqslant \omega_1$, where $\omega_1{}^2 \ll \omega_o{}^2$, (2) the region $\omega_1 \leqslant \omega \leqslant \omega_2$, where $\omega_2{}^2 \gg \omega_o{}^2$ and (3) the region $\omega_2 \leqslant \omega \leqslant \infty$. We write from (3), then, since $x_o(t)$ is real

$$r(t) = -\frac{1}{2\pi\omega_o{}^2} \int_o^{\omega_1} \left[\frac{\bar{a}_o(\omega)\,e^{j\omega t}}{1 - (\omega/\omega_o)^2 + 2j\zeta\omega/\omega_o} + \frac{\bar{a}_o^*(\omega)\,e^{-j\omega t}}{1 - (\omega/\omega_o)^2 - 2j\zeta\omega/\omega_o} \right] d\omega$$

$$- \frac{1}{2\pi\omega_o} \int_{\omega_1}^{\omega_2} \left[\frac{\bar{v}_o(\omega)\,e^{j\omega t}}{2\zeta - j\,(1+\omega_o/\omega)\,(1-\omega/\omega_o)} + \frac{\bar{v}_o^*(\omega)\,e^{-j\omega t}}{2\zeta + j\,(1+\omega_o/\omega)(1-\omega/\omega_o)} \right] d\omega$$

$$- \frac{1}{2\pi} \int_{\omega_2}^{\infty} \left[\frac{\bar{x}_o(\omega)\,e^{j\omega t}}{1 - (\omega_o/\omega)^2 - 2j\zeta\omega_o/\omega} + \frac{\bar{x}_o^*(\omega)\,e^{-j\omega t}}{1 - (\omega_o/\omega)^2 + 2j\zeta\omega_o/\omega} \right] d\omega \tag{4}$$

where an asterisk denotes complex conjugate and where $v_o(t)$ and $a_o(t)$ are the velocity and acceleration of the attachment point, respectively. Examination of the three integrals on the right side of (4) leads us to conclude that r(t) varies roughly as $a_o(t)$, $v_o(t)$ and $x_o(t)$ for low-frequency ($\omega^2 \ll \omega_o{}^2$), intermediate-frequency ($\omega \sim \omega_o$), and high-frequency ($\omega^2 \gg \omega_o{}^2$) input motions, respectively.

From the above development we conclude that, with respect to the "highest natural frequency" of a small attached system, we need not be concerned with upper intermediate- and high-frequency shell acceleration components or with high-frequency shell velocity components at the system's attachment point. This is fortunate, since high-frequency inaccuracies appear in computed acceleration histories before they appear in the corresponding velocity histories; high-frequency inaccuracies rarely appear in computed displacement histories.

In connection with the second use, that of predicting failure or survival of the shell itself, the quantities of interest are usually stresses or strains. These quantities are significant only to the extent that they combine in such a way as to reach a failure criterion, and, in almost all problems, a few of them greatly exceed the others in magnitude. Hence judgments regarding the accuracy of stress/strain computations should be based more upon considerations regarding peak values of significant stresses/strains and times of occurrence of the peak values than upon response details.

It is not always the case that stresses/strains are the quantities of interest in shell survivability studies. For example, in a nonlinear dynamic buckling investigation, one generally seeks the level at which a given type of excitation rather suddenly produces a dramatic increase in shell response. The selection of the response variable to be used is somewhat arbitrary, although the variable chosen is usually displacement.

FINITE DIFFERENCE AND FINITE ELEMENT METHODS

Although the finite difference method has been employed in the solution of two-dimensional continuum problems for over sixty years [3], the advent of large-scale digital computers has brought about a tremendous expansion in its application. The finite element method is truly a product of the computer age, its original applications to two-dimensional continuum problems dating back less than twenty years [4]. In static shell analysis both methods impose a gridwork on the shell, formulate a discrete model based on the gridwork, and solve a set of algebraic equations. The two methods differ in that the finite difference method discretizes the governing partial differential equations or energy functional for the continuous shell, while the finite element method discretizes the shell itself, representing it as an assemblage of plate or shell elements, each of which is described by a finite number of dependent variables.

In dynamic shell analysis, each method leads to a set of ordinary differential equations in time. In linear problems, these equations may be solved by modal superposition, following solution of the associated eigenvalue problem. A more direct approach that applies to both linear and nonlinear problems is solution by step-by-step numerical integration techniques. With this approach, computations march steadily outward from the initial time along a discretized time axis. Comprehensive treatments of the finite difference and finite element methods in structural mechanics are presented in [5] and [6]. A fairly recent treatment of step-by-step numerical integration techniques is found in [7]; [8] and [9] provide recent information regarding current numerical shell analysis capability.

Although we must necessarily refer the reader to the above references for an adequate review of the finite difference, finite element and step-by-step numerical integration methods, we would like to offer a few brief observations. The primary advantage of the finite difference method has derived from its simplicity of implementation with respect to either energy functionals or differential equations. The analyst uses the method with reasonable confidence that the numerical solutions will in fact converge to the true solution appropriate to the governing differential equations everywhere in the domain as the mesh size is reduced (strong convergence). The primary limitation of the finite difference method has been that it lacks the versatility to treat geometrically complex structures. The existence of this difficulty has naturally prompted research work in the development of variable grid capability [10].

The primary advantage of the finite element method has been precisely its ability to treat geometrically complex structures. There currently exists a vast array of beam, bar, plate and shell elements for use by the analyst. The primary limitation of the finite element method has been the lack of satisfactory criteria to assure convergence of the numerical solutions with decreasing mesh size.* Because very few finite element formulations reduce to the appropriate governing differential equations in the limit of vanishing mesh size, strong convergence is generally not to be expected. Recent work [11, 12] has produced certain criteria sufficient for weak convergence (in which appropriately smoothed numerical results converge with decreasing mesh size), which is all that is really required. A number of questions remain, however, and research is actively continuing in this area.

With regard to dynamic aspects, a critical need exists for increasingly rapid and accurate eigensolvers for application in solutions by modal superposition. The fastest eigensolvers for large systems at the present time employ iterative methods with modal sweepout and spectral shifting [13 - 15]. Recent work [16, 17] on step-by-step numerical integration techniques indicates that, even for nonlinear response problems, implicit methods (in which simultaneous equations must be solved at each time step) are becoming generally superior to explicit methods (in which solutions at a given time step may be computed directly from previously computed solutions at earlier times). This is because explicit methods are unstable if the time step exceeds a critical value that is often unacceptably small for shell structure response problems.

CONVERGENCE OF THIN SHELL RESPONSE COMPUTATIONS

We would now like to discuss in some detail a problem that arises in computations of the response of thin shells to rapidly applied loads. While the information presented is based on the finite difference method, it pertains to the general behavior of discrete models of continuous shell structures. As one looks for convergence of shell response computations with decreasing mesh size, one hopes at the very worst to find convergence as the mesh dimensions become comparable to or less than the thickness of the shell. This, in turn, implies convergence in the short structural wavelength region, where the effects of shell curvature are negligible. Hence, with regard to convergence questions in the short wavelength region, it is permissible to consider the simpler case of a plate instead of a shell.

---

*This pertains to the use of stationary as opposed to minimum functionals and to non-conforming elements.

The free in-plane and transverse motions of a thin isotropic plate are governed by the (elementary theory) partial differential equations [18]

in-plane:
$$
\frac{\partial^2 u}{\partial x^2} + \frac{1}{2}(1\text{-}\nu)\frac{\partial^2 u}{\partial y^2} + \frac{1}{2}(1+\nu)\frac{\partial^2 v}{\partial x \partial y} = \frac{1}{c^2}\frac{\partial^2 u}{\partial t^2}
$$

$$
\frac{\partial^2 v}{\partial y^2} + \frac{1}{2}(1\text{-}\nu)\frac{\partial^2 v}{\partial x^2} + \frac{1}{2}(1+\nu)\frac{\partial^2 u}{\partial x \partial y} = \frac{1}{c^2}\frac{\partial^2 v}{\partial t^2}
$$
(5)

transverse:
$$
\frac{\partial^4 w}{\partial x^4} + 2\frac{\partial^4 w}{\partial x^2 \partial y^2} + \frac{\partial^4 w}{\partial y^4} = -\left(\frac{1}{c\gamma}\right)^2 \frac{\partial^2 w}{\partial t^2}
$$

in which x and y are Cartesian coordinates, t is time, u (x,y,t), v (x,y,t) and w (x,y,t) are displacements of the plate's middle surface in the x-direction, y-direction and direction normal to the surface of the plate, respectively, $c = [E/\rho\,(1\text{-}\nu^2)]^{1/2}$ is the plate velocity, E, ρ and ν are Young's modulus, mass density and Poisson's ratio for the plate material, respectively, and $\gamma = h/\sqrt{12}$ is the plate's cross-sectional radius of gyration, where h is the plate thickness. These equations are readily converted to ordinary differential equations in time by application of the usual techniques involving second order $[0\,(\Delta x^2)\,]$ difference expressions. Assuming that we have written down these equations, we look for steady-state wave solutions of the form*

$$
\begin{Bmatrix} u_{mn} \\ v_{mn} \\ w_{mn} \end{Bmatrix} = \begin{Bmatrix} U \\ V \\ W \end{Bmatrix} \quad \exp\,[\,j\,(\omega t - k_x\,m\Delta x - k_y\,n\Delta y)\,]
$$
(6)

where ω is angular frequency, k_x and k_y are trace wavenumbers relative to the x- and y-axes, respectively, and m and n are mesh point indices along the x- and y- axes, respectively. Substitution of (6) into the ordinary differential equations for the finite difference plate yields homogeneous algebraic equations in U, V, and W. For non-trivial solutions, the determinant of the coefficient matrix must vanish, which leads to the frequency equations

$$
\frac{h}{c}\,\omega_{1,2}^{(i)}\,(k_x, \Delta x, k_y, \Delta y) = \left\{\,[\,(3\text{-}\nu) \pm (1+\nu)\sqrt{1\text{-}\epsilon}\,\,]\right.
$$

$$
\times \left.\left[\frac{1}{(\Delta x/h)^2}\sin^2\frac{k_x\Delta x}{2} + \frac{1}{(\Delta y/h)^2}\sin^2\frac{k_y\Delta y}{2}\right]\right\}^{1/2}
$$

---

*This approach differs slightly from that used in [19], which considers the analogous one-dimensional problem.

$$\frac{h}{c} \; \omega^{(t)} (k_x, \Delta x, k_y, \Delta y) = \frac{2}{3} \sqrt{3} \left[\frac{1}{(\Delta x/h)^2} \sin^2 \frac{k_x \Delta x}{2} + \frac{1}{(\Delta y/h)^2} \sin^2 \frac{k_y \Delta y}{2} \right]$$

(7)

where the superscripts (i) and (t) denote in-plane and transverse motion, respectively, and where

$$\epsilon = 4 \; \frac{\sin^2 \dfrac{k_x \Delta x}{2} \; \sin^2 \dfrac{k_y \Delta y}{2}}{(\Delta x/h)^2 \; (\Delta y/h)^2} \times \frac{\sin^2 \dfrac{k_x \Delta x}{2} + \sin^2 \dfrac{k_y \Delta y}{2} - \sin^2 \dfrac{k_x \Delta x}{2} \; \sin^2 \dfrac{k_y \Delta y}{2}}{\left[\dfrac{1}{(\Delta x/h)^2} \sin^2 \dfrac{k_x \Delta x}{2} + \dfrac{1}{(\Delta y/h)^2} \sin^2 \dfrac{k_y \Delta y}{2} \right]^2}$$

(8)

It is easily shown that these expressions reduce to the corresponding frequency equations for a continuous plate as $k_x \Delta x$ and $k_y \Delta y$ approach zero.

Frequency curves from (7), along with corresponding curves appropriate to (5) and to the exact theory of elasticity [18], are plotted in Figures 1-4 for various values of $\Delta x/h$ and $\Delta y/h$ and for the cases $k_x = k$, $k_y = 0$ and $k_x = k_y = k/\sqrt{2}$. In the first case, we see that $\epsilon = 0$, so that $\omega_2^{(i)} (k_x, \Delta x) = [\frac{1}{2} (1-\nu)]^{\frac{1}{2}} \; \omega_1^{(i)} (k_x, \Delta x)$; in the limit $k_x \Delta x \to 0$,

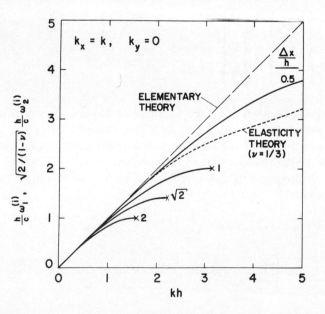

Fig. 1. Frequency Curves for Propagation of In-Plane Waves along a Mesh Axis (Elementary theory curve coincides with elasticity theory curve in the case of $\omega_2^{(i)}$)

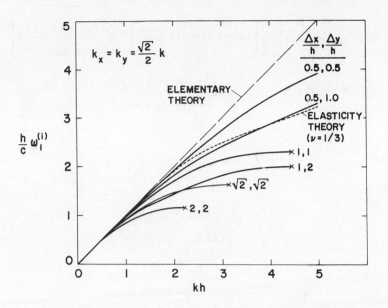

Fig. 2. Frequency Curves for Propagation of In-Plane Waves at 45° Angle with respect to Mesh Axes

$k_y \Delta y \to 0$, ϵ also vanishes, and $\omega_2^{(i)} (k_x, k_y) = [\frac{1}{2} (1-\nu)]^{\frac{1}{2}} \omega_1^{(i)} (k_x, k_y)$. The figures show that the frequency curves for the elementary theory discretized plate coincide with the corresponding curves for the continuous plate for small values of kh; as kh increases, the former deviate from the latter in a downward direction.

At "optimum" values of $\Delta x/h$, $\Delta y/h$, some elementary theory discretized plate curves lie close to the curves from elasticity theory over a wide kh range, one that considerably exceeds the range over which the elementary theory continuous plate curves lie close to the elasticity theory curves. This implies that a thin shell numerical solution with optimum discretization may in some cases be more accurate than even an exact solution of the corresponding thin shell differential equations! Figures 1 and 2 indicate that, for predominantly extensional motions with arbitrary k_x/k_y, the optimum mesh dimensions are $\Delta x/h = \Delta y/h \approx 0.7$. Figures 3 and 4 indicate that, for predominantly transverse motions with arbitrary k_x/k_y, the optimum mesh dimensions are $\Delta x/h = \Delta y/h \approx 1.2$. A comparison of Figures 1 and 2 with Figures 3 and 4 reveals that, for an arbitrary mix of in-plane and transverse motions, and for arbitrary k_x/k_y, the mesh dimensions $\Delta x/h = \Delta y/h \approx 1.2$ yield the grestest frequency range for accurate computations.

Each of the frequencies $\omega_1^{(i)}$, $\omega_2^{(i)}$ and $\omega^{(t)}$ reaches its maximum at $k_x \Delta x = k_y \Delta y = \pi$, beyond which the group velocity $c_g = d\omega/dk$ becomes negative. It has been shown [20] that, beyond such "cutoff" points, the wavenumber k is no longer real and no travelling wave exists. Hence we conclude that the maximum frequencies of propagating waves in the discretized plate are

$$\frac{h}{c}\,\omega_{M1,2}^{(i)}\,(\Delta x, \Delta y) = \left\{\left[(3\text{-}\nu) \pm (1+\nu)\sqrt{1\text{-}\epsilon_M}\,\right]\frac{(\Delta x/h)^2 + (\Delta y/h)^2}{(\Delta x/h)^2\,(\Delta y/h)^2}\right\}^{1/2}$$

$$\frac{h}{c}\,\omega_M^{(t)}\,(\Delta x, \Delta y) = \frac{2}{3}\,\sqrt{3}\,\frac{(\Delta x/h)^2 + (\Delta y/h)^2}{(\Delta x/h)^2\,(\Delta y/h)^2}$$

$$(9)$$

where

$$\epsilon_M = 4\,\frac{(\Delta x/h)^2\,(\Delta y/h)^2}{\left[(\Delta x/h)^2 + (\Delta y/h)^2\right]^2} \qquad (10)$$

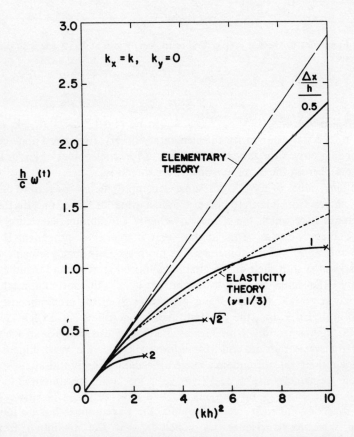

Fig. 3. Frequency Curves for Propagation of Transverse Waves Along a Mesh Axis

Actually, cutoff can occur for waves travelling in a prescribed direction at frequencies below those given by (9). For example, if $\Delta y > \Delta x$, flexural waves travelling along the y-axis will reach cutoff at a frequency lower than either the cutoff frequency for waves travelling along the x-axis or the maximum frequency given by the second of (9). The minimum cutoff frequencies are those appropriate to propagation along the axis (say the y-axis) with the coarser mesh ($\Delta y > \Delta x$), and are given by

$$\frac{h}{c} \; \omega^{(i)}_{m1,2} \; (\Delta y) = \left[\, (3\text{-}\nu) \pm (1\text{+}\nu) \, \right]^{1/2} / (\Delta y / h)$$

$$, \Delta y > \Delta x \qquad (11)$$

$$\frac{h}{c} \; \omega^{(t)}_{m} \; (\Delta y) = \frac{2}{3} \sqrt{3} / (\Delta y / h)^2$$

It is important to remember that (7) − (11) pertain only to second order finite difference expressions. The use of higher order difference expressions produces curves in

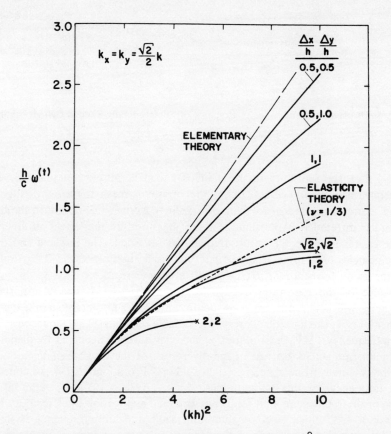

Fig. 4. Frequency Curves for Propagation of Transverse Waves at 45° Angle with respect to Mesh Axes

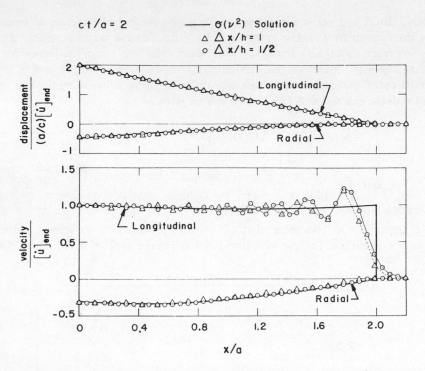

Fig. 5. Displacement and Velocity Response of Longitudinally Excited Cylindrical Shell

Figures 1-4 that, for a given mesh size, lie closer than the discretized plate curves shown to the elementary theory curves. Hence the optimum mesh for a set of higher order difference expressions is coarser than the $\Delta x/h = \Delta y/h = 1.2$ optimum mesh for the second order difference expressions. Simple measures of the effect of higher order difference expressions are the cutoff frequencies obtained. The in-plane and transverse cutoff frequencies of (11), for example, are increased about 15% and 29%, respectively, through the use of fourth order $[0\,(\Delta x^4)]$ difference expressions.

Knowledge of the maximum frequencies for a given mesh size is important since spurious noise will occur in the response computations if a discrete system is significantly excited (with respect to a particular response variable of interest) at frequencies above the maximum frequency [21]. Deterioration in accuracy actually occurs at frequencies below the maximum frequencies because of the divergence of the discrete plate frequency and group velocity curves from those of elasticity theory. From Figures 1-4, we conclude that the limit frequency for the optimum mesh $\Delta x/h = \Delta y/h \approx 1.2$ is $\omega_l \approx 2c/3h$, which is about two or three times the limit frequency for valid application of elementary thin shell theory. For a 1/8" thick steel or aluminum shell, the limit frequency is thus about 175,000 Hz., which is well beyond the frequency range of interest in most structural dynamics analyses.

TRANSIENT RESPONSE COMPUTATIONS

We have observed that a discrete model of a continuous shell is generally accurate only for response Fourier components with frequencies below a limit frequency that is somewhat lower than the minimum cutoff frequency for the mesh. Let us now observe some linear response computations that exhibit the effects of this limitation. These computations are all performed with a finite difference code based on the thin shell equations of [22]. Second order finite difference expressions are applied to these equations and the (explicit) central finite difference numerical time integration scheme is used [23].

Figure 5 shows displacement and velocity response computations for a long cylindrical shell that is subjected to an axisymmetric, longitudinal step-end-velocity excitation. Because the effects of dispersion are small, response "snapshots" (response as a function of axial coordinate) yield essentially the same information as response "histories" (response as a function of time). Hence only snapshots are shown. We see that the discontinuity in the longitudinal velocity response of the continuous shell causes spurious noise, or "ringing" of the finite difference mesh at approximately the mesh cutoff frequency $2c/\Delta x$ [see (11)]. Because of the absence of discontinuities, or alternately, because Fourier components with frequencies near or above the mesh cutoff frequency do not contribute significantly to longitudinal displacement response, radial displacement response or radial velocity response, the finite difference computations for these quantities are quite satisfactory. The $0(\nu^2)$ solutions given in Figure 5 are approximate, but accurate, analytical solutions obtained for a membrane shell by a perturbation technique based on the smallness of ν^2 [24].

Acceleration and stress snapshots for the same shell are shown in Figure 6. It is rather interesting to observe the attempt made by the code to deal with the Dirac delta-function that characterizes longitudinal acceleration response. We note that the spurious oscillations in the finite difference computations decay with increasing distance from the delta-function. As one would expect, the stress computations exhibit the same behavior as that which characterizes the longitudinal velocity computations.

Let us now examine the response of a cylindrical shell to a uniform radial impulse excitation. Figure 7 shows radial displacement and velocity snapshots at a time when a flexural disturbance generated at either end of the shell and travelling at the plate velocity would first reach the shell's mid-station ($x = L/2$). The displacement computations indicate that very little flexural energy travels this fast, suggesting instead that the front of the boundary-generated disturbance travels at approximately the shear velocity $c_s = [\frac{1}{2}(1-\nu)]^{\frac{1}{2}}c$. This is in agreement with the results of [25], which applies the method of characteristics to shell equations appropriate to improved (Timoshenko) theory. The velocity computations, however, show signs of convergence difficulties that appear to be unrelated to the velocity discontinuity at the shear wave front [25]. In fact, the $\Delta x/h = 1/2$ velocity computations predict the arrival of a boundary-generated disturbance that travels faster than even the plate velocity.

Because of the severe dispersion that characterizes the propagation of flexural waves, it

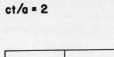

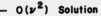

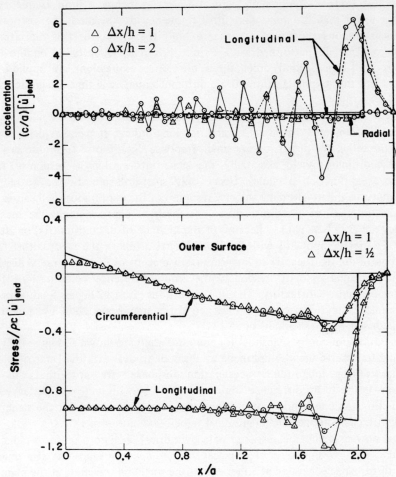

Fig. 6. Acceleration and Stress Response of Longitudinally Excited Cylindrical Shell

is generally easier to interpret computed response histories than response snapshots, especially at later times. Figure 8 shows radial displacement and velocity histories at the mid-station of the impulsively excited cylindrical shell; these display the same convergence behavior as that observed in Figure 7. Acceleration and strain histories at the mid-station are shown in Figure 9. We observe that the former are so completely

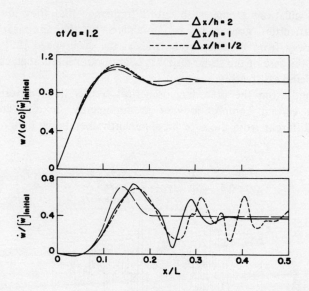

Fig. 7. Radial Displacement and Velocity Response of Radially Excited Cylindrical Shell

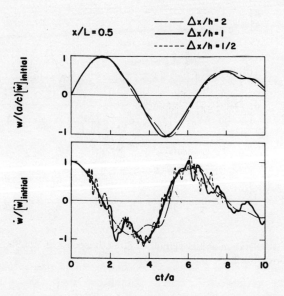

Fig. 8. Radial Displacement and Velocity Response at Mid-Station of Radially Excited Cylindrical Shell

dominated by oscillations at the mesh cutoff frequency that they are unusable.* The convergence difficulties appear to be unrelated to the propagation of a Dirac delta-function away from the shell boundaries at the shear speed [25]. The problem is not so severe in the case of the strain response computations, although convergence could hardly be considered satisfactory.

We have found from the preceding numerical results that convergence difficulties associated with certain computations of predominantly flexural shell response are fundamentally different from those of predominantly membrane shell response. Whereas

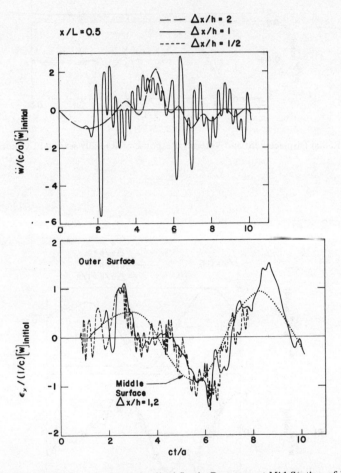

Fig. 9. Radial Acceleration and Longitudinal Strain Response at Mid-Station of Radially Excited Cylindrical Shell

---

*Acceleration computations for $\Delta x/h$ = 1/2, which are not shown in Figure9, oscillate at four times the frequency and with twice the amplitude of the $\Delta x/h$ = 1 computations.

the membrane response computations exhibit relatively mild difficulties in the vicinity of response discontinuities, flexural response computations exhibit rather severe difficulties throughout the temporal and spatial domains. This is not readily explained by Figures 1-4, because the membrane and flexural finite difference frequency curves exhibit essentially the same behavior (deviation from the elementary theory curves, cutoff, etc.). It seems likely that the difference in convergence behavior is connected with the fact that short wavelength membrane response is governed by hyperbolic equations (second order in time, second order in space) whereas short wavelength flexural response is governed by a parabolic equation (second order in time, fourth order in space). This implies that numerical computations based on improved (Timoshenko) shell theory will not exhibit the severe convergence difficulties just observed. A definitive statement in this regard must, however, await further study. In the meantime, it is useful to know when to expect convergence problems in thin shell response computations, and what one might do to alleviate them.

SMOOTHING

From an examination of the numerical results presented here and given in [26], we conclude that short wavelength convergence difficulties may be expected in thin shell computations for a certain kinematic response variable whenever the improved (Timoshenko) shell equations indicate that the prescribed excitation produces discontinuities of some order in that variable. With regard to a stress response variable, convergence difficulties may be expected in thin shell computations whenever the improved shell equations predict discontinuities in any of the displacement gradients to which that stress (as determined from elementary theory) is proportional. An alternate interpretation is that convergence difficulties may be expected in a response history if the Fourier transform of that history, as determined from elementary theory, does not decay faster than ω^{-1} for ω greater than the limit frequency for the mesh. For example, the computations for the step-end-velocity-excited cylindrical shell exhibit moderate convergence difficulties with respect to longitudinal velocity and stress response and appreciable convergence difficulties with respect to longitudinal acceleration response; the computations for the radial-impulse-excited cylindrical shell demonstrate moderate convergence difficulties with respect to radial velocity and longitudinal strain response, and severe difficulties with respect to radial acceleration response.

Perhaps the simplest method for alleviating the above convergence difficulties is temporal smoothing of the excitation. From Figures 1-4, we see that, for $\omega \lesssim \omega_l = 2c/3h$, all of the finite difference curves with $\Delta x/h \leqslant 1.2$, $\Delta y/h \leqslant 1.2$ lie close to the elementary theory and elasticity theory curves, which are themselves reasonably close. Hence, if all Fourier components of the excitation with frequencies above ω_l (i.e., with periods less than about ten times the extensional wave transit time through the thickness, $t_e = h/c$) are completely filtered out, we would expect to find satisfactory convergence in all quantities for $\Delta x/h \lesssim 1.2$, $\Delta y/h \lesssim 1.2$. A digital filter that does, in fact, completely filter out all Fourier components with frequencies above a prescribed filter cutoff frequency, while introducing no amplitude change or phase shift into the remaining components, can be constructed [27]. Application of an ideal filter is not really required, however, as we see in the following example.

Consider the radial-impulse-excited cylindrical shell discussed above; let us replace the impulse loading with the triangular pressure pulse

$$p\,(x,\theta,t) \;=\; \frac{2I}{t_w} \left\{ \begin{array}{l} 2\,t/t_w,\, 0 \leqslant t \leqslant \dfrac{1}{2}\,t_w \\[2mm] 2\,(1\text{-}t/t_w),\, \dfrac{1}{2}\,t_w \leqslant t \leqslant t_w \\[2mm] 0,\, \text{otherwise} \end{array} \right. \tag{12}$$

where I is the magnitude of the impulse. The Fourier transform of this excitation is

$$\bar{p}\,(x,\theta,\omega) = I \cdot \frac{\sin^2 \left(\dfrac{\omega t_w}{4} \right)}{\left(\dfrac{\omega t_w}{4} \right)^2}\, e^{-\frac{1}{2}\,j\omega t_w} \tag{13}$$

The spectrum $|\bar{p}\,(x,\theta,\omega)|$ is shown in Figure 10. It is clearly a compact function with a reasonably well defined excitation cutoff frequency

$$\omega_x \approx \frac{4\pi}{t_w} \tag{14}$$

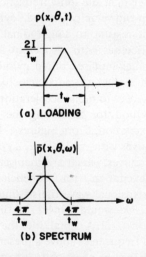

(a) LOADING

(b) SPECTRUM

Fig. 10. History and Spectrum of Smoothed Radial Excitation

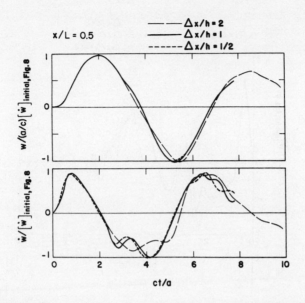

Fig. 11. Radial Displacement and Velocity Response of Cylindrical Shell to Smoothed
Radial Excitation

Thus we conclude that the replacement of the impulse excitation by the triangular
pressure pulse constitutes an effective smoothing operation.

Figure 11 shows radial displacement and velocity histories at the mid-station of the
cylindrical shell for a triangular pressure loading of width t_w = 16.4 h/c. This corresponds
to an excitation cutoff frequency of $\omega_x \approx 0.77$ c/h, which is slightly higher than ω_I.
Even so, the smoothing process has produced convergence behavior in the velocity
computations that is clearly superior to that shown in Figure 8. The improvement is even
more pronounced in the longitudinal acceleration and strain response computations
shown in Figure 12, which are to be compared with those of Figure 9.

Temporal smoothing of the excitation is not the only method for alleviating
convergence difficulties in computed responses. Other methods, for example, involve the
introduction of artificial damping into the shell or the use of a step-by-step numerical
integration scheme that artifically damps out high-frequency response. The advantage of
temporal smoothing of the excitation over these methods is that the effects of the former
on the original problem are rather easily evaluated, whereas the latter introduce
modifications into the problem whose effects are somewhat difficult to assess.

CONCLUSION

Following some brief comments on 1) common applications of shell response
computations, 2) the primary advantages and limitations of the finite difference and
finite element methods at the present time, and 3) recent efforts in the development and

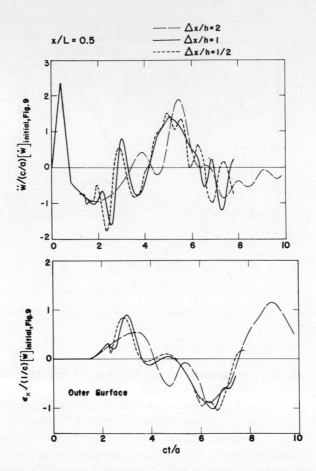

Fig. 12. Radial Acceleration and Longitudinal Strain Response of Cylindrical Shell to Smoothed Radial Excitation

use of efficient, large-scale eigensolvers and step-by-step numerical integration techniques, we have examined in some detail the convergence of thin shell response computations appropriate to rapidly applied loads. The results of this examination pertain in general to situations in which a continuous shell structure is modeled as a discrete system with a maximum natural frequency.

Although we have described the nature of convergence problems associated with the replacement of a continuous shell structure with a discrete analog and have proposed temporal smoothing of the excitation as the simplest remedy, we have left a number of questions unanswered. For example, why are convergence difficulties associated with the computation of flexural shell response significantly more severe than those associated with the computation of membrane shell response? Will the use of improved (Timoshenko) shell theory satisfactorily resolve the problem? What smoothing criteria does one use for nonlinear response problems? And finally, what about computations appropriate to an irregular mesh that is constructed to handle a configuration of complex geometry? We suggest that answers to these and related questions be high priority items for researchers in this area.

ACKNOWLEDGMENT

The authors are indebted to their colleagues at the Lockheed Palo Alto Research Laboratory, especially to Dr. C. A. Felippa and Mr. P. S. Jensen for their contributions to the discussion of the finite difference, finite element and step-by-step numerical integration methods. The work reported herein was carried out under Contract NAS 1-9111 with the NASA Langley Research Center and under the Lockheed Missiles & Space Company's Independent Research Program.

REFERENCES

1. S. Timoshenko and D. H. Young, *Vibration Problems in Engineering,* 3rd Ed. Van Nostrand, New York (1955).
2. J. A. Aseltine, *Transform Methods in Linear Systems Analysis,* McGraw-Hill, New York (1968).
3. C. Runge, Uber eine methode die partielle differentialgleichung Δu = constans numerish zu integrieren, *Z. Math, u. Phys.* **56**, 225 (1908).
4. M. J. Turner, R. W. Clough, H. C. Martin and L. J. Topp, Stiffness and deflection analysis of complex structures, *J. Aero. Sci.* **23**, 805 (1956).
5. G. E. Forsythe and W. R. Wasow, *Finite-Difference Methods for Partial Differential Equations,* Wiley, New York (1960).
6. O. C. Zienkiewicz and Y. K. Cheung, *The Finite Element Method in Structural and Continuum Mechanics,* McGraw-Hill, New York (1967).
7. P. Henrici, *Discrete Variable Methods in Ordinary Differential Equations,* Wiley, New York (1962).
8. R. F. Hartung, An assessment of current capability for computer analysis of shell structures, *J. Comput. & Struct.* **1**, 3 (1971); also AFFDL-TR-71-54.
9. R. F. Hartung (Ed), *Proc. Conf. on Computer Oriented Analysis of Shell Structures,* Lockheed Missiles & Space Co., Palo Alto, California (1970) AFFDL-TR-71-79.
10. P. S. Jensen, Finite difference techniques for variable grids, *Proc. Conf. on Computer Oriented Analysis of Shell Structures,* **9**.
11. C. A. Felippa and R. W. Clough, The finite element in solid mechanics, *Proc. Sym. Appl. Math.* **21**, Amer. Math. Soc., Providence, R.I. (1969).
12. E. R. de Arantes e Oliveira, Completeness and convergence in the finite element method, *Proc. of the Second Conf. on Matrix Methods in Structural Mechanics,* Wright-Patterson Air Force Base, Ohio (1968), AFFDL-TR-68-150.
13. J. H. Wilkinson, *The Algebraic Eigenvalue Problem,* Clarendon Press, Oxford (1965).
14. F. Brogen, K. Forsberg and S. Smith, Experimental and analytical investigations of the dynamic behavior of a cylinder with a cutout, *AIAA J.* **7**, 903 (1969).
15. W. D. Whetstone and C. E. Jones, Vibrational characteristics of linear space frames, *J. Struc. Div.,* ASCE **9**, 2077 (1969).
16. I. W. Sandberg and H. Schickman, Numerical integration of systems of stiff nonlinear differential equations, *Bell Systems Tech. J.* **47**, 511 (1968).

17. W. Liniger and R. A. Willoughby, Efficient integration methods for stiff systems of ordinary differential equations, SIAM *J. Num. Anal.* **7**, 47 (1970).
18. R. D. Mindlin, *An introduction to the mathematical theory of vibrations of elastic plates,* U.S. Army Signal Corps Engr. Labs, Fort Monmouth, N.J. (1955).
19. R. D. Krieg, *Phase velocities of elastic waves in structural computer programs,* SC-DR-67-816, Sandia Laboratories, Albuquerque, New Mexico (1967).
20. T. von Karman and M. A. Biot, *Mathematical Methods in Engineering,* McGraw-Hill, New York (1940).
21. R. S. Dunham, R. E. Nickell and D. C. Stickler, Integration operators for transit structural response, *Proc. Conf. on Computer Oriented Analysis of Structures,* **9**.
22. J. Kempner, *Unified thin shell theory,* Polytechnical Institute of Brooklyn, PIBAL Report No. 566 (1960).
23. L. H. Sobel, W. Silsby and B. G. Wrenn, *Computer user's manual for the STAR (Shell Transient Asymmetric Response) code,* Space & Missile Systems Organization Report SAMSO-TR-70-240 (1970).
24. T. L. Geers, Analysis of wave propagation in elastic cylindrical shells by the perturbation method, *J. Appl. Mech.* (to appear).
25. P. C. Chou, Analysis of axisymmetrical motions of cylindrical shells by the method of characteristics, *AIAA J.* **6**, 1492 (1968).
26. T. L. Geers and L. H. Sobel, *Analysis of transient, linear wave propagation in shells by the finite difference method,* NASA CR-1885.
27. T. L. Geers and S. A. Denenberg, *A digital filter for separating high- and low-frequency components of a transient signal,* David Taylor Model Basin (now NSRDC) Rpt 1795 (1964) AD432277.

DYNAMIC PROPERTIES OF FULL-SCALE STRUCTURES DETERMINED FROM NATURAL EXCITATIONS

DONALD E. HUDSON

Professor of Mechanical Engineering and Applied Mechanics
California Institute of Technology

Abstract – Most buildings and engineering structures are being constantly excited at low vibration levels by wind forces or by microtremor motions of the ground. In addition, larger earthquakes occasionally subject structures to higher dynamic force levels. Such natural excitations afford a convenient way to conduct certain types of dynamic tests of full-scale structures. By direct measurement of structural response to natural ambient excitations, natural frequencies, mode shapes, and damping may be determined with relative ease. As examples of such investigations, information is given on: (1) measurement and analysis techniques for the determination of structural dynamic properties from ambient tests; (2) the comparison of natural frequencies, mode shapes and damping as determined from low-level wind and microtremor tests with values obtained at higher force levels from steady-state sinusoidal forced vibration tests; (3) the alterations in natural frequencies of buildings caused by strong earthquake shaking as determined from before-and-after low-level ambient tests; and (4) the determination of structural dynamic properties at high force levels by the measurement of input and response motions during a strong earthquake.

INTRODUCTION

The difficulties of carrying out dynamic tests of full-scale civil engineering structures are manifold. Such structures are usually physically large, and require large forces for their excitation. The physical size usually precludes loading from a fixed point, which means that some form of inertia force loading or reaction force device is required. Each structure is likely to be unique, and no testing that could cause a degradation of the structure can be tolerated. The testing of a prototype to destruction, as can often be done in aircraft development, is an unlikely possibility for buildings, dams, etc. Furthermore, the structure is likely to be put into limited use even before it is finished, so that the experimenter cannot count on even a brief period of undisturbed non-destructive testing, nor can a time be found during which building occupants are not likely to be disturbed.

The economics of such test programs are also distinctly unfavorable as compared with other fields. Since civil engineering structures vary so much, it is difficult to transfer test information from one structure to another, and the value of the test program may often be limited to the specific system to which it was applied.

The above situation suggests that a considerable effort is needed to develop simple dynamic tests, and, in particular, to exploit to the utmost dynamic loadings which may arise naturally during the life of a structure. Most of these incidental dynamic loads, such as traffic on a bridge, or microtremor earth excitations of buildings, represent very low load levels compared with the design loads of the structure. Some of them, however, such

159

as strong winds or earthquakes, can indeed subject a structure to significant or even damaging dynamic loads. If suitable instrumentation can be available at the right place and the right time, such larger natural excitations can form a testing technique of great importance.

TYPES OF NATURAL EXCITATION

Perhaps the most important low force level excitations are wind loads. For structures such as tall buildings and bridges, it is seldom that the atmosphere is so quiet that easily measurable motions of the structure are not being continuously excited. At less frequent intervals, gusty winds of a higher velocity will cause larger motions, and will in many cases excite clear patterns of vibration in simple modes. Such loads are sufficiently common in most localities so that it is feasible to plan test programs in advance, with confidence that after the temporary installation of response measuring instruments, one will not need to wait more than a few minutes or at most an hour or so for a suitable excitation.

All structures are also being continually subjected to small motions transmitted from the ground to the foundations or arising from disturbances within the structure. These microseismic or microtremor motions are the result of microearthquakes, ground vibrations caused by traffic, machinery, wind in trees, surf on a coast, etc. In addition, the operation of machinery within a structure, such as elevators, air conditioning fans, or the movement of people, will result in a more-or-less random vibration at measurable levels. Such microtremor oscillations are sometimes larger than wind oscillations at some sites.

For higher levels of excitation, strong winds or earthquakes constitute the main possibilities. Since such events are fortunately rare, and are also unpredictable, this type of testing cannot often be expected, and must be carefully prepared for in advance of the event. Nevertheless, some notable results have been obtained from such excitations, as will be later described.

WIND EXCITATIONS

The simplest type of wind excitation test involves the installation of a seismograph having a magnification in the range of 200-800 in an upper story location [1]. A seismic transducer having a natural period in the range of one second is appropriate for most applications in high-rise buildings, since this relatively low-frequency device will not show response to machinery induced vibrations. If an accelerometer type transducer is used, a sharp high-frequency cut-off filter will be needed to suppress high-frequency excitations which would otherwise dominate the record. It will usually be found that within a few minutes of instrument installation and adjustment, a suitable small gust of wind will excite a clear sinusoidal oscillation at the fundamental natural frequency of the structure. In some cases, gusts having spatially distributed force patterns suitable for exciting higher

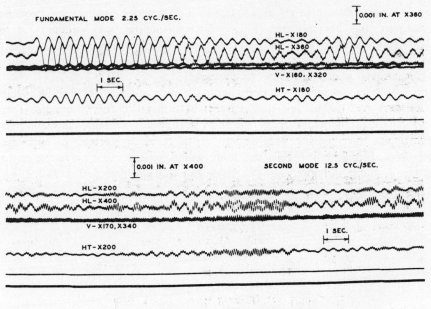

SAMPLE RECORDS OF WIND-EXCITED TOWER VIBRATIONS

Fig. 1. Wind-excited vibrations at the top of a concrete intake tower of a dam.

modes will occur. Figure 1 shows typical results of such wind excitations as measured on top of a concrete intake tower of a dam [2]. In this example a visual inspection of the record and a simple measurement of the number of cycles per unit time is all that is required to complete the analysis. It will be noted that a clear second mode motion has been excited as well as the fundamental.

Measurements of this type have been used to determine the fundamental natural frequencies of several hundred buildings in various cities [3,4]. The idea of these tests is to build up a body of "pre-earthquake" information on building frequencies for comparisons made after strong earthquake shaking. It would be expected that structural changes induced by strong earthquake loadings might manifest themselves as significant changes in such building frequencies, and such changes might in fact be a clue to structural damage not otherwise evident. Results related to this general idea will be presented in a later section.

It may happen that no wind gust loadings occur to excite clear natural modes, and that the record obtained from the total ambient excitation including atmospheric disturbances, microtremor excitations through the foundations, and machinery and personnel excitations from within the building, may have the appearance of a random response at a very low response level. In such cases, no natural frequencies are clearly evident to the eye, but a more involved spectral analysis may in fact reveal them.

One suggestion for the analysis of such quasi-random responses is that the autocorrelation function be calculated, from which both the fundamental frequency and the damping may be calculated [5,6,7]. Such autocorrelation functions are easy to calculate numerically with relatively simple techniques, but the method has been found to be limited in practical use to a determination of the lowest natural frequency and to give only a rough approximation to the damping [6].

A more direct and a more complete approach is to calculate the Fourier spectrum of the response [8,9]. This spectrum curve will indicate clearly the major frequency components, and the width of the spectrum peaks will yield reasonably good damping values for the lower modes. With modern digital computer techniques, such spectrum calculations are relatively simple, and there would appear to be no longer any special advantage to the more limited autocorrelation function approach [10, 11, 12, 13].

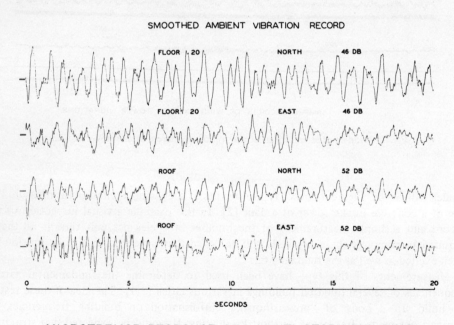

SMOOTHED AMBIENT VIBRATION RECORD

MICROTREMOR RESPONSE OF A MULTI-STORY BUILDING

Fig. 2. Microtremor response at the top of a 22-story steel-frame building.

In Figure 2 is shown a sample microtremor response measured on the top floor of a 22-story steel-frame building. Such response records are obtained by recording in the field on magnetic tape which is then played back through a laboratory analog-to-digital convertor which produces a computer-compatible digital tape input. The magnetic tapes were digitized at 200 points per second, and were then digitally filtered to remove a 30 cps noise signal generated by machinery operating within the building. A digital smoothing process was then carried out, and every 5th point was retained for analysis. The Fourier amplitude spectra shown in Figure 3 were then calculated using the Cooley-Tukey algorithm. It should be noted that the method becomes of practical importance only if the full resources of the above type of automatic data-processing can be employed.

It is important to note that the amplitudes of the motions involved in the above tests are of the order of 10^{-3} cm or 10 microns. A natural question is if motions at these low

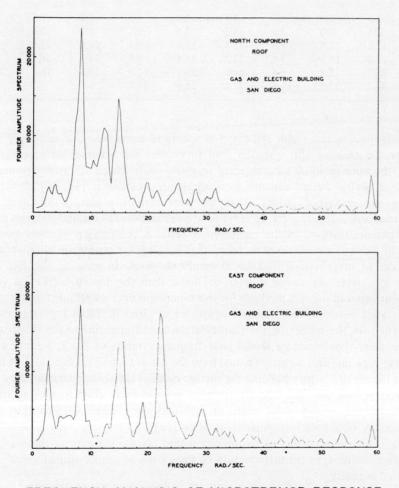

FREQUENCY ANALYSIS OF MICROTREMOR RESPONSE
OF A MULTI-STORY BUILDING

Fig. 3. Fourier spectra of microtremor response at the top of a 22-story steel-frame building.

TABLE I

Resonance Tests and Ambient Frequencies (2π f rad/sec) Compared [14]

Mode	NS Translation			EW Translation			Torsion		
	Ambient	Resonance	f_i/f_1 ratio ambient	Ambient	Resonance	f_i/f_1 ratio ambient	Ambient	Resonance	f_i/f_1 ratio ambient
1	2.7	2.5	1.0	2.5	2.4	1.0	2.7	2.6	1.0
2	7.5	7.0	2.8	8.2	7.5	3.3	8.1	7.8	3.0
3	12.5	12.5	4.6	15.4	14.3	6.2	14.6	13.5	5.4
4	19.0	18.9	7.0	22.2	21.4	8.9	21.3	21.1	7.9
5	25.3	26.8	9.4	29.1	26.9	11.7	29.0	28.1	10.8
6	31.7	32.1	11.8	37.8	31.0	15.1	–	–	–

levels will involve the major structural elements in essentially the same way as larger motions. To examine this question, and to explore the accuracy of the microtremor results, the same building was subjected to steady-state forced vibration resonance tests using an eccentric weight variable speed mechanical oscillator [14, 15]. By accurately defining the resonance curves associated with various modes of vibration, natural frequencies and damping were determined at peak amplitudes about 100 times as large as the amplitudes involved in the microtremor tests. A comparison of these two tests is shown in Table I, where it will be noted that very similar results are obtained over this two order of magnitude difference in amplitude level. In view of the fact that the microtremor tests are much simpler to make than the forced oscillation tests, the agreement between the two methods for this building is of considerable interest.

The ratios between the various frequencies as given in Table I provide interesting information on the nature of the mathematical model appropriate for the structure. A uniform shear type structure would have frequency ratios of (1, 3, 5, . . .), whereas a cantilever type uniform structure would have the ratios (1, 6. 25, 17. 5, . . .). It is evident that for the above 22-story building the torsion mode is closely approximated by the pure shear system, and that the transverse modes are also fairly close to such a simple pure shear model. One of the useful results of such frequency tests is the light they throw on the adequacy of simple mathematical models [16]. To treat of even relatively simple building structures in their full complexity would tax the capacity of the most modern computing facilities, so the introduction of some form of simplified mathematical model is a virtual necessity.

By making simultaneous microtremor measurements at various locations in a structure, mode shapes can be determined as well as frequencies. Figure 4 shows a comparison of microtremor determined mode shapes with those measured during the 100 times greater amplitude forced vibration tests [14]. Again it is clear that the simple microtremor test has produced useful results.

TABLE II

Comparison of Damping Values for Resonance and Ambient Tests
(% Critical) [14]

Mode	NS		EW		Torsion	
	Ambient	Resonance	Ambient	Resonance	Ambient	Resonance
1	9.5	2.6	11.62	1.9	9.2	3.0
2	3.5	2.7	4.0	1.6	5.7	3.4
3	2.6	3.7	4.6	3.1	–	2.9
4	–	3.9	–	2.8	–	3.0
5	–	3.1	–	2.9	–	3.0
6	–	4.4	–	2.8	–	3.4

It is difficult to obtain accurate damping values from low-level ambient tests, and the strong nonlinearities of certain types of damping mechanism increases further the difficulty of properly evaluating such damping determinations. In Table II is shown a comparison of damping values determined for the 22-story building by the microtremor and the forced vibration resonance test [14]. At the higher amplitudes of the forced

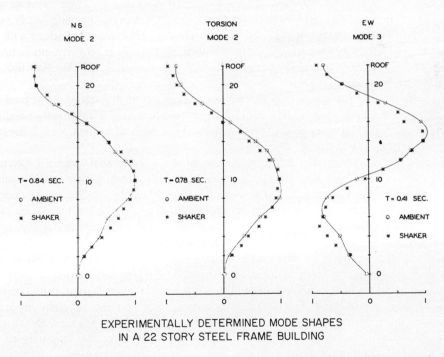

EXPERIMENTALLY DETERMINED MODE SHAPES
IN A 22 STORY STEEL FRAME BUILDING

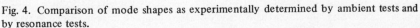

Fig. 4. Comparison of mode shapes as experimentally determined by ambient tests and by resonance tests.

vibration test, higher values of the equivalent viscous damping factors would certainly be expected. This was not in fact the case, and this can only mean that the damping values obtained for the microtremor test where somehow considerably in error. For this building this inaccuracy is probably caused by the fact that the frequency peaks are relatively close together, thus tending to superimpose response curves and broaden out the peaks [17]. For the more elaborate resonance tests, this situation could be more easily identified and corrected for [15, 17]. In general, it should probably not be expected that very useful damping information would be obtained from such low-level excitations.

MAN-EXCITED VIBRATIONS

For some types of structures an improvement in low-level wind excitation tests can be achieved by the use of resonant applied forces generated by the motion of people within the structure [18]. It has often been observed that the motion of people within buildings can excite measurable motions during tests with sensitive vibration pickups. During some wind-excited tests on a concrete intake tower, for example, it was noted that a considerable deflection on the sensitive seismograph at the top of the tower could be set up by the back-and-forth movements of the instrument operator. This suggested that the operator by synchronizing his movements with the natural frequency of the building might be able to build up resonant amplitudes greater than those excited by small wind gusts. It was in fact found that with a little practice the operator, by keeping one eye on a visual seismograph recorder display, could effectively synchronize his motions with the structure and produce a surprisingly large amplitude [19]. Such observations naturally recall old stories about bridges being destroyed by marching armies that did not "break-step" to avoid resonances.

At first thought it seems highly unlikely that a large enough exciting force could be generated in this simple way to be useful for large structures such as multistory buildings. In fact, it is just for such large structures with their low natural frequencies and damping that the method is most useful.

One important advantage of the method over the wind is that once a measurable amplitude has been built up, the operator can stop his motion to permit free damped vibrations to occur from which the structural damping can be determined with some accuracy. This requires, of course, that an amplitude of motion be built up which much exceeds that produced by random wind excitations, which would otherwise distort the free vibration decay curve.

To test this method of man-excited vibrations, measurements were made in a nine-story steel-frame building under construction. By carrying out the man-excited test at various locations in the building depending on the mode to be excited, it was possible to accurately determine eight natural frequencies corresponding to eight modes of vibration, and to get reasonably accurate values of damping [20]. Figure 5 shows some sample records made in this way, and indicates the clarity and the simplicity of the results. Complete vibration generator tests were also made on this particular building, and the more accurate data obtained from steady-state resonance curves agreed very well with

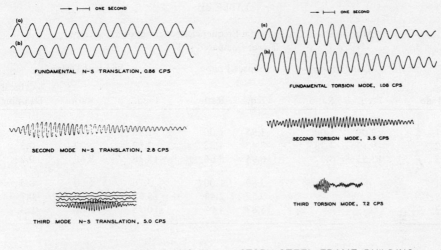

MAN-EXCITED VIBRATIONS IN A 9-STORY STEEL FRAME BUILDING

Fig. 5. Man-excited vibration modes in a nine-story steel-frame building.

the above man-excited tests [20].

Although such man-excited tests are of course not universally applicable, and are limited to relatively low excitation levels, it is so simple a procedure when it is appropriate that it can make a valuable contribution to many dynamic tests.

COMPARISON OF CALCULATED AND MEASURED STRUCTURAL PROPERTIES

There have been relatively few structures for which a direct comparison can be made between calculated natural frequencies and mode shapes, and those obtained from tests. One example is that of a reinforced concrete clock tower on the campus of the University of British Columbia [21]. This is a 121 ft. high tower having an octagonal section of approximately 13 ft. diameter. The natural frequencies were calculated using two different lumped parameter models for comparison. The first model was a simplified plane-frame system using 10 lumped masses. A more complicated space frame model using 36 masses was also investigated. Ambient vibration tests of the tower were of the low-level wind excitation type, with responses recorded on magnetic tape and Fourier analyzed by standard computer techniques. Table III shows the main results of these studies. The limitations involved in using a small number of degrees of freedom in the model are clearly indicated by the increasing inaccuracies of the higher torsional modes. These results are probably typical of the accuracies to be expected for relatively simple structures.

TABLE III

Analytical and Experimental Frequencies (cyc/sec) of a Concrete
Tower Compared [21]

	Mode	Plane Frame		Space Frame		Ambient Tests		
		Freq.	Ratio	Freq.	Ratio	Freq.	Ratio	% Critical Damping
Translation	1	2.09	1.00	1.95	1.00	1.78	1.00	2.7
	2	9.17	4.38	8.47	4.33	7.52	4.22	0.3
	3	17.55	8.37	16.95	8.66	15.38	8.66	0.2
Torsion	1	–	–	3.89	1.00	3.76	1.00	0.5
	2	–	–	9.52	2.44	5.65	1.50	0.7
	3	–	–	15.87	4.08	10.75	2.86	0.4

PERIOD CHANGES IN BUILDINGS

As mentioned above, a measurement of the natural periods of structures at relatively low excitations levels may reveal significant patterns of structural behavior. In particular, changes in these periods during larger excitations such as damaging earthquakes may indicate significant permanent structural alterations.

One of the first examples of earthquake excitation in which such period changes were clearly measured was the Peruvian earthquake of October 17, 1966. Prior to the earthquake the fundamental natural periods of a number of tall buildings in Lima had been measured by low-level wind excitation tests such as are illustrated in Figure 1 [22]. An accelerograph recorded in Lima during the earthquake peak accelerations of about 25% g, so that it is known that the buildings in Lima were subjected to a strong ground shaking. After the earthquake the low-level periods were again measured [23]. The most striking change in period occurred in an 11-story building with pre-earthquake periods in two lateral directions of 0.32 and 0.38 seconds, and post-earthquake periods of 1.2 seconds in both directions. This major change is believed to be the result of cracking of in-filled brick walls during the earthquake. No significant structural damage was believed to have been involved. During this same earthquake, other buildings also showed significant period increases: e.g. from 0.67 to 0.80 seconds for a 15-story building; from 0.71 to 1.04 in a 22-story building; and from 0.65 to 1.29 in a 22-story building [23]. In no case did significant structural damage appear to have occurred, so the full meaning of such period changes is open to question.

Another interesting example occurred during the Tashkent earthquake of 1966. For this earthquake the ground motions were believed to have been moderately strong, but no quantitative measurements are available. Period measurements were made in a number of structures befor and after the earthquake. In one case a 4-story brick building changed in fundamental period from 0.27 to 0.48 seconds [24]. In this building, structural damage was sufficient that repairs were required, after which the period was found to be 0.51 seconds. It is clear that the repairs did not restore the original stiffness to the structure.

A more extensive test, although at low excitation levels, was provided by the Borrego Mountain, California, earthquake of April 8, 1968. Although this magnitude 6.5 earthquake was over 100 miles from Los Angeles, it was well-recorded there on strong-motion accelerographs at many building sites. Ground accelerations were only of the order of 1-2% g, and the maximum building accelerations were not greater than about 8% g. The periods of many buildings had been measured before the earthquake by the Seismological Field Survey, NOA, NOAA [4]. Even though the structural motions were small, period changes of as much as 35% over the low-level wind excitation values were observed [25, 26].

The most complete tests of the possibilities of such period measurements will be afforded by analysis of the data from the San Fernando, California, earthquake of February 9, 1971 [27, 28]. In this case some 50 fully instrumented multi-story buildings were subjected to ground motions in the order of 10-20% g, and to building responses in the 30-40% g range [28]. Many of these buildings have had low-level period measurements made before and after the earthquake, as well as the acceleration response

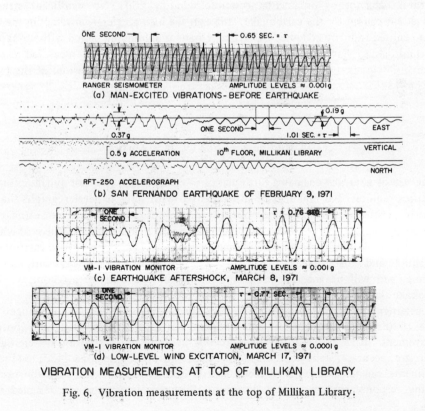

VIBRATION MEASUREMENTS AT TOP OF MILLIKAN LIBRARY

Fig. 6. Vibration measurements at the top of Millikan Library.

measured during the earthquake. Analysis of this data is now proceeding, and when complete should give conclusive results as to the practical usefulness of the technique.

Of special interest in this connection is the behavior of the 9-story reinforced concrete shear wall Millikan Library on the campus of the California Institute of Technology. The building has been subjected to complete resonance vibration testing during and after construction as well as before and after the earthquake [29]. Before the earthquake, a typical man-excited low-level vibration test gave a fundamental period in the EW direction of 0.65 seconds, as shown in the sample record in Figure 6a. This agreed well with forced vibration resonance tests at considerably higher excitation levels. During the earthquake, the upper portion of the buildind was subjected to a peak acceleration of about 37% g, as shown in the strong-motion accelerograph record of Figure 6b, which also indicates clearly that the EW vibration period at this excitation level was 1.01 seconds [30]. The dominant period during the strong earthquake motion is evidently considerably longer than for the low-level tests.

After the main shock of the earthquake, a smaller excitation caused by an aftershock gave the period of 0.76 seconds shown in Figure 6c, and a low-level wind excitation test illustrated in Figure 6d also showed a period of 0.77 seconds. These altered periods after the earthquake were also confirmed by a more elaborate ambient vibration study [31]. The permanent increase of period is belived to have been the result of alterations in the structural behavior of ornamental concrete window grills. No significant structural damage was caused by the earthquake, although the high acceleration values in the upper floors resulted in many collapsed bookcases. Changes in building period without apparent structural damage have also been commented on in connection with measured responses of tall buildings in Las Vegas caused by underground nuclear explosions at the Nevada Test Site [32].

HIGH-LEVEL TESTS OF BUILDINGS

The use of naturally occurring earthquakes to excite higher levels of dynamic forces in structures requires that buildings be instrumented with suitable accelerographs that will patiently await for many years the hopefully rare event, and will then be automatically triggered by the earthquake. Since no information is available as to the where or when of the next suitable event, numerous buildings in seismic regions must be so instrumented with simple and reliable devices that will be able to survive for many years with minimum servicing, and will be able to spring into action in one-tenth of a second when the earthquake finally arrives.

Accelerographs suitable for the above very special requirements were developed early in the 1930's by the U. S. Coast and Geodetic Survey [1]. Since that time continued improvements have resulted in commercially available strong-motion accelerographs which are accurate, reliable, compact, and relatively inexpensive [33, 34]. Such instruments can be used both to record earthquake ground motion, and the resulting building responses in upper story locations. Such accelerographs are designed to be relatively insensitive, so that the largest motions associated with destructive earthquakes will remain on scale.

The first important record of building response with such instrumentation was obtained during the San Francisco earthquake of 1957. Three accelerographs had been installed for a number of years in the 17-story Alexander Building in downtown San Francisco, and during the 1957 earthquake excellent records were obtained of the input ground acceleration as measured in the basement, and of the responses of the 11th and 17th floor. By comparing calculated and measured responses it was possible to achieve a fairly complete understanding of the structural dynamic characteristics of the building at intermediate force levels. [35].

Since that time, similar records were obtained and similar studies were made of a building in Akita, Japan, excited by the Niigata earthquake of 1964 [36], and of a building in Nagano which was shaken by one of the strong earthquakes of the Matsushiro, Japan, sequence [37].

The potential value of such investigations was one of the motivating factors behind the adoption by the City of Los Angeles in 1965 of an amendment to the Building Code requiring that three accelerographs be supplied by the building owner in all new buildings

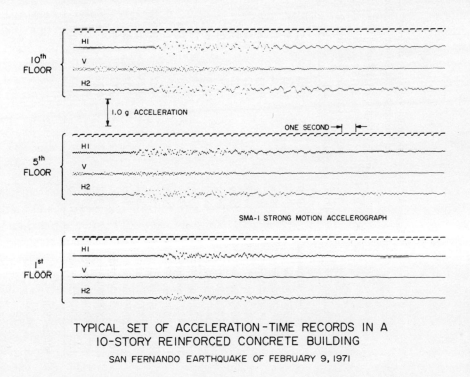

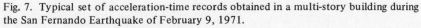

Fig. 7. Typical set of acceleration-time records obtained in a multi-story building during the San Fernando Earthquake of February 9, 1971.

TABLE IVa

Typical Peak Accelerations (g's) in Multi-Story Buildings: Reinforced Concrete
San Fernando Earthquake of February 9, 1971

H1, H2 = Orthogonal Horizontal Components; V = Vertical

No.	Building	Date	No. Stories	Ground, Basement or 1st Floor			Intermediate Level			Roof or Top Floor		
				H1	H2	V	H1	H2	V	H1	H2	V
1	15107 Vanowen	1970	7	0.11	0.11	0.11	0.24	0.23	0.20	0.26	0.40	0.17
2	8244 Orion	1967	7	0.25	0.15	0.19	0.21	0.25	0.24	0.40	0.34	0.26
3	1640 Marengo	1966	7	0.14	0.15	0.09	0.15	0.27	0.13	0.25	0.44	0.15
4	4680 Wilshire	1967	7	0.14	0.10	0.09	0.23	0.19	0.12	0.24	0.30	0.15
5	646 Olive*	1967	7	0.22	0.26	0.09	0.25	0.25	0.13	0.39	0.48	0.26
6	4687 Sunset	1966	8	0.20	0.18	0.15	0.31	0.24	0.15	0.45	0.47	0.22
7	2011 Zonal	1966	9	0.08	0.08	0.07	0.20	0.17	0.10	0.23	0.21	0.11
8	433 Oakhurst	1970	10	0.09	0.06	0.03	0.14	0.14	0.04	0.27	0.27	0.10
9	120 Robertson	1966	10	0.10	0.10	0.04	0.18	0.19	0.10	0.33	0.28	0.12
10	420 Roxbury	1969	10	0.21	0.17	0.05	0.21	0.24	0.11	0.30	0.22	0.14
11	7080 Hollywood	1966	11	0.11	0.11	0.08	0.21	0.13	0.16	0.21	0.13	0.22
12	3710 Wilshire*	1966	11	0.17	0.16	0.09	0.29	0.17	0.11	0.22	0.38	0.17
13	3470 Wilshire	1966	12	0.15	0.12	0.06	0.21	0.22	0.11	0.22	0.25	0.15
14	8639 Lincoln	1969	12	0.04	0.03	0.03	0.09	0.08	0.09	0.13	0.13	0.06
15	15250 Ventura	1971	12	0.17	0.24	0.11	0.25	0.28	0.16	0.18	0.30	0.18
16	2500 Wilshire	1969	13	0.11	0.13	0.06	0.14	0.16	0.07	0.20	0.20	0.15
17	6200 Wilshire	1970	16	0.12	0.13	0.03	0.29	0.17	0.05	0.28	0.26	0.08
18	4000 Chapman	1970	19	0.02	0.02	0.02	0.05	0.04	0.03	0.06	0.06	0.04

*Shear Wall

TABLE IVb

Typical Peak Accelerations (g's) in Multi-Story Buildings: Steel-Frame
San Fernando Earthquake of February 9, 1971

H1, H2 = Orthogonal Horizontal Components; V = Vertical

No.	Building	Date	No. Stories	Ground, Basement or 1st Floor			Intermediate Level			Roof or Top Floor		
				H1	H2	V	H1	H2	V	H1	H2	V
1	5260 Century*	1968	7	0.06	0.05	0.02	0.05	0.07	0.04	0.07	0.05	0.09
2	3407 Sixth	1966	8	0.17	0.20	0.06	0.22	0.22	0.10	0.28	0.22	0.27
3	1150 Hill	1970	10	0.12	0.09	0.05	0.10	0.12	0.09	0.12	0.12	0.15
4	900 Fremont	1971	12	0.14	0.14	0.07	0.14	0.15	0.12	0.18	0.15	0.17
5	L.A. Water & Power	1969	15	0.15	0.20	0.08	0.19	0.14	0.09	0.17	0.13	0.16
6	250 First	1967	15	0.10	0.14	0.06	0.21	0.17	0.09	0.17	0.18	0.20
7	1800 Century Park East	1970	16	0.08	0.11	0.08	0.23	0.25	0.16	0.28	0.28	0.33
8	800 First	1969	33	0.09	0.14	0.06	0.13	0.19	0.17	0.17	0.27	0.25

*Shear Wall

above a six-story height. At the time of the San Fernando earthquake, about 50 high-rise buildings in Los Angeles, had been so equipped, and the resulting remarkable set of ground motion and building response measurements may be said to constitute a structural dynamic test of unparalleled scope [28].

Table IVa, b summarizes a number of peak acceleration readings taken from these building accelerographs. It will be noted that upper floor accelerations exceeding 40% g occurred in numerous cases. When it is considered that the highest acceleration value which it has been possible to produce in a full scale building by vibration generators is of the order of 3-5% g, the unusual interest of these results will be apparent.

In Figure 7 is shown a typical set of acceleration-time records obtained in a tall building. Considering the quality of the traces, and the fact that the accelerographs were carefully calibrated before and after the earthquake, it is evident that a relatively high degree of accuracy can be expected for the analysis.

COMPARISON OF CALCULATED AND MEASURED EARTHQUAKE RESPONSE

As an example of the use to which building acceleration data of the above type can be

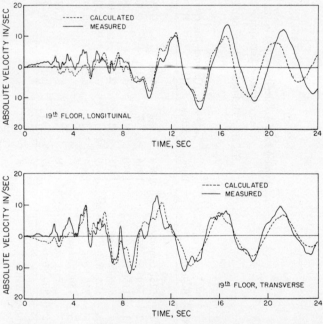

COMPARISON OF MEASURED AND CALCULATED BUILDING RESPONSES
SAN FERNANDO EARTHQUAKE OF FEBRUARY 9, 1971

Fig. 8. Comparison of measured and calculated building response to the San Fernando Earthquake [38].

put, some results of a preliminary investigation of a 42-story steel-frame building will be cited [38]. In the course of the design of the building, two computer models had been developed and studied in some detail. One model involved a flexible joint hypothesis, the other assumed infinitely rigid joints. These models had been used in earthquake design studies in which past recorded earthquakes were used as exciting forces.

During the San Fernando earthquake, input ground accelerations were obtained by an accelerograph in the 2nd basement, and building response was measured by an accelerograph at the 19th floor level. The measured ground acceleration was then used as the input excitation to the rigid joint structural model, assuming 5% damping in each of the eight modes included in the analysis, and the theoretical response of the 19th floor was calculated. In Figure 8 the comparison of the calculated and measured results are shown. The curve marked measured is the velocity as obtained from a direct integration of the recorded acceleration-time curve. It is evident that the main features of the earthquake response are being represented by the mathematical model with a useful accuracy, and that such questions as the likely response of the structure to earthquakes of even larger size can be answered with a high degree of confidence.

When similar analyses have been completed for the additional dozens of diverse types of structures for which such measurements are available, it is clear that a very considerable increase in basic knowledge of building dynamics will result.

CONCLUSIONS

It is evident that dynamic tests of the kind outlined above can contribute much to an increased knowledge of structural mechanics. Although the relatively simple low-level excitation tests have only limited applicability to the more practical problems of structural response, even such tests can contribute significantly to an understanding of the appropriateness of mathematical models. In view of the virtual impossibility of carrying out high-level load tests on actual buildings by artificial means, the use of natural earthquakes for this purpose should be carefully considered. Although the expense of instrumenting large numbers of buildings may seem large, it may not prove to be excessive when compared with alternative methods, and when the value of the measurements for other purposes is also considered.

REFERENCES

1. W. K. Cloud Instruments for earthquake investigations, in *Earthquake Investigations in the Western United States*, 1931-1964, (ed. D. S. Carder) Publication 41-2, U.S. Coast and Geodetic Survey, Dept. of Commerce, Washington, D. C., 1964.
2. W. O. Keightley, G. W. Housner and D. E. Hudson, *Vibration tests of the Encino Dam intake tower*, Earthquake Engineering Research Laboratory, California Institute of Technology, Pasadena, 1961.
3. D. S. Carder, Vibration observations, in *Earthquake Investigations in the Western United States*, 1931-1964, (ed. D. S. Carder) Publication 41-2, U. S. Coast and Geodetic Survey, Dept. of Commerce, Washington, D. C., 1964.

4. Building Period Reports, by Seismological Field Survey, U.S. Coast and Geodetic Survey, Dept. of Commerce, Washington, D.C.

5. K. Takahasi and K. Husimi, Method to determine frequency and attenuation constant from the irregular motion of an oscillating body, *Inst. Phys. and Chem. Res.* (Japan), **14**, No. 4.

6. S. Cherry and A. G. Brady, determination of structural dynamic properties by statistical analysis of random vibrations, *Proc. Third World Conference on Earthquake Engineering,* New Zealand, 1965.

7. E. Shima, T. Tanaka, and N. Den, Some new instruments used in earthquake engineering in Japan," *Proc. Second World Conference on Earthquake Engineering,* **II**, Tokyo and Kyoto, Japan, 1960.

8. R. Crawford, and H. S. Ward, Determinations of natural periods of buildings, *Bull. Seis. Soc. Amer.,* **54**, No. 6, December, 1964.

9. H. S. Ward and R. Crawford, Wind-induced vibrations and building modes, *Bull. Seis. Soc. Amer.,* **56**, No. 4, August, 1966.

10. R. R. Blandford, V. R. McLamore and J. Aunon, *Structural analysis of Millikan Library from ambient vibrations,"* Report No. 616-0268-2107, Teledyne Earth Sciences, February 1968.

11. M. D. Trifunac, *Ambient vibration test of a thirty-nine story steel frame building,* report No. EERL 70-02, Earthquake Engineering Research Laboratory, California Institute of Technology, Pasadena, 1970.

12. V. R. McLamore, *Ambient vibration survey of Chesapeake Bay Bridge,* Report No. 0370-2152 for the Bureau of Public Roads, U.S. Dept. of Transportation, Teledyne Geotronics, Long Beach, California, March 1970.

13. V. R. McLamore, *Ambient vibration survey of Newport Bridge,* Report No. 0370-2150 for the Bureau of Public Roads, U.S. Dept. of Transportation, Teledyne Geotronics, Long Beach, California, March 1970.

14. M. D. Trifunac, *Wind and microtremor induced vibrations of a twenty-two story steel frame building,* Earthquake Engineering Research Laboratory, California Institute of Technology, Pasadena, 1970.

15. P. C. Jennings, R. B. Matthiesen, and J. B. Hoerner, *Forced vibration of a 22-story steel frame building,* Report EERL 71-01, Earthquake Engineering Research Laboratory, California Institute of Technology, and Earthquake Engineering and Structures Laboratory, Univ. of California at Los Angeles, 1971.

16. G. W. Housner, The significance of the natural periods of vibration of structures, *Primeras Jornadas Argentinas de Ingenieria Anisismica,* San Juan-Mendoza, Argentina, 1962.

17. J. B. Hoerner, and P. C. Jennings, Modal interference in vibration tests, *Eng. Mech., Proc. ASCE,* **EM4**, August 1969.

18. D. E. Hudson, W. O. Keightley, and N. N. Nielsen, A new method for the measurement of the natural periods of buildings, *Bull. Seis. Soc. Amer.,* **54**, No. 1, February 1964.

19. W. O. Keightley, *Vibration tests of structures,* Earthquake Engineering Research Laboratory, California Institute of Technology, Pasadena, 1963.

20. N. N. Nielsen, *Dynamic response of multistory buildings,* Earthquake Engineering Research Laboratory, California Institute of Technology, Pasadena, 1964.

21. S. Cherry, and U. A. Topf, Determination of the dynamic properties of a tower structure from ambient vibrations, *Proc. Conf. on Earthquake Analysis of Structures,* Iasi, Romania, 1970.

22. J. Kuroiwa, *Periodos de vibracion y caracteristicas estructurales de los edificios en Lima y sus Alrededores,* Boletin No. 9, Institute de Estructuras, Universidad Nacional de Ingenieria, Lima, Peru, 1964.

23. L. Esteva y J. A. Nieto, *El Temblor de Lima, Peru, Octubre 17, 1966*, **XXXVII**, Revista Ingenieria, Mexico, 1967.
24. V. T. Rasskazovsky and K. S. Abdurashidov, Restoration of stone buildings after earthquake, *Proc. 4th World Conf. on Earthquake Engineering*, **III**, Santiago, Chile, 1969.
25. W. K. Cloud and R. P. Maley, Building-period measurements during and earthquake with comments on instruments, *Proc. Conf. on Earthquake Anal. of Structures*, Iasi, Romania, 1970.
26. R. B. Matthesen, P. Ibanez, L. G. Selna and C. B. Smith, *San Onofre nuclear generating station supplementary vibration tests*, UCLA-ENG-7095, December 1970.
27. W. K. Cloud and R. P. Maley, San Fernando earthquake of February 9, 1971, Preliminary Report on Strong-Motion Data, *Bull. Seis. Soc. Amer.*, **61**, No. 2, April, 1971.
28. R. P. Maley and W. P. Cloud, Preliminary strong motion results from the San Fernando earthquake of February 9, 1971, Professional Paper 733, U.S. Geological Survey and NOAA, Washington, D.C. 1971.
29. J. H. Kuroiwa, *Vibration test of a multistory building*, Earthquake Engineering Research Laboratory, California Institute of Technology, Pasadena, 1967.
30. G. W. Housner and D. E. Hudson, Preliminary report on the engineering aspects of the San Fernando earthquake of February 9, 1971, *Bull. Seis. Soc. Amer.*, **61**, No. 2, April 1971.
31. V. R. McLamore, *Ambient vibration tests of the Millikan Library*, Teledyne Geotronics, 1971.
32. J. A. Blume, Response of highrise buildings to ground motion from underground nuclear detonations, *Bull. Seis. Soc. Amer.*, **59**, No. 6, December, 1969.
33. D. E. Hudson, Ground motion measurements, in *Earthquake Engineering*, (ed. R. W. Wiegel), Prentice-Hall, Englewood Cliffs, N.J., 1970.
34. H. T. Halverson, Modern trends in strong movement (strong motion) instrumentation, *Proc. of Conf. on Dynamic Waves in Civil Engineering*, Swansea, Wales, 1970, Wiley-Interscience, 1971.
35. D. E. Hudson, A comparison of theoretical and experimental determinations of building response to earthquakes, *Proc. of the Second World Conf. on Earthquake Engineering*, Tokyo and Kyoto, Japan, 1960.
36. Y. Osawa and M. Murakami, Response analysis of tall buildings to strong earthquake motions," *Bull. Earthquake Res. Inst.*, Tokyo, Vol. 44, 1966.
37. H. Tajimi, Dynamic behaviour of multistory building during the Matsushiro earthquakes, in *Recent Researches of Structural Mechanics* (ed. H. Tanaka and S. Kawamata) Uno Shoten, Tokyo, 1968.
38. J. Lord, *Post-earthquake analysis of a 42-story tower*, Report 71811-15, P-394, A. C. Martin & Associates, Los Angeles, April, 1971.

A SURVEY OF ROTARY-WING AIRCRAFT CRASHWORTHINESS

GEORGE T. SINGLEY, III

Safety & Survivability Division, Eustis Directorate
US Army Air Mobility Research & Development Laboratory
Fort Eustis, Virginia

Abstract—Within military aviation the "-ilities" (reliability, maintainability, and survivability) have been the recent recipients of much attention. It is the purpose of this paper to survey one area of survivability — crashworthiness. The scope of the survey is restricted primarily to US Army rotary-wing aircraft crashworthiness, although many of the principles and techniques presented are applicable to other aircraft types.

Justification for improved crashworthiness design requirements is presented. Substantiation is presented in the form of a review and analysis of both Army operational experience and the 11-year crash survival research and development program conducted by the Eustis Directorate, US Army Air Mobile R & D Laboratory (formerly the US Army Aviation Materiel Laboratories). The review and analysis of the Eustis Directorate's crash survival R & D program includes a discussion of both the methods employed and those areas that have exhibited potential for significant crashworthiness improvement. Crash survival R & D methods discussed include: Aircraft accident investigation; crash testing of full-scale, fully instrumented aircraft; crash injury and crashworthiness evaluation of aircraft and aircraft mockups; selected component and subassembly testing.

A review of rotary-wing aircraft crashworthiness design principles and techniques is presented. This presentation is segmented into six design areas: Airframe, seat/restraint systems, occupant environment, ancillary equipment stowage, post-crash emergency escape, and fire prevention.

INTRODUCTION

Man's concern about aviation crash hazards was evidenced as long ago as the Greek myth concerning the tragic flight of Daedalus and Icarus. Unfortunately, man's fears have been well founded. Historically, aircraft creashes have proven to be tragic, chaotic events that extract a high price in terms of human suffering and economic loss. Man's earlier concern for aircraft occupant crash safety yielded such devices as the lap belt, parachutes, ejection seats, etc. Perhaps nowhere is this concern so prevalent, or well founded, than in military avaiation.

The purpose of this survey paper is threefold: (1) to present justification for the necessity of aircraft crashworthiness design, (2) to describe the aircraft crash environment and consequent hazards, and (3) to review recent advances in aircraft crashworthiness design. The scope of this paper will be restricted to rotary-wing aircraft crashworthiness; however, the reader will quickly realize that much of the content of this paper is applicable to fixed-wing aircraft as well.

A multitude of synonyms for, and definitions of, crashworthiness have evolved. For the purpose of this paper, aircraft crashworthiness is defined as the capacity of the aircraft to perform as a protective container in a manner which permits the occupants to

179

survive the impact and safely evacuate the wreckage in a survivable accident.

Investigation and analysis of past accidents have shown that occupant crash survival could be greatly enhanced if the following five general survivability factors were considered during the preliminary design stages of aircraft.[1]

1. Crashworthiness of the aircraft structure — the ability of the aircraft to maintain living space for occupants throughout the crash sequence.
2. Tiedown chain strength — the strength of the linkage retaining occupants, cargo, and equipment during the crash sequence.
3. Occupant acceleration environment — the rate of onset, magnitude, duration, and direction of accelerations experienced by seated occupants as a result of a crash.
4. Occupant environment hazards — barriers, projections, and loose equipment in the occupant's immediate vicinity which during the crash sequence may cause contact injuries.
5. Postcrash hazard — the threat to occupant survival posed by fire, drowning, exposure, etc., following the impact sequence.

CRASHWORTHINESS JUSTIFIED

In recent years there has been an increasing emphasis on the "-ilities" (reliability, maintainability, and survivability) within military aviation. Certainly the concept of aircraft crashworthiness is not new: what is new is an increased realization of the resources that can be saved through proper aircraft crashworthiness design. Considering the nature of the missions flown, the low level of flight, and the hazardous terrain that rotary-wing aircraft must fly over and land on, it is not difficult to understand why the accident rate per flying hour for rotary-wing aircraft is twice that of its Army light fixed-wing counterpart [2] and why renewed emphasis by the military has been placed on improving the crashworthiness of its rotary-wing aircraft.

Often those persons arguing for improved aircraft crashworthiness seek justification solely in the number of accidents and the resultant injuries and fatalities. The argument most often heard against improved crashworthiness is that it is synonymous with prohibitive increases in weight and cost and/or reduction in performance. As a result of recent investigations, studies, and crashworthiness research and development efforts, the argument for crashworthiness has been significantly enhanced.

A recently concluded crash injury economics study by the U.S. Army Aeromedical Research Laboratory (USAARL), Fort Rucker, Alabama, provides promoters of improved rotary-wing aircraft crashworthiness with much-soughtafter data relative to what costs the Army incurs as a result of an aviator's death.[3] Table I summarizes the total cost to the Government of the accidental death of an Army aviator by aviator grade level. These replacement costs represent the total of training, maintenance, direct support, and indirect support costs for the respective Army aviator grade level. As Table I illustrates, the cost to the Government to replace an Army aviator lost due to accidental death with an aviator of equal training, rank, and experience can range from $102,670 (2LT) to $759,954 (CWO 4). The cost to the Government to replace one WO1 killed by

TABLE I

Total Cost of Aviator Death to Government [9]

Rank	Total Cost to Death	Total Cost After Death	Total Cost
WO1	$144,076	$ 4,073	$148,149
WO2	137,375	47,068	184,443
WO3	422,911	61,183	484,094
WO4	648,043	111,911	759,954
2LT	98,551	4,118	102,670
1LT	128,004	42,894	170,898
CPT	181,893	69,324	251,217
MAJ	353,167	58,238	411,405
LTC	525,871	51,526	577,396
COL	672,354	33,559	705,913

postcrash fire is equal to the cost to furnish approximately 30 UH-1D/H aircraft during production, and 18 UH-1D/H aircraft on a retrofit basis, with new crash-resistant self-sealing fuel systems.

Based on these economic findings alone, it will be difficult to preclude future rotary-wing aircraft systems from being designed for improved crashworthiness; however, there is additional justification for crashworthiness design. Improved system crashworthiness generates improved mission effectiveness as a consequence of both higher aviator confidence in their aircraft and increased aviator/aircraft availability. Furthermore, improved aircraft crashworthiness can reduce material losses; e.g., by improving the energy-absorbing capacity of the landing gear, airframe damage can be reduced. In light of the estimated cost of hardware losses resulting from all 1969 Army rotary-wing aircraft accidents not due to hostile fire, $144.017 million, the potential savings in this area should not be taken lightly. Finally, past efforts have clearly demonstrated not only that improved crashworthiness design is feasible, but also that it is not synonymous with prohibitive cost and weight increases.[5, 6, 7, 8].

The optimum approach to eliminating these tragic and costly losses is to prevent aircraft accidents from occurring. Recent efforts in aviation accident prevention have made noteworthy advances toward this goal, as shown by the continuous downward trend in the rotary-wing accident rate;* however, common sense dictates that as long as the human element exists in aviation, aircraft must be designed for the crash environment if losses due to aircraft crashes are to be minimized.

---

*The Army rotary-wing aircraft accident rate, exclusive of RVN operations, has dropped from approximately 37/100,000 flying hours in 1962 to 9.9/100,000 flying hours in 1969; however, the level of flying hours has increased from 603,905 to 2,264,957 so that total accidents were identical for both years, 224.[13, 16]

EUSTIS DIRECTORATE CRASHWORTHINESS PROGRAM BACKGROUND

Although aircraft accidents will occur in the future, past injury rates and severity need not, and should not, persist. Based on this premise and recognizing the growing role of Army aviation, the Army began contributing in 1955 to Aviation Crash Injury Research (AvCIR), then associated with Cornell University. During the four years that followed, the Army's program support was in the form of financial grants. In 1959, the U.S. Army Transportation Research Command (USATRECOM)* awarded a contract to AvCIR,† then a division of the Flight Safety Foundation, for the conduct of research in fields related to Army aviation safety with the emphasis on crash injury and crashworthiness programs.

The crashworthiness R & D program conducted by the Eustis Directorate since 1959 has had as its objective the development of design criteria and engineering data to improve the crashworthiness of Army aircraft through:
1. Investigation and analysis of aircraft accidents
2. Dynamic crash testing of instrumented full-scale aircraft
3. Component static and dynamic testing
4. Crashworthiness evaluation of Army aircraft and aircraft mockups

Fig. 1. Droned UH-1 Crash Test.

*Now the Eustis Directorate, U.S. Army Air Mobility Research and Development Laboratory, Fort Eustis, Virginia; formerly, U.S. Army Aviation Materiel Laboratories.

†Now the AvSER Facility of Dynamic Science, a Division of Marshall Industries, Inc., Phoenix, Arizona.

Release of H-25 During Crash Test

Fig. 2. H-25 Drop Test.

These methods have been used successfully to achieve a better understanding of the crash environment, to determine the causes of crash casualties, and to develop design solutions to minimize the severity and occurrence of casualties in potentially survivable Army aircraft accidents.

Considerable knowledge about the crash environment and accident casualty causes has been acquired. Between 1959 and 1966, thousands of civilian and military accidents were reviewed and 21 aircraft accidents were investigated on site. Twenty-four instrumented full-scale aircraft dynamic crash tests were conducted, with the emphasis on re-creating actual crashes in detail. Full-scale aircraft dynamic test methods used included remote control, crane drop, and monorail tests (Figures 1, 2, and 3). During these tests, instrumented anthropomorphic dummies were used to simulate aircraft occupants. Test instrumentation and photographic coverage provided a detailed record of the response of the aircraft and occupants to the crash impact. Additional insight was achieved by conducting selected static and dynamic tests on aircraft components, subsystems, and equipment. Both drop tower and linear accelerator dynamic testing techniques were used. Items tested included: troop and crew seats, restraint systems, litter systems, fuel cells, fire inerting systems, fuel emulsions, and helmets. Also, during this period a method was developed to evaluate the crashworthiness potential of aircraft, under which the following six weighted factors are considered:
1. Crew retention system
2. Troop retention system
3. Postcrash fire potential
4. Basic airframe crashworthiness

Fig. 3. C-45 Monorail Crash Test.

5. Evacuation

6. Injurious environment

Since its establishment, this aircraft crashworthiness evaluation system has been used successfully to determine the crashworthiness of all operational Army rotary-wing aircraft, and many crashworthiness design improvements have been made to Army aircraft as a result of these evaluations. Significantly, history has proven that this method is valid.

In addition to effecting crashworthiness improvements in existing aircraft [5] during this period, all three of the Light Observation Helicopter (LOH) competitors incorporated crashworthiness design principles in their candidate aircraft and, significantly, without degrading airworthiness; [6] moreover, it was during this period that the data base was established for perhaps the most important crashworthiness document to date: the *Crash Survival Design Guide*. [1] This document was prepared initially by the Flight Safety Foundation under a Eustis Directorate contract. Published in 1967 as USAAVLABS TR 67-22, this document is periodically revised to reflect the most recent crashworthiness R & D findings in eight areas:*

1. Aircraft crash kinematic and survival envelopes
2. Airframe crashworthiness
3. Occupant environment
4. Seats (crew and troop/passenger)

---

*The current edition of this document is USAAMRDL TR-22.

5. Restraint systems (crew, troop/passenger and cargo)
6. Postcrash fire
7. Ancillary equipment
8. Emergency escape provisions

Since the initial publishing of the *Crash Survival Design Guide*, the Eustis Directorate crashworthiness program has been devoted to: (1) monitoring aircraft accident data for the purpose of noting any new aircraft crashworthiness or crash kinematics trends; (2) through exploratory and advanced development efforts, developing and testing prototype designs that demonstrate the feasibility of preliminary crashworthiness design criteria; (3) developing new and improved crashworthiness design criteria; and (4) incorporating improved crashworthiness design criteria into pertinent military specifications and systems requirements documents for both existing Army aircraft systems, when retrofit is practical, and future Army aircraft.

During the past 12 years, 36 full-scale aircraft crash tests have been conducted, 16 Army aircraft models have been evaluated with respect to crashworthiness, 22 aircraft accidents have been investigated, and 90 technical reports have been prepared by AvSER under contracts with the Eustis Directorate. Contractual efforts have also been carried out for the Eustis Directorate by several other firms and academic institutions. Moreover, a large number of the Eustis Directorate crashworthiness efforts would not have been possible had it not been for the assistance of aircraft manufacturers and other Government agencies.

THE ROTARY-WING AIRCRAFT CRASH ENVIRONMENT

The primary U.S. Army source for aircraft accident data pertaining to accidents not due to hostile fire is the U.S. Army Board for Aviation Accident Research (USABAAR)*,

TABLE II

Rotary-Wing Aircraft Accident Occurrences by Operational
Phase (From July 1968 through June 1969) [9]

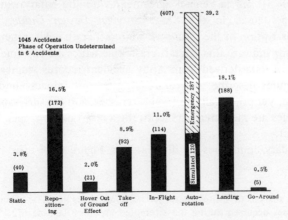

*Presently the U.S. Army Agency for Aviation Safety

TABLE III

Airspeed, Sinkspeed, and Impact Terrain for 185 Light Observation Helicopter (LOH),
Utility Helicopter (UH), and Cargo Helicopter (CH) Accidents [9]

		LOH	UH	CH
Average airspeed at impact (ft/sec)	Low*	45.1	35.6	9.5
Highest impact airspeed attained (ft/sec)	High	67.7	53.4	21.3
Average sink speed at impact (ft/sec)	Low	11.1	11.8	11.5
Highest sink speed attained (ft/sec)	High	19.4	20.5	19.5
Impact terrain	Land	67%	47.4%	76%
	Water	24%	13.3%	0%
	Trees & Other†	9%	39.3%	24%

*Each investigator made both a high and a low speed estimate.
†"Other" includes tree stumps, rocks, fuel drums, pierced steel plank, and asphalt.

Fort Rucker, Alabama. USABAAR has recently completed a comprehensive study of occupant injury experience in Army rotary-wing aircraft for the period January 1967 through December 1969. [9] Additional publications describing the aircraft crash environment include References 1, 2, 4, and 10 through 13.

As Table II indicates, two out of every three Army rotary-wing aircraft accidents occur during attempted autorotation (39.2%), landing (18.1%), and takeoff (8.9%) phases of operation.* Insight relative to representative airspeed, sink speed, and impact terrain for three of the aircraft models studied in Reference 9 is presented in Table III. One surprising aspect of the contents of Table III is the high percentage of water impacts for the Army LOH (24%) and UH (13.3%) rotary-wing aircraft studied. The velocities contained in Table III are in general agreement with the rotary-wing aircraft accident kinematic data contained in the *Crash Survival Design Guide,* which presents a comprehensive description of the rotary-wing aircraft crash environment. [1]

Accident records indicate that aircraft crash kinematics are extremely variable. Figures 4, 5, and 6 present data derived from Army accident records studies from 1960 to the present. These figures describe: (1) rotary-wing aircraft impact conditions that can be expected in severe, yet potentially survivable, crashes, and (2) survivable crash limits. The ΔV and G_{avg} values are measured parallel to the aircraft axes irrespective of the aircraft's attitude at impact.

Velocity boundary conditions are illustrated in Figure 6 for three rotary-wing aircraft

---

*A recent statistical analysis of United States jet transport losses by phase of operation indicates that four out of five losses occurred during the attempted approach/landing (52.1%) and takeoff/climb (33.3%) phases of operation. [14]

crash categories: survivable, marginally survivable, and unsurvivable. The survivable accident velocity region represents accidents which at least one occupant survived, or could have survived, had it not been for other factors; e.g., postcrash fire and failure to use restraint system. Although some might equate the relatively small size of this area with insignificance, a review of all Army rotary-wing aircraft accidents during the period January 1967 through December 1969 indicates that 2388 accidents of the total 2546 accidents reported, or 93.8% were considered survivable; moreover, as Table IV illustrates, 41.1% of the fatalities and 98.7% of the injuries during this period occurred in accidents classified survivable. There have been aircraft accidents with survivors which have crashed at velocity levels within the marginally survivable region of Figure 6; however, survival in this region is a function of several factors which are not typical for existing aircraft; e.g., improved fuselage energy absorption capacity. Probability of occupant survival is extremely questionable within the unsurvivable region of combined crash velocities due to the catastrophic nature of the accidents included in this regime.

Figure 4 presents distribution curves for ΔV_x and ΔV_z that were incurred during survivable accidents involving substantial structural damage and/or occupant injury. Although insufficient data has been collated and analyzed to accurately plot a distribution curve for ΔV_y, experience indicates that ΔV_y generally does not exceed 30 ft/sec. Analysis of crash kinematics data to date also indicates that the resultant velocity change for combined ΔV_x, ΔV_y, and ΔV_z components of the 95th percentile survivable

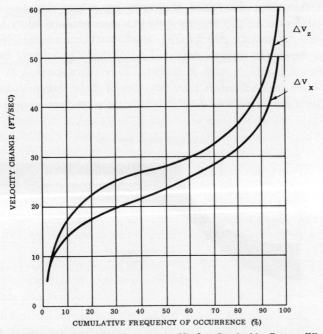

Fig. 4. Distribution of Velocity Changes (Δ V) for Survivable Rotary-Wing Aircraft Accidents. [1]

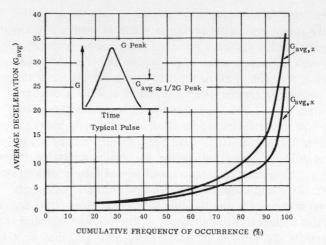

Fig. 5. Distribution of Average Impact Decelerations (G_{avg}) for Rotary-Wing Aircraft.[1]

rotary-wing aircraft accident does not exceed 50 ft/sec, nor do the ΔV_x, ΔV_y, and ΔV_z components of the resultant ΔV exceed 50 ft/sec, 30 ft/sec, and 42 ft/sec respectively. [1]

Analysis of accident records shows that the average G_x and G_z decelerations measured at the aircraft floor near the center of gravity for rotary-wing aircraft in survivable accidents involving substantial structural damage and/or personal injury are distributed as shown in Figure 5. As was the case for ΔV_y, insufficient data exists to plot a distribution curve for G_y; however, present data indicates that G_y seldom exceeds 18G for rotary-wing aircraft, and G_{peak} can be expected to be twice the G_{avg} values presented in Figure 5. Obviously, these G values are a function of many design characteristics of the aircraft and therefore are presented only to indicate what deceleration levels have been

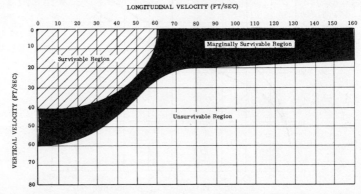

Fig. 6. Initial Impact Velocities (Based on Accident Case Histories of Military and Civilian Aircraft). [1]

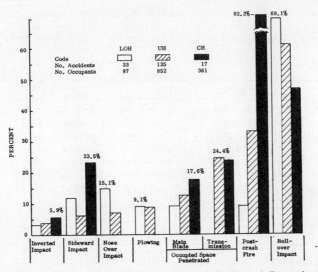

Fig. 7. Helicopter Kinematics Experience, January 1967 through December 1969. [9]

typical with past aircraft types; moreover, decelerations at points on the aircraft other than at the floor level near the aircraft c.g. can differ significantly from those values in Figure 5.

Graphical descriptions of how three rotary-wing aircraft models — LOH, UH, and CH — and their occupants have fared during the crash conditions described above are provided in Figures 7 and 8. Figure 7 emphasizes the statistical significance of the dreaded postcrash fire. In the case of the CH model, a fire broke out upon impact or

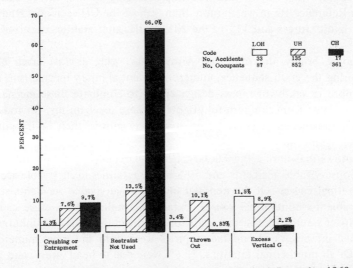

Fig. 8. Occupant Injury Experience, January 1967 through December 1969. [9]

TABLE IV

Rotary-Wing Aircraft Injury Data (January 1967 – December 1969) [9]

	NUMBER			%†	
	S*	NS	TOT	S	NS
Accidents	2388	158	2546	93.8	6.2
Occupants	10599	735	11334	93.5	6.5
Fatalities	439	655	1094	41.1	59.9
Thermal	181	147	328	55.2	44.8
Non-thermal	258	508	766	33.7	66.3
Injured (non-fatal)	2663	36††	2699	98.7	1.3
Thermal	161				
Non-thermal	2502				
No injury	7497	44††	7541	99.4	0.6

*S – survivable accident, NS – nonsurvivable accident.
†Survivable or nonsurvivable number for respective item as a % of the item total, e.g., 93.8% represents the percentage of survivable accidents.
††Some occupants escaped fatality as a result of being ejected from the aircraft during the crash sequence.

within 3 minutes after impact in 82.2% of the accidents for the period studied. The Army has been concerned about postcrash fire hazards for years, and noteworthy advances have been made, particularly in the area of crashworthy fuel systems, which will be discussed later in this paper. What is particularly significant about the contents of Figure 7 is the high incidence of roll-over and sideward impacts. This finding agrees with the study conducted in Reference 10, in which more than 50% of the UH accidents studied resulted in significant lateral forces and 111 of the 201 injuries and fatalities occurred in roll-over accidents.

The foregoing briefly describes the Army rotary-wing aircraft crash environment dynamics. During the crash sequence, numerous potential injury mechanisms are created. It is the purpose of crashworthiness design either to eliminate these mechanisms or to isolate the occupants from the harmful effects of these mechanisms. Obviously, in order to do this, the designer must be aware of these mechanisms, their injury potential, and their statistical significance.

Table IV shows that during the January 1967 – December 1969 period, 77.5% of the fatalities in nonsurvivable accidents and 59% of the fatalities in survivable accidents were due to nonthermal causes. Of the nonfatal injuries in survivable accidents studied, 94% were attributable to nonthermal causes. Fatalities, totaling 1094, were caused by the following: impact trauma (61%), thermal (30%), drowning (5%), and unknown (4%). Of the nonthermal injuries, head and face injuries were both the most numerous and the most lethal. Of all nonthermal fatalities, 22.6% were due to head injury; and 23.2% of all personnel receiving nonthermal injuries, both fatal and nonfatal, received head and face injuries. These head and face injuries are attributable to a number of injury mechanisms; e.g., body flailing, failure of seat/restraint system, structural collapse, loose ojbect strikes, shoulder harness not available, and failure to use restraint system provided. Second only to the head in terms of nonthermal, nonfatal injury frequency was the back. Although back injuries are not as lethal as head injuries, they are often incapacitating. Of all personnel receiving nonthermal injuries, both fatal and nonfatal, 15.7% incurred back injuries; 34.3% of these injuries were vertebra compressive fractures, 27.8% were strain/sprain injuries, and 37.9% were contusions, lacerations, or unknown. [9]

It is evident from the above that numerous injuries and fatalities are being incurred in accidents in which the occupants could tolerate the abrupt decelerations created during the crash sequence. Causes of these injuries include airframe collapse, seat/restraint system inadequacies, fire hazards, postcrash hazards, lethal projections in the occupant's immediate environment, main rotor blade and transmission penetration of occupied space, inadequate crash force attenuation, and operational problems such as failure to properly wear available restraint systems and helmets. This is attested to by the fact that 41% of the fatalities in Table IV occurred in accidents that were survivable based on impact conditions alone. Consequently, it is the goal of crashworthiness design criteria to reduce injuries and fatalities by: (1) minimizing the hazards that are injuring occupants in accidents in which the occupants can tolerate the abrupt decelerations created by the impact, and (2) providing the occupants with an aircraft system having adequate energy absorption capacity to attenuate crash impact forces to humanly tolerable levels. Pursuit of these goals not only will minimize the frequency of injuries and fatalities in accidents presently considered survivable, but also will expand to practical limits the population of impact conditions in which occupant protection can be provided. The remainder of this paper surveys design criteria and solutions that promise to significantly enhance the crashworthiness of future rotary-wing aircraft. This material is segmented into the following crashworthiness design areas: seating systems, fuselage (including landing gear), and postcrash considerations.

SEATING SYSTEMS

An Army rotary-wing aircraft seating system* is many things to many people: to the vulnerability engineer, it provides a structure to which he can attach armor to protect the pilot; to the pilot, it is too often uncomfortable and restricting, and if armored, it is protection; to the human factors engineer, it represents a task possessing many seemingly incompatible uncompromising human engineering requirements; to the crew chief, troop seats are often so bothersome that frequently they are removed from the aircraft, or the restraint systems are stowed away; but to the engineer charged with crashworthiness design responsibilities, it not only represents a system whose design must comply with

---

*Seating system is defined as all seat and restraint system components; e.g., restraint system, seat bucket and support structure, seat cushion, and seat attachment hardware.

requirements from several engineering disciplines, but it presents an opportunity for him to significantly enhance the future occupants' probability of survival during a future potentially survivable accident.

Pioneers of aircraft crashworthiness were quick to recognize the significance of seating system design with respect to aircraft crash injury. [4, 5, 17, 18, 19, 20, 21] Early accident investigation and full-scale crash tests indicated that seat and restraint system tiedown strengths were considerably lower than the occupant inertia loads created in potentially survivable crashes and lower than the decelerative loading that could have been tolerated by the occupant. [18] Lack of, or poorly designed, upper torso restraint systems permitted injurious contact with surrounding structure. Occupants were "submarining" under their lap belts and receiving spinal and abdominal injuries. In rotary-wing and light fixed-wing aircraft accidents, a high incidence of spinal compressive fractures were noted. In many instances, injuries and fatalities caused by seating system failures were the result of simple design deficiencies; e.g., understrength fasteners or tracks. It became evident that aircraft seating systems were neither performing their intended role of occupant retention during moderate crashes, nor approaching their potential as a crash-force attenuating mechanism.

Fig. 9. RVN Operationally Expedient Seating System.

The adequacy of aircraft seating system crashworthiness is a function of three of the five aforementioned crash survivability factors; i.e., tiedown chain strength, occupant acceleration environment, and occupant environment hazards. Not only must the seating system be retained generally in its original position during the crash sequence, but it must restrain the occupant from injurious body flailing and contact with injurious surroundings; and it must protect the occupant from injurious decelerations during the crash sequence by providing him with an energy-absorption mechanism and a properly designed restraint system and seat cushion. Obviously, these are but a few of the seating system requirements that must be considered. Included among the most restrictive requirements are those that pertain to ballistic protection, human factors, volume, weight, cost, environment, crash force attenuation mechanism, stroking distance limits, and operational considerations. Operational considerations have been particularly defeating with respect to design efforts to provide the occupants with more crashworthy seating systems.

Although there certainly are combat situations that dictate a lessening of crash safety precautions in order to achieve a mission or crew safety (e.g., rapid ingress and egress of troops in a "hot" landing zone), experience provides too many examples in which the aircrew and passengers do not make proper use of hardware associated with occupant crash survival. Figures 9 and 10 are examples of operationally expedient and hazardous

Fig. 10. RVN Operationally Expedient Seating System.

seating configurations witnessed in RVN rotary-wing aircraft operations. The seating system designer can play a major role in alleviating the occurrence of these situations by providing the occupant with a comfortable, nonrestricting seating system that is not operationally troublesome.

The first line of crash protection that the occupant has is his restraint harness. A major portion of the Army's earlier crash survival research was devoted to identifying crash hazards associated with occupant restraint systems, determining the adequacy of existing restraint systems, and developing preliminary design concepts and criteria for improved restraint systems. It was noted that failure of the tiedown chain was the major cause of unnecessary injuries and fatalities in several aircraft crashes. This is unfortunate, because tiedown chain adequacy is the easiest to achieve of the previously mentioned five general crash survivability factors. [4] Adequacy of the occupant tiedown chain is perhaps the single most important crashworthiness design consideration because: (1) providing sufficient aircraft strength and crash energy dissipation to human tolerance levels is little solace to the occupants if their seating system fails and no longer performs its retention function, and (2) an improved occupant restraint system can mean life or death during a postcrash fire or ditching because escape from the burning or submerged aircraft is enhanced if the restraint system is easily, quickly doffed and if no injury or debilitation has been received by the occupant. The results of the Eustis Directorate's efforts pertaining to restraint systems are presented in References 1 and 17 through 23. Due to space limitations, only aircrew and troop/passenger seating systems will be discussed here. Advances in litter systems and cargo restraint crashworthiness design are summarized in References 1 and 24 through 28.

Restraint system design must be guided by human deceleration limits and by the decelerations anticipated in potentially survivable accidents. Moreover, the restraint system should be designed and tested as a portion of the complete seating system. Human tolerance to applied decelerative force is a function of restraint system characteristics and the rate of onset, magnitude, duration, and direction of the applied decelerative force. These factors are briefly summarized in the paragraphs that follow.

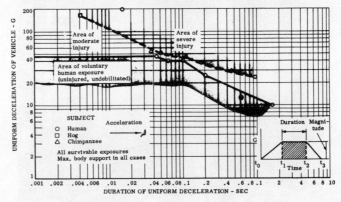

Fig. 11. Duration and Magnitude of $-G_x$ Acceleration Endured by Various Subjects. [1]

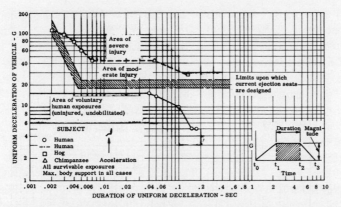

Fig. 12. Duration and Magnitude of $+G_z$ Deceleration Endured by Various Subjects. [1]

The effect of the magnitude of the applied decelerative force and the duration for which the force is applied is illustrated in Figures 11 and 12. Past studies have demonstrated that the rate of onset of the applied force has a definite effect on human tolerance, and under some impact conditions, the rate of onset appears to be a determining factor, as indicated in Figures 13 and 14, which indicate the significance of

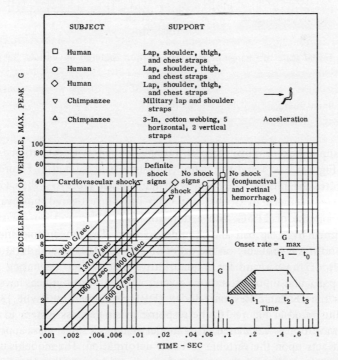

Fig. 13. Initial Rate of Change of $-G_x$ Deceleration Endured by Various Subjects. [1]

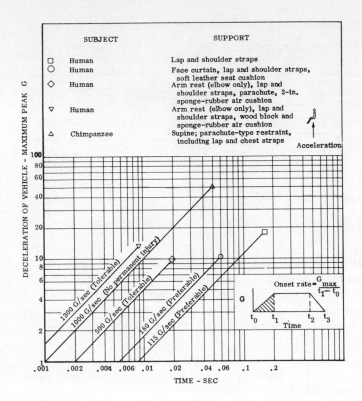

Fig. 14. Initial Rate of Change of $+G_z$ Deceleration Endured by Various Subjects. [1]

direction of applied decelerative force on human tolerance. Deceleration force terminology used in Figures 11-14 and throughout this report is defined in Figure 15. As these figures illustrate, the human body can withstand much more G_x than G_z decelerative loading. Human tolerance to G_y has not been fully investigated; however, two studies indicate that a pulse of 11.5G with a duration of 0.01 second can be sustained by a subject restrained by a lap belt and shoulder harness. The human tolerance limit to G_y abrupt decelerations is probably on the order of 20G for 0.10 second duration. [1]

As a result of the demand for a quantitative method for evaluating the degree of severity of a particular input excitation, several injury severity indices have evolved. One such index is the Dynamic Response Index (DRI) developed by Payne [31]. Using the DRI model illustrated in Figure 16, the response of single-mass system to the excitation acceleration waveform is calculated. This model assumes a single-mass approximation for the body that acts upon the vertebra to cause deformation. The model is used to predict the maximum deformation of the spine and associated force within the vertebral column

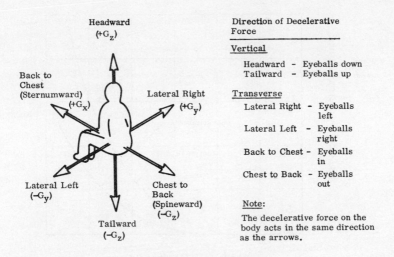

Fig. 15. Terminology for Decelerative Forces on the Body. [1]

for abrupt decelerations. This is achieved by solving the following equation through the use of a computer.

$$Z = \frac{d^2 \delta}{dt^2} + 2 c\omega_n \frac{d\delta}{dt} + \omega_n^2 \delta$$

The term $\omega_n^2 \delta$ represents the spinal deformation. Model spring stiffness was derived from

$$\text{Dynamic Response Index (DRI)} = \frac{\omega_n^2 \ \delta_{max}}{g}$$

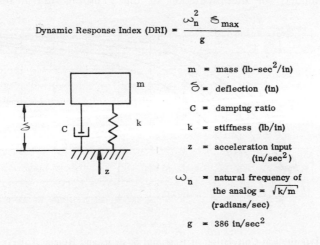

m = mass (lb-sec^2/in)

δ = deflection (in)

C = damping ratio

k = stiffness (lb/in)

z = acceleration input (in/sec^2)

ω_n = natural frequency of the analog = $\sqrt{k/m}$ (radians/sec)

g = 386 in/sec^2

Fig. 16. DRI Spinal-Injury Model. [1]

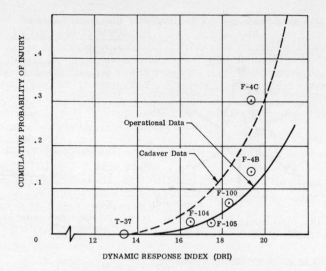

Fig. 17. Probability of Spinal Injury Predicted From Cadaver Data Compared to Operation Experience. [1]

tests of cadaver vertebral segments, and damping ratios were derived from measurements of mechanical impedance of human subjects during vibration and impact. The DRI is correlated with cumulative probability of spinal injury in Figure 17. The U.S. Air Force has adopted a system using a combination of acceleration as a function of duration and the DRI for establishing acceptable ejection seat acceleration environments. [32] The DRI has been shown to be effective in predicting spinal injury potential for $+G_z$ acceleration environments, and efforts are under way to apply the DRI to other directions. [1] More detailed discussions of human tolerance, the DRI, and other mathematical models for the prediction of injury during human body impact are

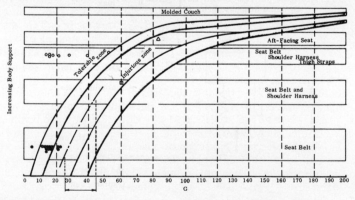

NOTE: The shaded areas indicate transition between the tolerable, injurious and fatal deceleration limits.

Fig. 18. Hypothetical Correlation of Restraint Systems and Human Tolerance to Short Duration Transverse Deceleration for Durations of the Order of 0.10 Second. [18]

presented in References 1, 11, 23 and 29 through 39.

The effectiveness of the restraint system is dependent upon the area over which the total force is distributed, the location on the body at which the restraint is applied, and the degree to which the restraint device limits residual freedom of movement. [1] The greater the contact area between the restraint system and the body, the more human tolerance is enhanced, as shown in Figure 18. Personnel restraint system research and development efforts pursued by the Eustis Directorate during the past decade have produced recommended design criteria and concepts for adequate personnel restraint systems. [1, 17, 18, 19] Concisely stated, any practical restraint system should:

1. Be light in weight and comfortable.
2. Be easy and quick to don and doff.
3. Preferably contain only one point of release, since a stunned or injured person might have difficulty releasing more than one buckle.
4. Provide freedom of movement to operate aircraft controls.
5. Provide restraint in the vertical, longitudinal, and lateral directions equal to known human tolerance to abrupt deceleration limits.

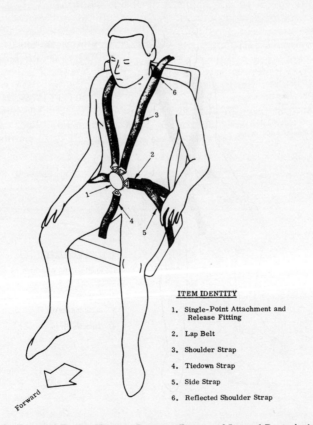

ITEM IDENTITY

1. Single-Point Attachment and
   Release Fitting

2. Lap Belt

3. Shoulder Strap

4. Tiedown Strap

5. Side Strap

6. Reflected Shoulder Strap

Fig. 19. Forward-Facing Harness Concept (Improved Lateral Restraint).[1]

6. Consist of webbing which covers the maximum area in the shoulder and pelvic body
   regions consistent with other items listed.

Minimum acceptable restraint system configurations for Army aircraft seating system
orientations are shown in Figures 19 and 22. Recognizing the inadequacies of present
restraint systems, the need to prove the recommended design criteria practical, and the
need for a detailed military specification to insure that future aviators will be provided
with a restraint system void of known deficiencies, the Eustis Directorate is presently
developing both a forward-facing aircrew restraint system which complies with the *Crash
Survival Design Guide* and a consequent detailed military specification. This program will
not be completed before the fall of 1971; however, the present preliminary design is
illustrated in Figure 23. This system is scheduled to be tested statically and dynamically
to the criteria specified in the *Crash Survival Design Guide*.

An adequate restraint system does little good if the seat to which it is attached is
understrength or permits decelerative loading of the occupant in excess of human
tolerance limits. In recognition of the undesirable performance of current aircraft seats in
moderate rotary-wing aircraft crashes, the Eustis Directorate has conducted a series of

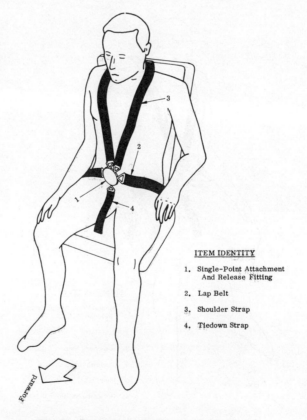

ITEM IDENTITY

1. Single-Point Attachment
   And Release Fitting

2. Lap Belt

3. Shoulder Strap

4. Tiedown Strap

Fig. 20. Forward-Facing Harness Concept. [1]

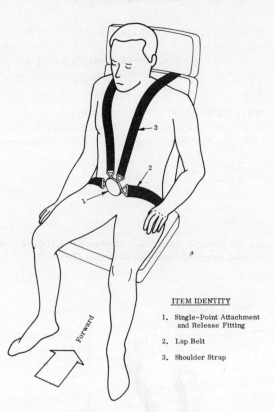

ITEM IDENTITY

1. Single-Point Attachment
   and Release Fitting

2. Lap Belt

3. Shoulder Strap

Fig. 21. Aft-Facing Harness Concept. [1]

seat crashworthiness studies and development programs, with the majority of the effort concentrated on crew and troop seats.

Initial analyses of military specifications governing the design and fabrication of non-ejection crew seats used in Army aircraft revealed that the strength requirements quoted in the specifications were considerably lower than (1) those which would be dictated by the upper limit of decelerations which could be tolerated by the seat occupants, and (2) the decelerations and forces representative of Army aircraft accidents. [19] Table V compares the equivalent ultimate static loads for crew seats required by MIL-S-5822 (1957) and MIL-S-81771 (10 June 1970) with the corresponding static design ultimate loads required by the *Crash Survival Design Guide.*

A seating system designed for crashworthiness must be designed in consideration of several basic principles. These principles are discussed in detail in Reference 1 and are briefly summarized below:

1. *Seat Orientation* — Because human tolerance is dependent upon the amount of contact area provided by the restraint system, the most desirable upright seat orientation is aft facing with its large contact area for forward impacts. This orientation is desirable in conjunction with the restraint system shown in Figure 21, mission requirements

TABLE V

Ultimate Static Load Factors for Crew Seats

Load Direction	MIL-S-5822* (1957)	MIL-S-81771 (10 June 1970)	Crash Survival Design Guide [1]
Forward	8.0 G	20.0 G	35.0 G
Aft	5.0 G	–	12.0 G
Down	15.0 G	20.0 G	25.0 G
Up	7.5 G	–	8.0 G
Side	10.0 G	10.0 G	20.0 G

*For comparison purposes, the ultimate static loads presented in MIL-S-5822 have been converted to load factors in terms of G based on a 95th percentile occupant (199.7 lb).

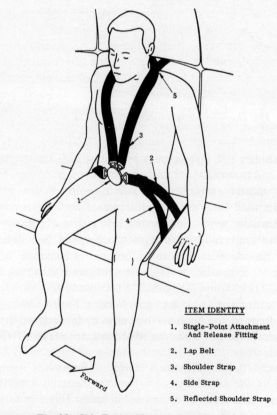

ITEM IDENTITY

1. Single-Point Attachment
   And Release Fitting

2. Lap Belt

3. Shoulder Strap

4. Side Strap

5. Reflected Shoulder Strap

Fig. 22. Side-Facing Harness Concept. [1]

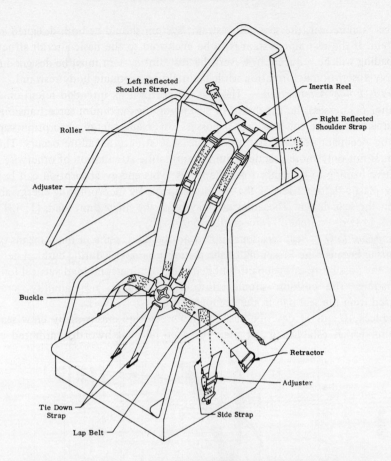

Fig. 23. Improved Forward-Facing Army Aircrew Restraint System Concept.

permitting. Forward-facing seats with restraint system configurations like those shown in Figures 19, 20, and 23 are a second choice to aft-facing seats. Side-facing seats are least desirable.

2. *Floor Attachment* — Two types of floor distortion are common in an aircraft crash: (1) floor surface "bulge" or "dishing" which can produce a rotation of the seat leg and possible attachment failure, and (2) twisting or warping of the floor surface which can cause seat structure distortion to the point of attachment failure. Consequently, seat attachment design should allow for these distortions and minimize their effect. Such techniques as a "yield hinge" and a "free rotation" joint are useful to this end. [1] Another means of alleviating floor distortion effects on tiedown is to attach the seat to a bulkhead.

3. *Restraint System* — The restraint system should be designed in accordance with the configurations presented in Figures 19, 20, 21, and 22 and the design criteria presented in

Reference 1; moreover, the seat and restraint system should be both designed and tested as a system. If the restraint system is to be anchored to the basic aircraft structure, seat frame loading will be reduced; however, the restraint system must be designed to permit seat energy-absorption deformation while maintaining adequate body restraint.

4. *Crash Force Attenuation* — The seat, to perform its intended retention function, must either (1) possess the capability of sustaining the maximum force transmitted from the occupant without collapse or (2) possess sufficient energy absorption capacity to reduce the occupant velocity to zero before seat structural failure occurs. This second approach is not only more practical, but it permits the attenuation of otherwise injurious decelerative loading to humanly tolerable levels. This energy absorption can be achieved either by plastic deformation of the seat structure or by incorporating energy-absorption devices in the seat design. These devices may be of the single limit load, [1, 43] multiple limit load, [44] or "notched" force-deflection [45] type.

5. *Seat Cushions* — Seat cushions should not be used as the primary means of vertical load limiting because the loosening of the restraint harnesses during cushion deformation increases the possibility of occupant "submarining" during combined vertical-longitudinal crash loading. The cushion should, however, attenuate, not amplify, crash loads transmitted from the seat pan to the occupant. [46]

References 40 and 41 describe past Eustis Directorate crashworthy crew seat designs and testing which established the basis for a current crashworthy, armored crew seat

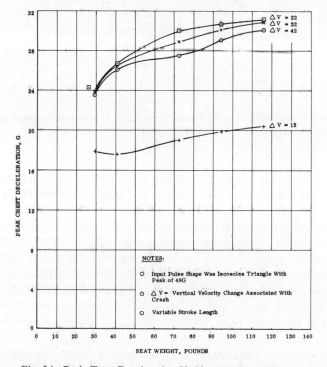

Fig. 24. Peak Chest Deceleration Vs Movable Seat Weight.

project. This project has two objectives: (1) design, fabricate, and test a prototype armored crew seat in accordance with a proposed military specification and (2) expand and improve crashworthy seat design technology. The crashworthiness requirements of this proposed military specification are essentially those recommended in the *Crash Survival Design Guide*. Although this project is not completed, several accomplishments to date are noteworthy.

Early armored crew seats were the result of attaching armor plates to seats already understrength from a crash loads standpoint. Gradually, seat strengths have been improved, with some complying with the crash loads cited in MIL-S-81771 as listed in Table V; however, none of the existing rotary-wing aircraft armored or unarmored crew seats comply with either the strength or the crash force attenuation requirements stated in the *Crash Survival Design Guide*. Numerous attempts have been made to develop a crashworthy armored crew seat; however, most have been plagued by inadequate load limiting mechanisms, restraint system inadequacies, and/or structural defects.

In addition to the increased strength requirement caused by its large mass, an armored crew seat poses an additional complication due to the large armor-to-man mass ratio. Because of the interaction between the occupant and the heavy bucket, the occupant can experience large decelerations during the crash force attenuation sequence. This relationship is shown in Figure 24. This data was obtained using a computer program of the 5-degree-of-freedom seat/occupant model shown in Figure 25. This damped,

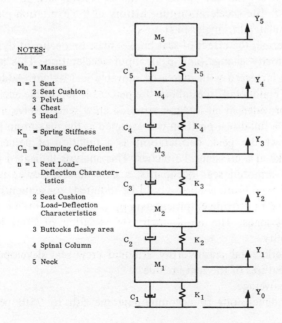

NOTES:

M_n = Masses

n = 1 Seat
2 Seat Cushion
3 Pelvis
4 Chest
5 Head

K_n = Spring Stiffness

C_n = Damping Coefficient

n = 1 Seat Load–
Deflection Character-
istics

2 Seat Cushion
Load–Deflection
Characteristics

3 Buttocks fleshy area

4 Spinal Column

5 Neck

Fig. 25. Lumped Parameter Model Representing a Seat, Seat Cushion, and Seat Occupant. [42]

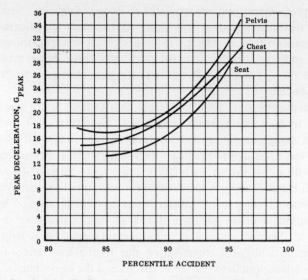

Fig. 26. Peak Deceleration Vs Percentile Accident (Constant Stroke Length of 12.0 In.).

spring-mass system models the occupant as a 3-degree-of-freedom system sitting erect on a seat cushion and seat so that the spinal column is vertical and so that only vertical translation is permitted. The acceleration-time history of the excitation pulse at the floor is assumed to be an equilateral triangle. [42]

The analysis to ascertain the effect of seat bucket mass on occupant dynamic response varied seat weight, velocity change, and peak input acceleration. This analysis showed that peak deceleration increased with increased movable seat weight. Additional analyses using this model were conducted to establish the percentile survivable accident for which protection could be provided in an integral armored crew seat. Twelve inches of stroke length with the seat in a full-down position was assumed as the maximum practical stroke length. Figure 26 shows the peak deceleration as a function of percentile survivable accident with the stroke at a constant 12 inches. This analysis indicated that to provide protection to integral armored seat occupants, a considerably lower limit load than calculated by standard rigid-body analysis could be required; consequently, the dynamic test matrix was designed to provide empirical data by varying percentile crash pulses, seat orientations, and limit loads. The requirement for variation of limit load demanded energy absorber flexibility.

The prototype experimental crashworthy armored crew seat developed is shown in Figure 27. Advanced features of the seat include:
1. 13 sq ft of ballistic coverage.
2. Configuration and dimensions to accommodate the 5th to 95th percentile Army aviator.
3. Three-point floor attachment which permits floor buckling while minimizing the forced racking of the seat structure.

4. Spherical rod-end floor attachments which permit floor rotation in any axis relative to the leg without bending the attachment structure.
5. A minimum of 12 inches of energy absorber stroke length.
6. A three-level load limiter which reduces deceleration imposed on lighter crewmembers in systems sized for heavier crewmembers.
7. Roller guide bearing assemblies to reduce frictional resistance to stroking.

The seat consisted basically of a simulated boron carbide-glass armored bucket, a guide frame assembly, and a support structure. The bucket was attached to the guide frame through four roller bearing assemblies and the energy-absorbing system. The energy absorber consisted of one stainless-steel tube and two energy-absorbing cables that were attached to the bottom bearing cross member and to the top of the guide frame assembly. The two cables could be selectively engaged to provide different limit loads for three different weight categories. The stainless-steel tube device was selected for its predictability, economy, and fast turnaround to provide test program flexibility. The upper yoke shown in Figure 27 was necessary to provide energy absorber and stroke flexibility during testing. This yoke would be eliminated in a flight-weight, preproduction prototype design. Figure 28 shows the same seat design modified to accommodate a rolling-torus type energy absorber instead of the tube and cable. Typical load-deflection characteristics for these two devices are shown in Figures 29 and 30.

Fig. 27. Prototype Crashworthy Armored Crew Seat With Tube and Cable, Variable Energy Absorber.

Fig. 28. Crashworthy Armored Crew Seat With Rolling-Torus Type Energy Absorber.

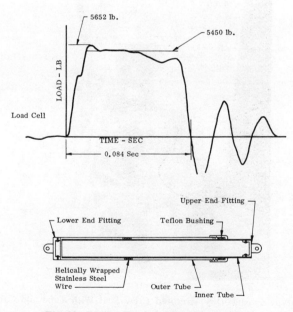

Fig. 29. Rolling Torus Load Limiter. [1]

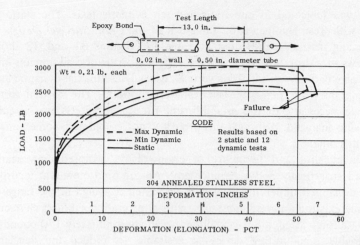

Fig. 30. Comparison of Dynamic and Static Load-Elongation Curves for Stainless Steel Tubes. [1]

TABLE VI

Crashworthy Armored Crew Seat Test Conditions

Test	Percentile Crash Pulse	Orientation of Seat (yaw, pitch, roll)	Peak Velocity Acceleration (G_{peak})	Time Change (ft/sec)	Duration (sec)
1	88*	0°, 0°, 0°	27	35	.081
2	93*	0°, 0°, 0°	38	39	.064
3	95*	0°, 0°, 0°	48	42	.054
4	93*	0°, 0°, 0°	38	39	.064
5	95*	0°, 0°, 0°	48	42	.054
6	99*	0°, 30°, 10°	32	41	.081
7	93*	0°, 30°, 10°	45	46	.064
8	95†	0°, 30°, 10°	48	50	.065
9	93	30°, 0°, 0°	23	47	.127
10	95††	30°, 0°, 0°	30	50	.103

*Vertical crash percentile.
†Test 1, Table 3-II, Crash Survival Design Guide. [1]
††Test 2, Table 3-III, Crash Survival Design Guide. [1]

The seat was subjected to one static and ten dynamic tests. The static test was in accordance with Test Number 6 of Table 3-II of the *Crash Survival Design Guide*. The dynamic testing matrix is presented in Table IV. Figures 31 and 32 show pre- and post-test views of a drop test and a sled test. The seat withstood all impacts imposed; i.e., dynamic crash environments involving vertical, triaxial, and biaxial dynamic loading up to and including the 95th percentile survivable crash pulse. The desired vertical energy absorber data necessary to the establishment of design criteria was acquired and is currently being analyzed. Although this project is not completed, some conclusions can be made at this time:

1. The practicality and feasibility of crashworthy, armored crew seats has been demonstrated analytically and experimentally. An armored crew seat incorporating the technology developed under this project is estimated to weigh in the range of 170-180 pounds. Some existing armored crew seats with poor crashworthiness characteristics and negligible crash-force attenuation capabilities weigh approximately 170 pounds.

2. A variable load-limiting device is desirable to reduce the deceleration level experienced by lighter seat occupants. Figure 33 illustrates the desirability of this feature for one sample calculation; e.g., the deceleration level experienced by the light-range occupant is 25% less than he would experience in a seat with a single limit energy absorber.

PRE-TEST   POST-TEST

Fig. 31. Crashworthy Armored Crew Seat Drop Test (Test Number 6 from Table VI).

PRE-TEST

POST-TEST

Fig. 32. Crashworthy Armored Crew Seat Sled Test (Test Number 9 from
Table VI).

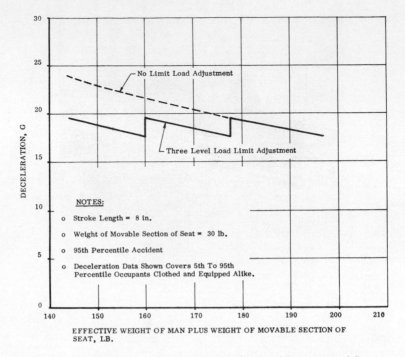

Fig. 33. Deceleration Vs Effective Movable Weight of Occupant and Seat.

Other recent crashworthy crew seats designed generally in accordance with Reference 1 include those shown in Figures 34 and 35. The Army-Sikorsky CH-54A crashworthy crew seat (Figure 34) weighs 192 pounds in the armored configuration and 34 pounds in the unarmored configuration. It achieves vertical crash force attenuation by stroking two rolling-torus type energy absorbers over an available vertical stroking distance of approximately 8 inches. This stroke restriction, imposed by the seat being designed for CH-54A retrofit, necessarily limits the crash force attenuation protection of the seat to a value less than the 95th percentile survivable accident. This seat was dynamically tested to the 95th percentile crash conditions required by Reference 1 and passed with respect to structural strength. The crash force attenuation performance is being analyzed by USAARL. Other active crashworthy crew seat projects include two projects recently initiated by the U.S. Air Force: one pertaining to fixed-wing transport aircrew seats and the other to a universal crew seat for three USAF helicopters.

In addition to possessing adequate strength and energy absorption capabilities like a crew seat, a troop seat must be designed to demanding operational considerations. Troop seats must be able to accommodate the lightly clothed and equipped 5th percentile soldier (approximately 134 pounds) as well as the heavily equipped 95th percentile soldier (approximately 255 pounds). [1] Also, troop seats must be easy to store in the aircraft, light in weight, not cumbersome, and durable. If the seat is not operationally acceptable, situations like those shown in Figures 9 and 10 arise.

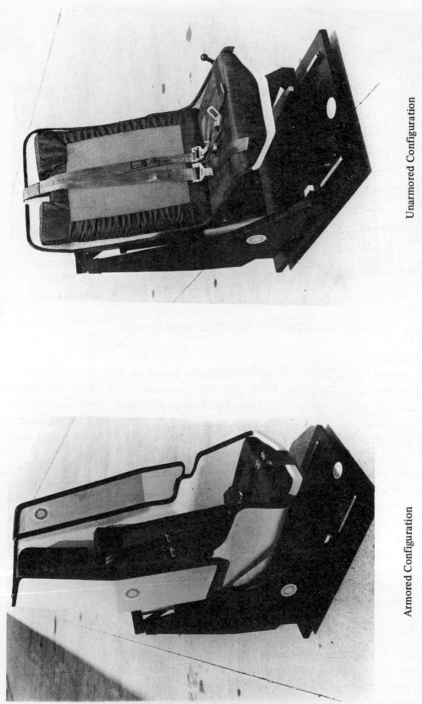

Unarmored Configuration

Armored Configuration

Fig. 34. Army-Sikorsky CH-54A Crashworthy Crew Seat.

Fig. 35. Boeing-Vertol Crashworthy Armored Crew Seat.

One experimental troop seat concept dynamically tested to the design criteria in Reference 18 is shown in Figure 36. [47] The seat consisted of two basic functional units: (1) a seat base incorporating an energy-absorbing strut to provide vertical support, and (2) a wrap-around-type seat back designed to provide the occupant with restraint in the lateral and longitudinal directions in addition to that provided by the lap and chest belts.

Boeing-Vertol, under a contract with the Aerospace Crew Equipment Department, Naval Air Development Center, has developed a lightweight, crashworthy troop seat weighing 6.6 pounds for a single-unit seat and approximately 5.6 pounds per seat in units of three. The seat consists of a fabric sling suspended from the aircraft overhead structure by webbing straps located at the front and back of the seat. The suspension straps are attached to the overhead structure by a compact, wire bending-type energy absorber. By suspending it from the overhead structure rather than supporting it on the floor structure, the crash input pulse to the seat is reduced up to 60% as a result of the crushing and deformation of structure between the floor and overhead structure. [48]

Fuselage (Including Landing Gear)

Initial efforts to improve the crashworthiness of the fuselage were stifled by the prevading belief that increased crashworthiness means increased costs, in terms of weight, money, and performance. However, experience shows that if crashworthiness design criteria are considered during the initial design stages of the aircraft, its crashworthiness can be significantly improved without intolerable costs being incurred. A case in point is the OH-6A and its crash performance record. [6, 49]

Basically, the foremost criterion which the aircraft structure must meet to be considered crashworthy is that the structure surrounding occupiable areas must remain reasonably intact, without dangerously reducing occupant living space; i.e., the occupants must be provided a "protective shell". The structure should be allowed to deform in a controlled, predictable manner, with crash forces and decelerations imposed on the occupants being minimized while still maintaining the protective shell. This protective shell should be delethalized with respect to the occupant's immediate environment; i.e., the requirements of Reference 1 pertaining to occupant environment and ancillary

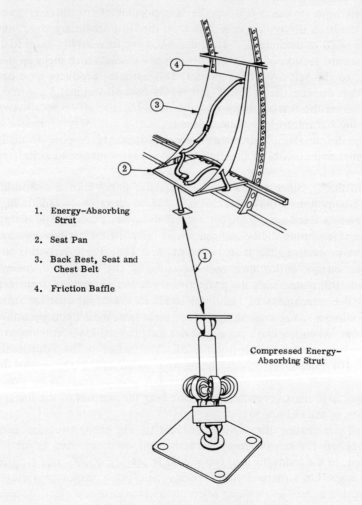

1. Energy–Absorbing Strut

2. Seat Pan

3. Back Rest, Seat and Chest Belt

4. Friction Baffle

Compressed Energy–Absorbing Strut

Fig. 36. Experimental Troop Seat. [47]

equipment stowage design should be met. This would prevent injurious body contact with sharp, nonpadded projections and with loose objects which become dangerous projectiles during the crash sequence.

The most frequent modes of structural failure resulting in occupant injury include: [1]

1. Longitudinal (crushing) loads on the cockpit structure
2. Vertical (crushing) loads on the fuselage shell
3. Lateral (crushing) loads on fuselage shell
4. Transverse (bending) loads on the fuselage
5. Deformation (buckling) of floor structure
6. Landing gear penetration of fuselage shell
7. Rupture of flammable fluid containers

The aircraft nose structure is often the first portion of the fuselage to contact the ground in a crash; in the typical nose structure, the light secondary structural members are rapidly crushed or discharged, causing the remaining forward bulkhead to dig into the earth. The forward section of the aircraft becomes a scoop, which picks up the earth and accelerates it to the velocity of the aircraft. This scooping produces large decelerations and loads which displace the bulkhead and reduce cockpit volume. A continuous keel or slide surface along the fuselage belly which permits the aircraft to slide over the terrain without causing hazardous longitudinal decelerations is one way to reduce crash force severity, cabin deformation, and occupant decelerations. [1] A more detailed discussion of the dynamics and kinetics of the longitudinal impact is presented in References 1, 12, and 50.

Collapse of the "protective shell" and compressive spinal injuries due to high vertical loading are common to rotary-wing aircraft due to their characteristic high-sink-rate impacts. Whereas a fixed-wing transport aircraft dissipates crash kinetic energy primarily by ground contact during sliding and can absorb very little kinetic energy by structural collapse if the protective shell is to be maintained, [12] the rotary-wing aircraft must absorb kinetic energy during the crash primarily by the collapse of energy-absorbing structure while still maintaining the protective shell. Maintenance of a protective shell is hindered by the attachment of high-mass items to the upper fuselage structure; e.g., engines, transmissions, and rotor masts. Often these large-mass items penetrate occupied areas; moreover, when the crash pulse includes longitudinal loads in addition to vertical loads, the resulting pitching of the rotor mast often causes the rotor blades to strike the fuselage. [9, 10] Means of alleviating hazards associated with vertical impacts include: [1]

1. Locating large mass concentrations at or near the bottom of the fuselage; e.g., the engine location of the OH-6A. [6]

2. Generally increasing the vertical stiffness of the cabin structure and designing structural supports for massive components located overhead, such as the transmission and rotor mast, to the following load factors: $N_y = 18G$, $N_x = 20G$, and $N_z = 20G$.

3. Increasing cabin structural elastic energy absorption and/or providing for plastic energy absorption at load levels which permit maintenance of a protective shell.

4. Attenuating vertical decelerations to tolerable levels in conjunction with seat

load-limiting capabilities by increasing subfloor structural energy-absorption capability and by providing energy-absorbing landing gear to reduce the severity of cabin decelerations for minor impacts.

A recent structural crashworthiness study [10] illustrated analytically the potential for reduction of floor vertical decelerations through proper crashworthiness design. This was achieved through the use of a 23-degree-of-freedom, non-linear lumped-mass mathematical model programmed for computer solution. Figure 37 shows the reduction in peak floor vertical decelerations achieved by the analytical redesign of landing gear, fuselage belly, and both items for a particular aircraft model. Improvement of both landing gear and fuselage belly energy-absorption characteristics reduced floor vertical decelerations by 65 percent for a 30 ft/sec vertical impact. The practicality of crash energy absorbing landing gear has been demonstrated by such operational crashworthy landing gear as those on the OH-6A [10] and the CH-54A. [8, 51] Care should be taken, however, to prevent landing gear from penetrating the protective shell or any part of a flammable fluid system.

Lateral impact of rotary-wing aircraft occurs more frequently and produces a more hazardous environment than previously believed. The accident data survey conducted in Reference 10 showed that over half of the severe rotary-wing aircraft crashes studied resulted in a roll-over or side impact. Because fuselage sides are not usually designed for crash protection, severe injuries have been caused in relatively minor crashes. Specific lateral impact hazards include: occupants being located close to the fuselage sides, inadequate occupant restraint, large open doorways, inadequate structural stiffness, and insufficient crushable structure in the fuselage walls. As a result of the significance of lateral impacts indicated by the study contained in Reference 10, the Eustic Directorate

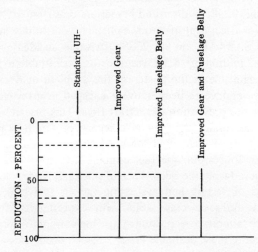

Fig. 37. Floor Acceleration Comparisons (30-ft/sec Vertical Impact) for UH-. [10]

is scheduled to begin in July 1971 a rotary-wing structural crashworthiness study concentrating on impacts containing both vertical and lateral crash forces.

When an aircraft impacts in an attitude not parallel or perpendicular to the impact surface, the angular acceleration of the fuselage often creates bending loads which rupture the fuselage at occupiable cabin sections; consequently, every attempt should be made to move the failure point to fuselage locations which pose a lesser threat to occupants.

Finally, the structure surrounding flammable fluid containers should not deform or fracture in a manner which causes container penetration. A more detailed discussion of crash-resistant flammable fluid containers is presented in the next section.

POSTCRASH HAZARDS

Of all possible threats to the survivability of an aircraft and its occupants, the one which creates more panic, causes the greatest number of fatalities, and is the most difficult to reduce or control is fire. Flammable fluid spillage is the most frequent cause of aircraft fire and is caused by either ruptured or torn fuel tanks, severed oil or hydraulic lines, or separation of fuel tank fittings.

In an effort to develop new concepts and criteria to improve the overall crash survivability of Army aircraft, the Eustis Directorate is devoting a major effort to the investigation of new concepts and techniques to reduce or eliminate postcrash fires.

An examination of USABAAR Army aircraft accident statistical data for survivable accidents for a 3-year period (CY 67 − CY 69) shows that there were 2388 survivable accidents, with fire occurring in 251 or 10.5% of the total. At first glance, 10.5% isn't too awesome; however, of the 439 fatalities occurring in these survivable accidents, 181 or 41% were the direct result of postcrash fire. Thermal injuries on the other hand accounted for only 161 or 6% of the 2663 injuries which occurred in these survivable accidents. The fires and fatalities discussed herein are crash impact fires only and do not include in-flight fires or fires resulting from ballistic hits or other causes. The total dollar loss as reported by USABAAR for the 2388 survivable accidents amounte; to $295.2 million, of which $76.2 million or 24.5% was directly attributable to postcrash fires.

Recognizing the magnitude of the postcrash fire problem in Army aircraft, the Eustis Directorate, has for several years been actively engaged in investigating and developing various techniques for the prevention of aircraft fires. This research has been conducted in four areas which fall under the general category of "built-in protection": (1) fire inerting systems, (2) breakaway fuel tanks, (3) fuel modification, and (4) fuel containment. Of these four areas, the fuel containment approach offered the greatest immediate potential in reducing the occurrence of postcrash fire.

After several years of testing and evaluating various fuel tank materials, including approximately twenty full-scale crash tests with experimental fuel tanks and other experiments related to aircraft fire prevention, a high-strength nylon laminate material was found to provide the most desirable crash-resistant properties and greatly exceeded the performance of other materials tested.

Using the results of these tests and concurrent test programs, an effort was initiated to design, fabricate, and qualify a complete crashworthy fuel system for the UH-1D/H helicopter. During the design of this system, hundreds of tests were conducted on various breakaway valve combinations, frangible attachments, and hose and hose fitting combinations.

The crashworthy fuel system designed and qualified for the UH-1D/H incorporated the following improvements:

1. Replacement of present fuel tanks with new crash-resistant fuel cells made of ARM-061 material.

2. Replacement of present metal tank fittings with new high-strength fittings capable of remaining in the tank during loads ten times greater than previously possible.

3. Replacement of the thin metal access and sump plate covers with covers capable of carrying anticipated crash loads.

4. Replacement of the sump drains and defueling valves with redesigned fittings capable of surviving crashes without becoming snagged in displacing structure, or being wiped off by dragging on the ground.

5. Replacement of rigid fluid lines with flexible, steel-wire-covered hose.

6. Replacement of hose and fittings with an improved design capable of carrying considerably higher crash loads.

7. Attachment of all flexible hoses to the tanks with self-sealing breakaway valves.

8. Use of frangible fasteners between metal fittings in the tank and the basic airframe structure.

The development of a crashworthy fuel system resulted in Revision B to military specification MIL-T-27422, "Aircraft Crash Resistant Fuel Tanks". This revision incorporated several new crashworthiness requirements, among which were a puncture-resistance test, a constant-rate tear test, a fitting-strength test, and a 65-ft drop test of a tank filled with water. Also, a draft military standard has been prepared which establishes design requirements for aircraft flammable fluid and electrical systems which will substantially reduce or eliminate postcrash fires in survivable crashes of military aircraft.

One of the biggest problems to date in qualifying a crashworthy fuel system for use in Army aircraft has been the successful completion of the 65-ft drop test. The military specification requires that a production-model fuel tank be filled with water and dropped from a height of 65 ft with no leakage occurring upon impact. High pressures and stresses are developed from this test which act on the tank wall and create tank design problems which to date have been overcome only by trial and error methods of tank design. In an attempt to define these internal pressures and stresses, an effort is currently under way to drop-test instrumented fuel tanks of various sizes and configurations to record data on the various parameters which affect crashworthy fuel tank design. This data will be used hopefully, to allow fuel tank manufacturers to design fuel tanks to successfully pass the 65-ft drop test the first time, thereby reducing delays in the qualification of the tanks.

Qualification of the UH-1D/H crashworthy fuel system (CWFS) was completed in October 1969, incorporation of the system into production aircraft was initiated in April 1970, and retrofit installation began in December 1970. As of 30 June 1971, there will be 884 production aircraft with these systems installed and 439 aircraft which have been retrofitted. The cost of these crashworthy fuel systems is approximately $4800 each for production aircraft and approximately $6450 each for retrofit of existing aircraft.

Installation of the crashworthy fuel system results in an increase of approximately 89 pounds in aircraft net weight. The fuel capacity of the CWFS is only 72 pounds less than the existing fuel system capacity.

USABAAR, USAAMRDL, and the Project Manager are closely monitoring crashes of UH-1D/H helicopters equipped with the CWFS in order to determine the effectiveness of the CWFS in reducing the incidence of postcrash fire. As of 31 May 1971 there had been 186 mishaps recorded, 23 of which were classified as major accidents. All of the six postcrash fires resulting from these accidents occurred in accidents classified as nonsurvivable, and the 6 fatalities and 27 injuries were nonthermal.

In addition to the UH-1H, crashworthy fuel systems are currently being qualified for the UH-1B, UH-1C, AH-1G, CH-47, OH-58, and OH-6 aircraft. Design, development, and qualification of all systems other than the UH-1 aircraft are being handled by USAAVSCOM and the responsible Project Manager. All of these fuel systems are scheduled to complete qualification by April 1972, and all aircraft are scheduled to be equipped with the CWFS by FY 77.

As a result of the study described in Reference 9, the significance of occupant drowning as a result of rotary-wing aircraft water impacts has been exposed. Little is known about the postcrash hazards associated with water impacts; however, it is anticipated that the current Office of Naval Research study of the Navy rotary-wing aircraft crash environment will provide more detailed information pertaining to this postcrash hazard. It has been suggested that all rotary-wing aircraft use some of the features of rotary-wing aircraft designed specifically for water landings, e.g., S-61 and S-62. [8] One means of reducing the severity of vertical water impacts and the hull pressures created by water impact is to design the fuselage bottom with a dead-rise angle as shown in Figure 38; e.g., the S-61 and S-62 amphibious rotary-wing aircraft possess a hull dead-rise angle of 12 degrees. Reference 8 recommends that both the hull and the nose section possess sufficient strength to withstand the water pressures created during a

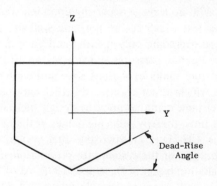

Fig. 38. Fuselage Bottom Dead-Rise Angle. [8]

50 ft/sec touchdown. If a deep, lower fuselage designed to restrict the entry of water during water impacts can be designed with energy-absorbing material (e.g., honeycomb) used as filler material in voids, both the aircraft's flotation potential and crash force attenuation capacity will be enhanced. Additional factors to be considered in occupant survival during water impacts and postcrash fire include the ease with which personnel restraint systems can be removed and the adequacy of postcrash evacuation provisions. Postcrash evacuation design criteria are presented in Reference 1.

CONCLUSION

This paper has presented justification for crashworthiness design, described the rotary-wing aircraft crash environment and associated hazards, and reviewed recent advances in rotary-wing aircraft crashworthiness design. As a result of the wealth of crashworthiness knowledge acquired during the past decade, the stature of crash-worthiness design in the design process has been significantly enhanced; consequently, improved crashworthiness features have been noted in recent rotary-wing aircraft designs. Although some retrofit efforts (e.g., the UH-1D/H crashworthy fuel system project) are feasible and promise a considerable return on investment, the design principles summarized herein and discussed in detail in the *Crash Survival Design Guide* must be incorporated in the initial design stages of future aircraft if the most effective mix of crashworthiness design techniques is to be obtained. The Army has adopted this philosophy as evidenced by (1) the extensive inclusion of the *Crash Survival Design Guide* design criteria in the Utility Tactical Transport Aircraft System (UTTAS) and Heavy Lift Helicopter (HLH) system requirements documents, and (2) the continuing program to revise or replace military specifications and standards not in accordance with verified aircraft crashworthiness design criteria (e.g., MIL-T-27422, "Aircraft Crash Resistant Fuel Tanks"). As a consequence of the adoption of this philosophy and continuing aircraft crashworthiness research and development efforts, a quantum reduction in the future Army rotary-wing aircraft crash injury and fatality rate is anticipated.

REFERENCES

1. J. W. Turnbow, et. al., *Crash survival design guide,* USAAMRDL Technical Report 71-22 (1971).
2. J. Haley, Helicopter structural design for impact, AHS Paper SW 70-16 (November 1970).
3. A. E. Zilioli, Capt., MC FS, *Crash injury economics: The costs of training and maintaining an army aviator,* USAARL Report No. 71-17 (April 1971).
4. R. G. Pearson, Ph.D. and M. H. Piazza, *Mechanisms of injury in modern lightplane crashes: A statistical summary of causative factors,* TCREC Technical Report 62-83 (November 1962).
5. H. F. Roegner, *Summary evaluation of U.S. Army HU-1A Bell Iroquois Helicopter,* TREC Technical Report 60-73 (December 1960).

6. H. G. Smith and J. M. McDermott, Designing for crashworthiness & survivability, AHS Paper 225 (May 1968).
7. N. B. Johnson, *Crashworthy fuel system design criteria and analysis,* USAAVLABS Technical Report 71-8 (March 1971).
8. M. J. Rich, Vulnerability & crashworthiness in the design of rotary-wing vehicle structures. SAE Paper 680 673 (October 1968).
9. J. Halley, *Analysis of existing helicopter structures to detect impact survival problems,* US Army Board for Aviation Accident Research, Fort Rucker, Alabama (1971).
10. C. I. Gatlin, D. E. Goebel and S. E. Larsen, *Analysis of helicopter structural crashworthiness,* USAAVLABS Technical Report 70-71 A & B (January 1971).
11. C. A. Yost and R. W. Oates, *Human survival in aircraft emergencies,* NASA CR-1262 (January 1969).
12. D. L. Greer, J. S. Breeden and T. L. Heid, *Crashworthy design principles,* FAA ADS-24 (September 1964).
13. K. L. Mattox, Capt., MC, *Injury experience in army helicopter accidents,* USABAAR HF 68-1 (1968).
14. *Flight Safety Foundation Accident Prevention Bulletin* 71-4 (April 1971).
15. *Crash Survival Investigation Textbook,* Dynamic Science, A Division of Marshall Industries, The "AvSER" Facility, 1800 W. Deer Valley Drive, Phoenix, Arizona (October 1968).
16. *U.S. Army aircraft accident data,* U.S. Army Board for Aviation Accident Research, Fort Rucker, Alabama (September 1970) (Unpublished).
17. J. W. Turnbow, Ph.D., et. al., *Military troop seat design criteria,* TREC Technical Report 62-79 (November 1962).
18. J. L. Haley, Jr. and J. P. Avery, Ph.D., *Personnel restraint systems study — basic concepts,* TREC Technical Report 62-94 (December 1962).
19. V. E. Rothe, et. al., *Crew seat design criteria for army aircraft,* TRECOM Technical Report 63-4 (February 1963).
20. J. L. Haley, Jr. and J. P. Avery, Ph.D., *Personnel restraint systems study UH-1A & UH-1B Bell Iroquois Helicopters,* TRECOM Technical Report 63-81 (March 1964).
21. J. L. Haley, Jr. and J. P. Avery, Ph.D., *Personnel restraint systems study, CH-47 Vertol Chinook,* TRECOM Technical Report 64-4 (April 1964).
22. L. W. T. Weinberg and J. W. Turnbow, Ph.D., *Survivability seat design dynamic test program.* USAAVLABS Technical Report 65-43 (July 1965).
23. G. M. Bruggink and D. J. Schneider, Capt., *Limits of seat-belt protection during crash decelerations,* TRECOM Technical Report 61-115 (September 1961).
24. L. W. T. Weinberg, et. al., *Dynamic test of an aircraft litter installation,* TRECOM Technical Report 63-3 (March 1963).
25. J. P. Avery, *Cargo restraint concepts for crash resistance,* USAAVLABS Technical Report 65-30 (June 1965).
26. L. W. T. Weinberg, *Crashworthiness evaluation of an energy absorption experimental troop seat concept,* USATRECOM Technical Report 65-6 (February 1965).
27. L. W. T. Weinberg, *Aircraft litter retention system design criteria,* USAAVLABS Technical Report 66-27 (April 1966).
28. J. Shefrin, et. al., *Integral helicopter cargo restraint system,* USAAVLABS Technical Report 69-68 (October 1969).
29. A. V. Zaborowski, Lateral impact studies lap belt shoulder harness investigations, *Ninth Stapp Car Crash Conf.,* Univ. of Missouri, Minneapolis, Minnesota (1966).
30. A. V. Zaborowski, Human tolerance to lateral impact with lap belt only, *Eighth Stapp Car Crash and Field Demonstration Conf.,* Wayne State Univ., Detroit, Michigan (1966).

31. E. L. Stech and P. R. Payne, *Dynamic models of the human body*, USAF AMRD-TR-66-157 (November 1969).
32. MIL-S-9479B (USAF), *Seat system, upward ejection, aircraft, general specification for*, (24 March 1971).
33. D. Orne and Y. K. Liu, A mathematical model of spinal response to impact, *J. of Basic Engineering*, ASME Paper No. 70-BHF-1 (1970).
34. G. H. Kydd and C. T. Reichwein, *Review of the dynamic response index* (D.R.I.), NADC-MR-6810 (August 1968).
35. P. Webb, M.D., *Bioastronautics data book*, NASA SP-3006 (1964).
36. A. B. Thompson, A proposed new concept for estimating the limit of human tolerance to impact acceleration, *Aerospace Medicine*, Vol. 33, No. 11 (November 1962).
37. A. M. Eiband, *Human tolerance to rapidly applied accelerations: A summary of the literature*, NASA Memorandum 5-19-59E (June 1959).
38. M. Kornhauser, Theoretical prediction of the effect of rate-of-onset on man's G-tolerance, *Aerospace Medicine*, Vol. 32, No. 5 (May 1961).
39. S. Ruff, Brief accelerations: Less than one second, *German Aviation Medicine in World War II*, Chapter IV-C, Vol. I, U.S. Government Printing Office, Washington, D.C. (1950).
40. L. W. T. Weinberg and J. W. Turnbow, *Survivability seat design dynamic test program*, USAAVLABS Technical Report 65-43 (July 1965).
41. L. R. Anderson, et. al., *Study & design of armored aircrew crash survival seat*, USAAVLABS Technical Report 67-2 (March 1967).
42. W. L. Auyer and J. W. Turnbow, *A study of the dynamic response of a damped, multi-degree of freedom, spring-mass system which simulates a seat, seat cushion, and seat occupant subjected to vertical impact acceleration*, AvSER 69-8, Dynamic Science (The AvSER Facility), A Division of Marshall Industries, Phoenix, Arizona (1969).
43. M. Schwartz, *Dynamic testing of energy attenuating devices*, Report No. NADC-AC-6903 (October 1969).
44. S. P. Desjardins and H. Harrison, *The design, fabrication, and performance characterization of an integrally armored crashworthy crew seat*, Dynamic Science, Division of Marshall Industries, USAAMRDL Technical Report (1971).
45. R. W. Carr and N. S. Phillips, *Definition of design criteria for energy absorption systems*, Report No. NADC-AC-7007 (June 1970).
46. P. R. Payne, Injury potential of ejection seat cushions, *J. of Aircraft*, **6**, No. 3 (May-June 1969).
47. L. W. T. Weinberg, *Crashworthiness evaluation of an energy-absorption experimental troop seat concept*, TRECOM Technical Report 65-6 (February 1965).
48. *Energy Attenuating Troop Seat Development Report*, Report No. D210-10244-1, Naval Air Development Center, Warminster, Pennsylvania (March 1971).
49. *Crash Survival Evaluation OH-6A Helicopter*, AvSER M67-3, Aviation Safety Engineering & Research, Phoenix, Arizona (April 1967).
50. W. H. Reed and J. P. Avery, Ph.D., *Principles for improving structural crashworthiness for STOL and CTOL aircraft*, USAAVLABS Technical Report 66-39 (June 1966).
51. M. J. Rich, *An energy absorption safety landing gear for helicopter & V/STOL aircraft*, IAS Paper 62-16 (1962).

AN ASSESSMENT OF ENERGY ABSORBING DEVICES FOR PROSPECTIVE USE IN AIRCRAFT IMPACT SITUATIONS

ARTHUR A. EZRA and RICHARD J. FAY

University of Denver, Denver, Colorado 80210

Abstract—This paper presents a survey of energy absorbing devices and systems for absorbing the kinetic energy of a moving body at impact, discusses their characteristics, and explores the use of these devices in aircraft crashworthiness design. Energy absorbing devices fall into three general categories according to the primary mechanism used for the absorption of energy. They are: material deformation, friction, and extrusion.

The important characteristics of energy absorbers are the specific energy absorption capacity per unit weight of device or system, the efficiency of the stroke, the stroke to length ratio, the reliability, the repeatability, the ability to sustain rebound loads, and the cost. In specific applications it is desirable to optimize the design in the sense that some desired combination of low cost, low weight, small size, and high performance is achieved. The understanding of the characteristics of particular energy absorbing devices is required to accomplish this.

Energy absorbers have an important role to play in the improvement of aircraft crashworthiness. In crashes of light fixed wing and rotary wing aircraft and transport aircraft impact, the loads are usually low enough to be survivable by the occupants if some measures are taken to provide improved protection through the use of energy absorbing devices. Areas in which these devices may be applied include the landing gear, the bottom of the fuselage, the seats, and the mountings for massive structures such as helicopter transmissions.

INTRODUCTION

Because of growing public sentiment concerning the need for improved crashworthiness in moving vehicles, a study was initiated in 1968, under the sponsorship of the NASA Office of University Affairs, to determine the possibility of exploiting unused NASA patents in the area of energy absorption. This study began with a comparative analysis of 62 U.S. patents and was extended to include the more general area of energy absorbers and systems. Later studies were made concerning actual applications in motor vehicle frames, vehicle seats, auto steering columns, auto bumpers, energy absorbing highway structures, elevators, and semi-trailers.

A study was recently begun under the sponsorship of the Office of Naval Research, Structural Mechanics Program, to study the application of energy absorbing devices to the improvement of aircraft crashworthiness.

SELECTION CRITERIA

Perhaps the greatest stimulus to the proliferation of such devices is the lack of widespread understanding of the criteria that determine the merits of one device over another for a particular application.

225

The first distinction that must be made is the difference between the performance required from a device and the prediction of the performance of a given device. To absorb energy and bring a moving body to rest, all devices must generate a resisting force over a certain stroke. The energy U absorbed is given by

$$U = \int_{0}^{S} F dx$$

where F is the resisting force and S is the total stroke.

The performance required of a device is limited by the maximum deceleration that may be applied to the moving vehicle and the space available to bring it to rest. The maximum deceleration that may be applied is determined by either the strength of the vehicle or the physical limitations of the occupants. The available stroke is governed by the configuration. Since the maximum possible amount of energy should be absorbed within these constraints, the ideal performance will be given by a constant force acting at the maximum permissible level over the entire stroke.

In actual practice, the instantaneous rise of the force level from zero to maximum value is neither possible nor desirable for the protection of the occupants of the decelerating vehicle. Some finite onset rate for the force is needed, but the desired onset rate is still a matter of opinion. The above relationship between force and distance represents an ideal from the energy absorption point of view.

While the ideal performance is therefore easy to specify, it is quite difficult to predict it accurately for a given device. This is where a considerable amount of theoretical work remains to be done.

The energy absorption per unit weight (specific energy) is an important measure of an energy absorbing device, particularly for aircraft where weight is important.

The ratio of stroke to length of device is another important parameter, because every inch of decelerating length in a collision is valuable.

Cost is always an important consideration. An individual energy absorbing device may be low in cost but the cost of the associated mechanisms required to adapt it to a particular use may be high. Therefore the energy absorbed per unit cost must include the cost of all the supporting systems in order to compare one type of energy absorber with another. Considerable savings can be made in safety applications where it is possible to design the elements of a structure to operate in an energy absorbing capacity if their design load is exceeded.

It can be seen therefore that the selection of the best energy absorber for a given application should not be made on the basis of specific energy alone, but must include consideration of all the other factors mentioned.

The stroke efficiency is a measure of how closely its performance approaches the best performance possible. It is measured by the ratio $U/(F_{max} S)$. A value of unity for this ratio represents the maximum energy absorption a device is capable of within the constraints of allowable force and maximum available stroke.

Repeatability and reliability are other desirable characteristics which govern the selection of an energy absorber. Devices which rely mostly on friction will give a wide range of performance due to the variability of the solid friction coefficient between the rubbing surfaces and due to change in viscosity with operating temperature.

Directionality of loading is another important consideration in the selection of an energy absorber for a given application. Devices which are only capable of uni-directional loadings have to be arranged in patterns to resist omni-directional loadings.

TYPES OF ENERGY ABSORBERS AND THEIR CHARACTERISTICS

Energy absorbing devices may be classified in one of three general categories; these are material deformation, extrusion, and friction. The classification is made on the basis of the primary energy absorbing mechanism. In many devices there is more than one energy absorbing mechanism. In many devices there is more than one energy absorbing mechanism operative but, in general, one is dominant. There have been a large number of energy absorbing devices conceived (as of November 1970 there were 62 U.S. patents on energy absorbers). However, the results of a 3 year study sponsored by NASA at the University of Denver show that a relatively small number are worthy of serious consideration. This paper will be limited to just these. Of this set there is an even smaller subset of energy absorbing devices that show the most promise for application to aircraft impact.

Material Deformation

The material deformation category includes a wide variety of energy absorbers which rely on the deformation of materials for the absorption of energy. A very simple example of this type of device is a rod or wire which is extended plastically by a tensile force. Other variations on this same theme include the plastic extension of cables or tubes. In all of these it is desirable to use a ductile material for the extensible element so that the energy absorption may be maximized. Other devices have been conceived which absorb energy through different types of material deformation including buckling, bending and

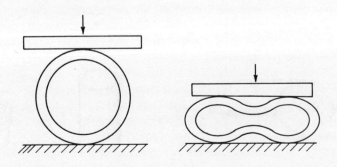

Fig. 1. Flattening Tube

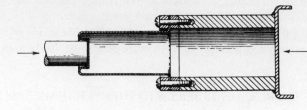

Fig. 2. Inside Out Tube

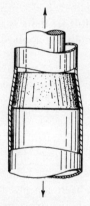

Fig. 3. Expanding Tube

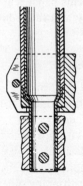

Fig. 4. Contracting Tube

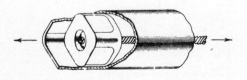

Fig. 5. Changing Shape Tube

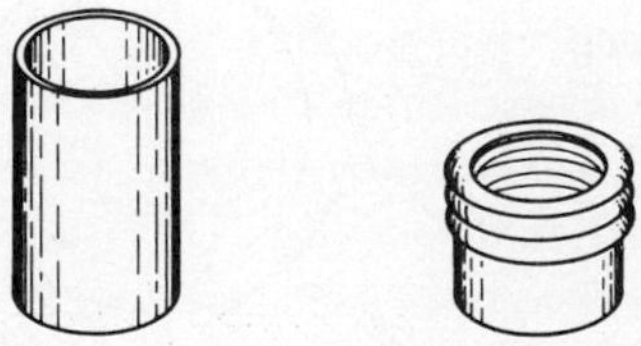

Fig. 6. Folding Tube

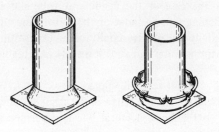

Fig. 7. Tube and Mandrel

shearing. Within the category of material deformation there are several well defined types of energy absorbers which deserve additional discussion.

Deforming Tubes—Tubes are deformable elements which lend themselves to a wide variety of uses as energy absorbers. In addition to plastic extension, they can be flattened [1]* as illustrated in Figure 1, made to turn inside out [2] as illustrated in Figure 2, made to expand [3] as illustrated in Figure 3, made to contract [4] as illustrated in Figure 4, made to change in cross-sectional shape [5] as illustrated in Figure 5, made to fold [6, 7, 8] as illustrated in Figure 6, or split and curl up (known also as the frangible tube) [7, 8, 9, 10, 11] as illustrated in Figure 7.

Of the deforming tube energy absorbers, plastic extension resists the highest sustained tensile force; however, the stroke of this mode of deformation is limited by the ductility of the material and over extension results in fracture. Buckling limits the axial compressive force which a tube can resist so tube folding provides the greatest resistance in the compressive mode. In the case of the tube and mandrel (splitting and curling) it may be desirable to use a less ductile tube material than that used for the other modes of deformation since splits must form to separate the tube into strips.

Of the tube energy absorbers, the flattening tube has the greatest potential stroke to length ratio since the solid height is only twice the thickness. However, in practice the tubes are usually used in an array so that complete flattening is not possible. The folding tube has a stroke to length ratio which is on the order of 0.7 or greater depending on the geometry of the original tube. For a simple plastic extension the stroke to length ratio is limited by the ductility of the tube material and at best is on the order of 0.5. For the tube and mandrel or frangible tube the stroke to length ratio is on the order of 0.8 or better depending on the geometry of both the tube and the mandrel.

The lowest stroke to length ratio of the axial deforming tube devices is found in those in which the tubes retain their original length while experiencing a change in cross-sectional geometry. These include the expanding and contracting tubes and the tube which experiences a change in cross-sectional shape.

The stroke to length ratio is low in these devices because the stroke is entirely due to

---

*Numbers in square brackets refer to references collected at the end of the paper.

overlap of the parts. In the case of the inside out tube the stroke to length ratio is somewhat better, approaching 0.5. In most of the deforming tube energy absorbers the load must be applied axially or nearly axially. The exception to this is the array of flattening tubes. The expanding tube can take some off axis loading if a generous mandrel overlap is used.

Bar and Wire Bending–Several tension energy absorbers have been developed which use the cyclic bending of metal bars and wires as the principal means of absorbing energy. In the device illustrated in Figure 8, a metal bar is deformed plastically as it is pulled

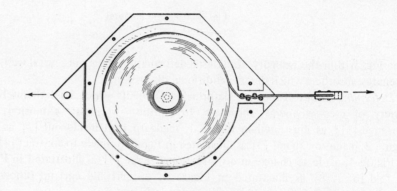

Fig. 8. Metal Strip Bending Device

between a set of pins.[12, 13] The stroke is limited only by the length of the bar provided in the coil. In a tension energy absorbing device directionality of loading is not a critical issue since proper end fittings will allow it to align itself with the load. Wire and bar bending energy absorbers are relatively expensive; however, the major parts are reuseable so the expense can be easily justified in applications involving frequent activation.

Cyclic Plastic Deformation–Several other energy absorbers have been developed which make use of the cyclic deformation of metal elements to absorb energy. One very well known energy absorber of this type makes use of a torus which is subjected to cyclic plastic deformation as it is rolled between two tubes;[14] this device is illustrated in Figure 9. Obviously this energy absorber depends on friction to effect the rolling of the

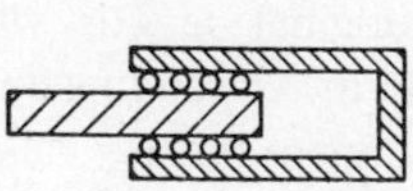

Fig. 9. Rolling Torus Device

torus as the two tubes are moved relative to each other. Therefore, the performance of the device is quite dependent on the fit of the torus and the tubes. In practice each device uses a wire helix rather than individual toroidal elements and is adjusted to give the desired resistive force by varying the amount of wire used.

Many cyclic deformation energy absorber concepts which have been patented involve complex schemes for inducing cyclic plastic deformation in metal elements. Figure 10

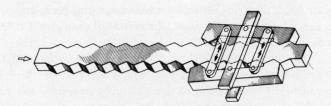

Fig. 10. Cyclic Deformation

illustrates how elaborate some of these schemes are.[15] The device uses a ratchet and linkage to cause the cyclic plastic deformation of four small elements. Obviously this device presents some design problems since the bearing area of the pins must be larger than the cross section of the elements where the deformation is desired. This device is composed of many parts and only a small amount of the material absorbs energy by deformation. In general, the specific energy of a device which relies on material deformation to absorb energy is directly related to how much of the device deforms during the energy absorption process.

Plastic Torsion of Shafts and the Plastic Hinge—Shafts made of ductile materials such as mild steel can be subjected to significant amounts of plastic torsion before rupture.

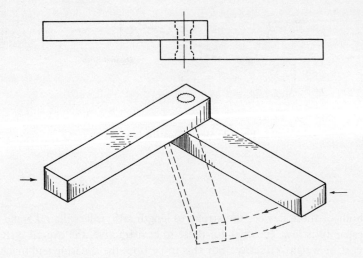

Fig. 11. Plastic Hinge

With mild steel it is possible to plastically rotate a shaft more than ½ turn per diameter of length. For one shot operation a device can be made which resembles a door hinge but has the pin fixed so that the rotation produces plastic torsion in the pin. This device, illustrated in Figure 11, has become known as the plastic hinge.

After elastic torsion is exceeded the shaft rotates with nearly constant torque. Plastic behavior begins at the outside and proceeds toward the center of the shaft as it is twisted, therefore, there is an angle of rotation in which the deformation is both elastic and plastic. The plastic hinge does not provide a constant resisting force even when the pin is fully plastic because of geometry. However, the effects of geometry can be minimized in some applications.

A variation of this device has been championed by the Department of Transportation for use in controlled collapsing automobile structures.[16] In this particular case structural shapes such as I sections and rectangular tubes are mitred and welded together in the shape of a "V". Forces on the legs of the device produce plastic deformation which is concentrated at the joint. This design has been plagued with problems of fracture at the joint.

Shearing and Machining—Since the machining of metals requires energy, this process has been considered as an energy absorbing mechanism. One such device absorbs energy by shearing the edges of a thin strip of metal.[17] This is illustrated in Figure 12. A device which absorbs energy by machining grooves in a metal tube[18] is illustrated in Figure 13. These processes will give nearly constant resistive forces over the stroke of the

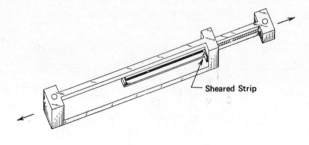

Fig. 12. Metal Shearing Device

devices. In both of these devices the stroke to length ratio is less than 0.5 and the energy absorption capacity is low. In order for these to be attractive, the overall system must be simple and cost or weight effective. For this to be true, the shearing and machining must involve parts of the mechanism which have an additional function(s) in the system. As

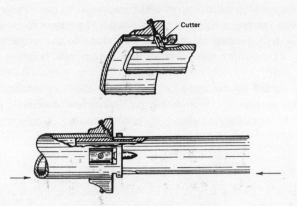

Fig. 13. Metal Machining Device

with the other devices which utilize metal deformation, the material properties and the dimensions must be consistent to achieve repeatable results.

Crushing Materials—Perhaps the simplest of all energy absorbers is crushable material. Wood, plastic foams, corrugated paper, and honeycombs of aluminum and paper are among the substances which have been used for this purpose. Except for the honeycomb type structures the force deflection curves for these materials tend to be exponential, which, while not being optimum in the sense of maximizing the energy absorbed for a given maximum resisting force, is very desirable for many applications. One very advantageous feature of crushing material energy absorber is that it is highly omni-directional. Crushing materials may be used by themselves as padding or as elements of a structure. Also, they may be used as energy absorbing cartridges in tube and piston energy absorbers.

Extrusion

A significant number of energy absorbers have been conceived which fall in the classification of extrusion devices. A familiar example is the automobile shock absorber (a similar device is used in aircraft landing gear). A variation on this theme [19] is illustrated in Figure 14. Here a solid material is extruded. One very successful device extrudes a

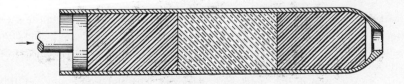

Fig. 14. Extrusion Device

viscoelastic substance which stores energy and later uses it to power the return stroke. An extrusion device can operate without a piston since it is possible to change the volume of a container simply by changing its geometry as is done in the extraction of toothpaste from a tube. A water filled energy absorbing automobile bumper operates on this principle.[20]

An important feature of extrusion devices is that they are velocity sensitive. This is useful in applications where it is desirable to have a low resistive force in low speed impacts and also the capability of absorbing larger amounts of energy in higher velocity impacts.

Friction

Each of us depends, for his safety, on an energy absorber which converts mechanical energy into heat by means of friction—the automobile brake. A variety of devices have been conceived which depend on friction as the primary mechanism for absorbing energy. In general, these devices consist of surfaces designed to rub together and a means of providing contact pressure between them. A remarkably simple friction energy absorber [21] is illustrated in Figure 15. This device consists of two partially overlapping

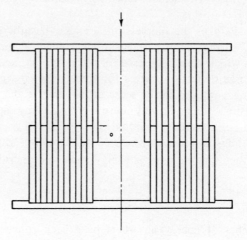

Fig. 15. Friction Device

sheets which have been wrapped over a cylindrical core. During wrapping the sheets are in lateral tension. When tension is removed the sheets contract, producing radial compressive stresses between the overlapping surfaces. These radial compressive stresses provide the necessary contact pressure to produce friction between the surfaces.

Due to the variation in the coefficient of friction with the nature and condition of the surfaces and the relative velocity between them, frictional energy absorbers require a means for controlling the contact pressure to produce a desired resistive force. In the automobile brake the driver presses on the pedal, observes the rate of deceleration of the vehicle, and then corrects the pedal pressure to achieve the desired result. Many energy absorber applications cannot afford such a complex control system. In the overlapping cylinder concept presented above, the device would be affected by the temperature and the surface condition of the plates; also, it has a low stroke to length ratio (probably on the order of 0.3) and a very low specific energy.

General Comments

All energy absorbing devices have this in common—they produce a resistive force over some stroke. Therefore it should be possible to adapt any energy absorbing device to virtually any application. However, in the real world it is highly desirable to optimize our designs. This means we must achieve some desired combination of low cost, low weight and small size for a given level of performance in the finished design. For safety applications the energy absorbers are designed to operate perhaps only once. In the case of an automobile or an aircraft the vehicle structure may be sacrificed for the purpose of human safety. This allows a certain freedom in the design for energy absorption since vehicle structure may be used for the purpose of absorbing energy in the crash situation.

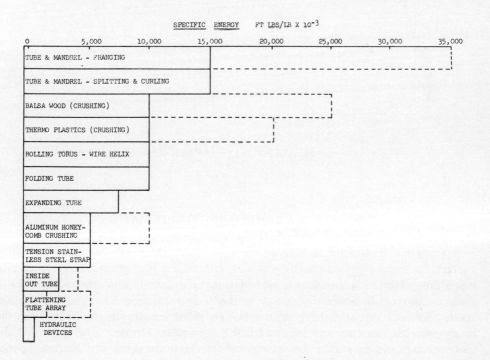

Fig. 16. Specific Energy Comparison of Energy Absorbers

A structural element which can also double as an energy absorbing element is a desirable item in such a design. Considerable discussion was devoted to deforming tubes earlier in this paper because some of these can be employed in this manner. The failure modes of other structural elements should also be considered so that the energy absorbing characteristics of structures can be optimized in the design of vehicles.

A comparison is made of a variety of energy absorbing devices on the basis of their length to stroke ratios in Figure 17. A summary of the characteristics, which influences the selection of an energy absorbing system for a given application, is given in Table 1.

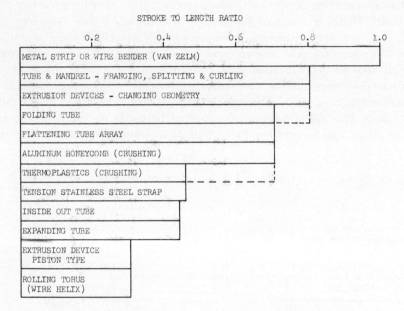

Fig. 17. Comparison of Stroke to Length Ratios of Energy Absorbers

AIRCRAFT CRASHWORTHINESS APPLICATIONS

A study has shown that in 95% of the crashes of light fixed-wing and rotary wing aircraft, the vertical impact velocity was 42 fps or less and the horizontal impact velocity was 50 fps or less.[22] This data is presented in Table 2 along with other data from this study concerning the acceleration pulses in the 95 percentile crash and allowable human acceleration tolerance. This data gives us cause to believe that the light fixed-wing and the rotary wing aircraft crash can be made highly survivable. The data also shows that the area needing the greatest improvement is the attenuation of the vertical impact acceleration on the occupant. In addition to having survivable accelerations, the living space in the aircraft must be preserved during the crash if survival is to be possible.

TABLE 1

Energy Absorber Data

Device	Stroke to Length Ratio (Approx.)	Force-Stroke Characteristics	Specific Energy (Approximately) Ft.lb/lb x 10^{-3}	Comments
Tension Stainless Steel Strap	0.5	Strain Dependent	5	Simple, Inexpensive, Fractures if over extended.
Flattening Tube Array	0.7	Nearly Constant for S/1* 0.7	1.5 – 5	Reliable.
Inside Out Tube	0.5	Constant	2 – 4	Reliable, Can Resist Rebound Loads.
Contracting Tube	0.5	Constant	–	–
Expanding Tube	0.5	Constant	8	Reliable, Accepts Some Off Axis Loading, Inexpensive.
Folding Tube	0.7 – 0.8	Nearly Constant (sinusoidal)	10	Reliable, Inexpensive, Can Accept Rebound Loads.
Tube and Mandrel 1. Franging	0.8	Large Variations (Average Nearly Constant)	15 – 35	Large Fluctuation in Load
2. Splitting and Curling	0.8	Nearly Constant	15	Reliable, Inexpensive if Conical Mandrel is Used.
Metal Strip or Wire Bender	1.0	Constant	–	Commercially Available, Reliable, Long Stroke, Works in Tension.

A. A. EZRA and R. J. FAY

TABLE 1 (Continued)

Energy Absorber Data

Device	Stroke to Length Ratio (Approx.–	Force-Stroke Characteristics	Specific Energy (Approximately) Ft.lb/lb x 10⁻³	Comments
Rolling Torus (wire helix)	0.3	Constant	10	Commercially Available, Worked in Tension or Compression, Each Device must be "Tuned" by Adding or Subtracting Wire.
Plastic Hinge (pin)	0.8	Function of Geometry	–	Very Simple, Reliable, Resists Rebound.
Extrusion Devices				
1. Piston Type	0.3	Velocity Sensitive	–	Hydraulic and Viscoelastic are Reuseable.
2. Changing Geometry	0.8	Velocity Sensitive	–	Very Simple, An Example of this is a Water Bumper.
Crushing Materials				
1. Balsa Wood	–	–	10 – 25	Omni-directional Load Capacity.
2. Aluminum Honeycomb	0.7	Constant for S/1* 0.7	5 – 10	Omni-directional Load Capability, Reliable.
3. Thermo Plastics	0.5 – 0.7	Exponential	10 – 20	Omni-directional Load Capability, Reliable.

*S/1 = Stroke to length ratio.

TABLE 2

Summary of Data for 95th Percentile Accident of Rotary and Fixed-Wing Aircraft
and Human Acceleration Tolerance

Impact Direction	Velocity Change fps	Peak G	Average G	Pulse Duration "T" Second	Human Tolerance G's	"T"
Longitudinal (Cockpit)	50	30	15	0.104	45	0.104
Longitudinal (Passenger Compartment)	50	24	12	0.130	35	0.130
Vertical	42	48	24	0.054	15	0.54
Lateral	25	16	8	0.097	11.5	0.1

Instances have been cited in which large masses such as wings, rotors, and transmissions
have exerted severe loads on the fuselage during impact resulting in collapse and a
decrease in living volume. Improvement is possible here through the use of frangible
connections which allow parts of the structure to break away at impact.[26] As
illustrated in Figure 18, the design of three areas of a rotary wing aircraft should receive
additional consideration with regard to crash performance. These are the landing gear, the
fuselage, and the seats. The same is true of light fixed-wing aircraft.

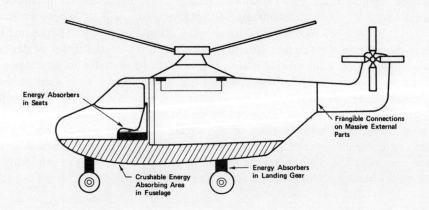

Fig. 18. Possible Improvements in Crashworthiness

In many crashes the occupants have survived the impact but were unable to survive the post-crash fire. This is an extremely important subject; however, it is outside the scope of this paper.

Landing Gear

Aircraft landing gear can be designed to absorb energy in a crash landing. The air-oil shock absorbers can be equipped with pressure relief valves so that they will not lock up in impacts at higher than normal velocities; some existing aircraft landing gear are made this way. Also, additional absorbers can be designed into the gear for use in harder than normal impacts.[26] Honeycomb energy absorbers are in use in the landing gear on some existing aircraft. These designs have been proven and should be used more extensively; other energy absorbers such as the folding tube should also be considered. The landing gear should also be studied for other possible ways of absorbing more energy; for instance, the wheels, and other parts might be designed to deform and absorb impact energy.

Fuselage

In crashworthiness design, the basic philosophy is to sacrifice the aircraft structure in favor of protecting the crew and passengers. Energy dissipation can be accomplished by controlled plastic deformation of fuselage structure. Space generally exists between cockpit and cabin floors and the bottom of the aircraft fuselage. This space should be designed as effectively as possible for the dissipation of energy in a crash. To accomplish this, structures must be included in the design which maximize the stroke efficiency and the specific energy of the structural system in this location while satisfying other aircraft design requirements. After a glance at Table 1 we are encouraged to try folding tubes or honeycomb. Naturally the energy absorbing structure must be integrated with the skeletal structure so that fuselage crushing is localized in the underfloor area at impact.

Controlled collapse of additional fuselage structure could also be useful in improving crashworthiness. For instance, mountings for massive parts such as helicopter transmissions could be designed to deform plastically thus dissipating energy and reducing the forces transmitted into the fuselage structure. This could reduce the threat of fuselage collapse without the need for making it stronger. It might be desirable in some cases to design for a certain amount of fuselage collapse; this would be the case if some decrease in living space could be permitted and if occupant seating were suspended from the upper fuselage structure rather than being mounted on the floor.

Frangible connections between parts of the aircraft may be a way of reducing the crash loads on occupant containing fuselage structure. In one helicopter designed for use as a crane, the cockpit separates from the rest of the aircraft in a crash. The possibility of designing for the break away of other parts has also been considered.

Seats

In addition to having adequate restraints and mounts sufficiently strong to prevent break away of the seats during a crash, seat structures should be designed to absorb energy, especially in the vertical direction. Both the Navy and the Army are working on this; however, there are problems which confront the design of energy absorbing seats. Of these, space limitation is the most perplexing. Traditionally crews in smaller aircraft are afforded a minimum of space in which to operate and seat adjustment usually consumes most of the room available for seat movement. One possible solution to this problem is to make the seats stationary and the controls adjustable. In future aircraft this promises to be easy since there is a trend toward "fly by wire" (electrical) controls. With this type of control system if would even be possible to make the controls move with the seat.

As one would suspect, it has been found that structure high in the fuselage experiences lower acceleration loads during a crash than the structure below due to energy absorption in the structure itself. This fact motivates the consideration of suspending the seats from the upper fuselage structure rather than mounting them on the floor. This might make it possible in some cases to provide ample crashworthiness without the specific use of energy absorbing devices. Also, this would have the added advantage of protecting the occupants from the problems associated with floor buckling.

Rebound due to elastic unloading of the deformed structure must be considered in the design of any energy absorbing seat. Energy absorbing devices used in this application must be capable of sustaining these loads after being stroked.

Passenger seating in most aircraft is more amenable to crashworthiness design than crew seating since seat adjustment is not required and seat height can be somewhat greater. The Navy and Army are both working on the development of energy absorbing passenger seats. An optimum design would use seat mounting structure in the energy absorption process so that a minimum of weight and size increase is required to provide the required force attenuation. Tubular energy absorbers and plastic hinges (plastic rotation of pins) are prime candidates since tubes and pins are normal parts of aircraft seats.

PREDICTION OF PERFORMANCE

The energy absorber is part of a dynamic system and the force that it transmits to the rest of the vehicle and finally to the passengers depends on many factors. Some of these are mass distribution, material properties in the plastic range, changes in geometry due to large deflections, solid and viscous friction, edge conditions, and other factors inherent in the particular system.

The large nonlinearities that appear in the equations of motion preclude exact analysis in closed form and detailed analysis of the system response requires numerical solutions that are costly and time-consuming.

For preliminary design, the force-displacement relation for the energy absorbing device is needed. This gross behavior can often be predicted satisfactorily by approximate methods. This section will summarize some estimates that have been used in predicting the specific energy of energy absorbers.

Extrusion Devices

Extrusion devices (Figure 14) utilize the principle of extruding a solid or a fluid through an orifice. Materials that exhibit solid as well as viscous type behavior can be treated as a Bingham material. The approximate analysis reduces to the equation of motion

$$\frac{dV}{dX} + \frac{1}{MV} \left[\alpha V + \beta V^2 + h(X) \right] = 0 \tag{1}$$

where V is the velocity of the mass M driving the device, X is the displacement, α a constant depending on viscosity and the ratio of the contact area to orifice area. The term BV^2 is related to the change in kinetic energy of the material extruded through the orifice. The constant B depends on the discharge coefficient for the orifice and the density of the material. The function h(x) describes the quasi-static behavior of the device including the force required to extrude the material as a solid.

Equation (1) can be solved approximately or numerically for any given set of parameters. The important results for a given initial velocity are the distance required to bring the system to rest and the maximum acceleration.

Figure 19 shows a comparison of theory and experiment for a patented device that extrudes a type of silicone rubber.

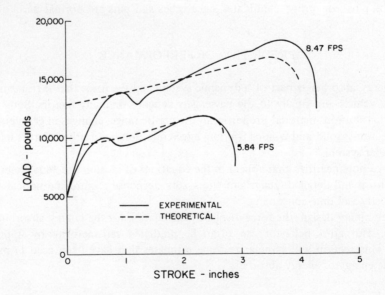

Fig. 19. Experiment and Theory for a Menasco Device

Folding Tube

The folding tube is simply a thin-walled right circular cylinder. Under the application of a sufficiently large axial load, the tube forms axisymmetric folds or the folds in thin tubes assume an asymetric pattern of nearly plane triangles with severe bending along their sides.

Timoshenko [27] notes that attempts at the static analysis of the axisymmetric problem goes back to Föppl and to Geckeler in 1926 and 1927. The numerical solution for the axisymmetric problem is relatively simple, however, to date, no one has made an exhaustive parametric study for this problem.

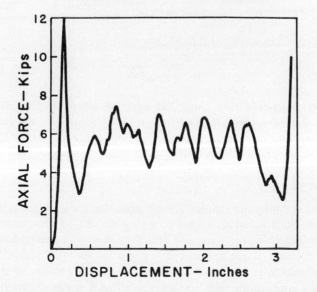

Fig. 20. Force-Displacement Curve for Thick Walled Collapsing Tube

Figure 20 is a typical force-deflection curve for a 3-inch diameter aluminum tube of 3003-H14 alloy. The diamater to thickness ratio was 44 so that it was thick enough to buckle in an axisymmetric mode. The first limit load on the curve at 12,000 pounds corresponds to the formation of the first fold at the end of the tube. The later peaks in load are related to subsequent folds which are not affected by edge conditions. By crimping the edge of the tube, the magnitude of the first peak can be lowered.

Further changes in the shape of the tube would reduce the magnitude of the oscillations in the load curve, but this would increase the cost of the device over straight commerical tubes.

An empirical formula for the mean folding force is

$$F = 8 \left(\sigma_y + .25 \, E_t\right) h \sqrt{2ah} \qquad (2)$$

where

$$
\begin{aligned}
\sigma_y &= \text{Nominal yield stress} \\
E_t &= \text{Tangent modulus} \\
a &= \text{Radius of tube} \\
h &= \text{Wall thickness}
\end{aligned}
$$

This can be compared to the classical expression for the buckling load

$$F = \frac{2\pi \, h^2 \, E_t}{\sqrt{3 \, (1-\mu^2)}} \qquad (3)$$

An empirical expression for length of shell generator in each fold is

$$L = 2.7\sqrt{ah} \qquad (4)$$

The half-wave length predicted by buckling theory is

$$L = 1.72\sqrt{ah} \qquad (5)$$

Once the static load-deflection curve has been obtained, its area can be used to estimate the amount of energy absorbed from a mass dropped on the end of the tube. Accelerometer data indicate that the dynamic load-deflection curves are higher than the static curves by amounts up to 25 per cent.

Tube and Mandrel Device

Figure 7 shows the tube and mandrel energy absorbing device developed by NASA for the space program. Such a device consists of a tube pressed upon a mandrel. The end of the tube on the mandrel fractures and deforms thus absorbing energy. Such a device has several excellent features. It can be designed to give a flat force deflection diagram. There may be an initial spike when the cracks are initiated, but if notches are cut into the end of the tube and the inside of the tube beveled, this effect is essentially eliminated.

The static axial force in the tube depends on the contact pressure between the tube and mandrel and the coefficient of solid friction between the two surfaces. If axial symmetry is assumed along with a rigid plastic material, a solution for the axial force F neglecting bending is

$$F = \frac{2\pi \sigma_y \rho}{1 + \mu^2} \left\{ (1-\mu^2) + e^{\mu d} \left[2\mu \sin d - (1-\mu^2) \cos d \right] \right\} \qquad (6)$$

where σ_y =   Yield stress

           μ =   Coefficient of friction between tube and mandrel

           ρ =   Radius of curvature for the mandrel

           d =   Polar angle from edge of mandrel to crack tip

To be consistent with axial symmetry, the angle d is given by

$$\frac{\epsilon a}{\rho} = 1 - \cos d \qquad (7)$$

where ϵ =   Rupture strain in tension

           a   =   Tube radius

However, once the crack is initiated it propagates along the tube destroying the axial symmetry. Experimental results give reasonable correlation with Equation (6) if d is computed from

$$\cos d = 1 - .15 \ \epsilon a/\rho \text{ and } \mu = 1.0.$$

CONCLUSIONS

A variety of energy absorbing structural elements are available for use in improving aircraft crashworthiness. The best one to use depends on the specific application, and the best choice depends not only on its specific energy (i.e., energy absorbed per unit mass) but also on its length to stroke ratio, reliability, and cost.

REFERENCES

1. J. A. DeRuntz and P. G. Hodge, Jr., Crushing of a tube between rigid plates, *J. of Appl. Mech.*, Sept., 1963.
2. C. K. Kroell, Energy absorbing vehicle bumper assembly, U.S. Patent No. 3,146,014, August 25, 1964.
3. T. O. Eddins, Space Craft Soft Landing System, U.S. Patent No. 3,181,821, May 4, 1965.

4. J. F. Spielman, Energy Absorbing Means, U.S. Patent No. 3,059,966, October 23, 1962.
5. G. Hendry, Energy Absorbing Seat Belt Attachment, U.S. Patent No. 3,016,972, March 27, 1962.
6. A. O. Coppa, *Collapsible shell structures for lunar landings*, General Electric Co., TIS Rep. No. R62509, 1962.
7. *Program for the exploitation of unused NASA patents,* First Annual Report, Univ. of Denver, 1969.
8. *Program for the exploitation of unused NASA patents,* Second Annual Report, Univ. of Denver, 1971.
9. J. R. McGehee, et al, Frangible Tube Energy Dissipation, U.S. Patent No. 3,143,321, August 4, 1964.
10. J. R. McGehee, *Experimental investigation of parameters and materials for fragmenting tube energy absorption process,* NASA TN-D-3268, February 1966.
11. G. F. Von Tusenhausen, Energy Absorbing Device, U.S. Patent No. 3,381,778, May 7, 1968.
12. M. A. Jackson, Energy Absorbing Device, U.S. Patent No. 3,211,260, October 12, 1965.
13. M. A. Jackson, et al, Cargo Tie Down Apparatus, U.S. Patent No. 3,377,044.
14. B. Mazelsky, Absorbing Device, U.S. Patent No. 3,369,634, February 20, 1968.
15. D. L. Platus, et al, Energy Absorbing Device, U.S. Patent No. 3,231,049, January 25, 1966.
16. P. M. Miller and K. N. Naab, *Basic research in automobile crashworthiness − Testing and evaluation of the forward structure modification concept,* CAL Report No. YB-2684-V-1, September 1969.
17. H. A. Son Moberg, Energy Absorbing Means, U.S. Patent No. 3,232,383, February 1, 1966.
18. J. F. Rayfield, et al, Deceleration Device, U.S. Patent No. 2,961,204, November 22, 1960.
19. G. H. Peterson, Variable Kinetic Energy Absorber, U.S. Patent No. 3,380,557, April 30, 1969.
20. J. W. Rich, Shock Absorbing Bumper, U.S. Patent No. 3,284,122, November 8, 1966.
21. E. W. Conrad, Non-Reuseable Kinetic Energy Absorber, U.S. Patent No. 3,164,222, January 5, 1965.
22. J. N. Turnbow, et al, *Crash survival design guide,* Dynamic Science, Phoenix, Arizona, USAAVLABS Technical Report 70-22, August, 1969.
23. N. Perrone, *Crashworthiness and biomechanics of vehicle impact,* NSF Frant No. GK 23747, The Catholic University of America, Washington, D.C., September, 1970.
24. J. Shefrin, *Integral helicopter cargo restraint system,* The Boeing Company, Vertol Division, Phila., Penn. USAAVLABS Technical Report 69-68, October, 1969.
25. A. P. Coppa, *New ways of shock absorption,* Machine Design, March 28, 1968.
26. M. J. Rich, *Vulnerability and crashworthiness in the design of rotary wing vehicle structures* SAE Paper 680673, Aeronautic and Space Engineering and Manufacturing Meeting, October, 1968.
27. S. P. Timoshenko and J. M. Gere, *Theory of elastic stability,* 2nd Edition, McGraw-Hill Book Co., New York, N.Y., 1961, pp. 457-461.

HUMAN TOLERANCE LIMITATIONS
RELATED TO AIRCRAFT CRASHWORTHINESS

ALBERT I. KING

Wayne State University, Detroit, Michigan

Abstract—A survey of available information regarding human tolerance to impact acceleration is made. The problems encountered in the scaling of animal data to the human level and in the correlation of single-directional tolerance data to actual multi-directional impacts in aircraft crashes are discussed. Examples of the effective use of restraint systems to raise the tolerance limits are given together with the injury potential of these systems. It is recommended that multi-directional impact studies be carried out along with the development and validation of mathematical models to generalize the experimental results.

INTRODUCTION

The crashworthiness of an aircraft structure or seat can only be evaluated on the basis of its ability to mitigate the injuries sustained by the occupants in a variety of crash configurations. Field accident data have shown that the lack of distortion of the cabin structure cannot be used as a measure of crashworthiness. In a survey of 913 light aircraft crashes, Hasbrook[1] ** found that 29.4% of the occupants were killed or seriously injured while 56.1% of the structures remained "intact to distorted". As a matter of fact, it may be desirable for interior surfaces to undergo large deformations to attenuate the impact of the occupant during a crash.

From the point of view of injury reduction or prevention, crashworthiness is meaningful if the limits of human tolerance can be first established. The study of human tolerance, however, is fraught with difficulties and subjected to various forms of definitions. A practical measure of tolerance is the acceleration (or deceleration) sustained by the human body or body segment, since it is a physical quantity that can be measured readily and is usually taken to be an input condition in an analysis of an impact event. Unfortunately, this simple measure is quite insufficient to describe fully the dynamic response of a biological structure. A variety of other factors have been added to supplement the acceleration information. These include duration of the impact, rate of onset, type of restraint system used and the forces generated in the restraint system. Empirical data given in this form have been obtained by many investigators over the past 25 years after some 220 research studies on man and other mammals [2].

---

*Associate Professor in Mechanical Engineering Sciences and Associate in Neurosurgery.

**Numbers in brackets refer to similarly numbered references at the end of the paper.

DEFINITIONS

Human tolerance to impact is classified as voluntary, injury threshold, minor injury and severe injury [3]. This classification came about as a result of a conglomeration of test data obtained by various investigators who used live human subjects, human cadavers and animals in their studies. Voluntary tolerance is obtained from tests on human volunteers. The acceleration level beyond which the volunteer is unwilling to exceed is defined as the voluntary tolerance limit. The level at which injury is imminent but does not occur is termed the injury threshold limit. It is generally difficult to achieve in tests using human subjects, and is obtained by interpolation between the voluntary limit and the minor injury limit. Human tolerance at the injury level is generally inferred from studies using cadavers or animals. Occasionally, injury data from human subjects have been obtained through over-exposure [4, 5]; but these are usually minor injuries which are reversible and do not constitute a serious threat to life. Severe injury includes all levels of trauma up to the fatal level.

We can immediately deduce from the above definitions of tolerance limits that there cannot be an exact quantitative description of tolerance. At each level, the limit can best be defined as some average value of a wide band of data points. There is very little one can do about the large variations encountered in biological systems, and it is because of this variation that there are no clear-cut answers to the establishment of satisfactory safety standards.

Patrick and Grime[3] are of the opinion that the minor injury category should be used as the best tolerance value for use in automotive safety programs. They reasoned that this level may cause severe injury to the weakest segment of the population, minor or no injury to the average individual and the strongest individuals will be in the sub-injury category. Whether this choice is equally applicable to aircraft crash environments is debatable. A strong reason for being more conservative is the requirement of mobility on the part of the occupant immediately post-crash. The hazards of fire and toxic fumes necessitate rapid exit from the aircraft in most circumstances. Furthermore, almost all of the available data on human impact tests were obtained from military volunteer subjects. These subjects are generally young males in good health and are hardly representative of the general population [2].

In addition to the various classifications of tolerance limits, there are two major subdivisions in the type of impact, namely whole-body impact and regional impact. When the force is distributed over a large part of the body, the impact is classified as whole-body. The distinction between the 2 classes of impact is primarily the result of the evolution of impact research. Early attempts to obtain human tolerance to impact consisted of investigations of survivability of falls from great heights [6, 7, 8]. These studies were followed by work in impact protection in crashes and escape systems in military aviation and aerospace studies, led by the pioneering work of Stapp [9, 10]. Eiband [11] summarized some of the results in the form of tolerance curves for whole-body impact for the seated human subjected to uni-directional accelerations. Biomechanical research in automotive safety had a different requirement. Restraint systems were either absent or so simple that whole-body tolerance limits cannot be used.

Various body segments were injured as the result of localized impact against inferior structures of the vehicle or the restraint system itself. This led to the establishment of regional tolerance limits. The implication is that injury is cause by forces and/or moments and that tolerance should be related to these loads. However, because of the impracticability of specifying and measuring such loads in many instances, we are forced to use acceleration as the principal measure for tolerance. Forces are used for certain skeletal structures such as the knee, facial bones and the pelvis.

In order to abbreviate the description of the acceleration direction, a set of conventions and notations was proposed by the Biodynamics (formerly, Acceleration) Committee of the Aerospace Medical Panel, AGARD [12]. Vehicle acceleration directions are defined along the rectangular cartesian coordinates shown in Table 1 (System 1) and are described in System 2. A different set of rectangular cartesian coordinate axes is used by physiologists to describe the reactive displacement of organs and tissues with respect to the skeletal structure, as shown in System 4. The "vernacular description" listed in Table B is helpful in visualizing the direction of the acceleration, particularly when decelerations are involved.

TABLE 1

CONVENTION OF SIGNS FOR LINEAR ACCELERATION
(REF. 12)

| SYSTEM I | SYSTEM 2 | SYSTEM 3 | SYSTEM 4 |

| Table A | | Table B | | |
| Direction of Acceleration | | Inertial Resultant of Body Acceleration | | |
Aircraft Computer Standard (System 1)	Acceleration Description (System 2)	Physiological Description (System 3)	Physiological Standard (System 4)	Vernacular Description
$+a_x$	Forward Accel.	Transverse A-P, Prone, or Chest to Back G	$-G_x$	Eyeballs in
$-a_x$	Backward Accel.	Transverse P-A, Supine or Back to Chest G	$+G_x$	Eyeballs out
$-a_z$	Headward Accel.	Positive G	$+G_z$	Eyeballs down
$+a_z$	Footward Accel.	Negative G	$-G_z$	Eyeballs up
$+a_y$	R. Lateral Accel.	Left Lateral G	$+G_y$	Eyeballs left
$-a_y$	L. Lateral Accel.	Right Lateral G	$-G_y$	Eyeballs right

AIRCRAFT IMPACT CONDITIONS

Aircraft crash kinematics have been estimated by Turnbow, et al. [13]. A total of 563 military aircraft accidents were reviewed. Of these, 373 cases were used to establish impact conditions. The data were divided into two categories. Rotary-wing and light, fixed-wing aircraft had similar impact conditions and the data from these crashes were put into one group. A second group consisted of 60 major accident of fixed-wing transport aircraft. All of the selected cases involved one or more of the following factors:
   (a) substantial structural damage
   (b) post-crash fire
   (c) personnel injuries
   (d) at least one person survived the crash
The information is thus representative of the type of crash which is survivable and is applicable to our discussion on tolerance.

The deceleration pulses measured at the floor of the aircraft were apprxoimated by a isosceles traingle, the peak being twice the average value. For each category of aircraft, the average vertical ($-a_z$) and longitudinal ($-a_x$) acceleration were plotted against the relative frequency of occurrence. These are shown in Figures 1 and 2 for the rotary- and light fixed-wing aircraft and fixed-wing transport aircraft respectively. Data for lateral acceleration were insufficient for a distribution plot but the magnitudes were found to be substantial, particularly for rotary-wing aircraft. Turnbow, et al. [13] estimated that lateral accelerations rarely exceed 16g. At the 95% level, the recommended design pulses for the two types of aircraft are summarized in Tables 2 and 3.

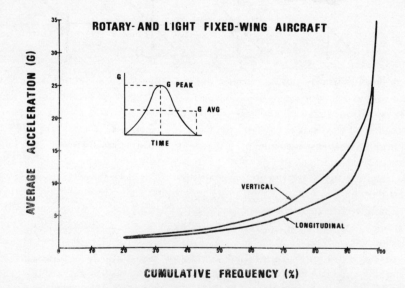

Fig. 1. Vertical and Longitudinal Accelerations for Rotary- and Light Fixed-Wing Aircraft (13)

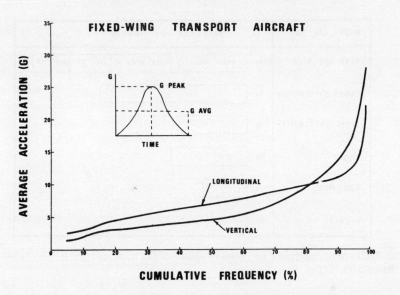

Fig. 2. Vertical and Longitudinal Accelerations for Fixed-Wing Transport Aircraft (13)

It can be seen from these data that aircraft crashes give rise to deceleration pulses along all 3 axes, with peaks that probably occur out-of-phase with each other. In addition there can be secondary impacts of lesser magnitude. It is within this framework of expected accelerations that we are to consider the tolerance limitations of human occupants.

AIRCRAFT ACCIDENT INJURIES

Having considered the input accelerations experienced by an occupant in an aircraft crash, we should briefly discuss the injury pattern resulting from such crashes. The source of this information is civilian and the most recent breakdown of incidences of injury to various regions of the body is given by the FAA [14] for 1965. There were 1019 business aircraft accidents that year out of a total of 5196 general aviation accidents. The frequency of head and neck injuries was the highest, followed by injuries to the lower and upper extremities, the chest, the abdomen and the pelvis. The distribution is shown in the form of a bar chart in Figure 3. Vertebral injuries are not listed but Snyder [15] states that they rank fourth in frequency of occurrence behind that of upper extremity injury. Earlier studies have also shown that trauma to the head is most common and most severe. These include the work of Marrow [16], DeHaven [17] and Hasbrook [1]. Although injuries to the extremities are seldom life-threatening, those to the lower extremities reduce mobility and can seriously affect the occupants chances of survival in the presence of fire and toxic fumes. Similarly, vertebral injuries can be considered to be serious in view of involvement of the spinal cord.

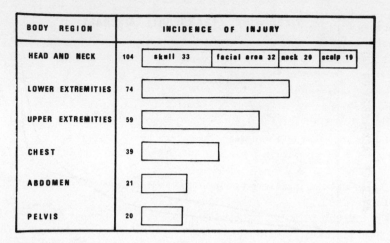

Fig. 3. Frequency of Injury to Various Regions of the Body in 1019 Business Aircraft Crashes in 1965 (14)

HUMAN TOLERANCE LIMITS

Efforts to compile available data on human impact tolerance were intensified in the last decade as a result of increasing emphasis on automotive safety. The recent work by Snyder [2] is an excellent summary of the state-of-the-art of our knowledge on this subject. It contains 446 references of impact studies and tables of data from all known investigations of animal and human deceleration tests. It is not the intention of this paper to duplicate this effort. Rather, some of the tolerance limitations regarding the head, the spine and the lower extremities will be reviewed and commented upon in the light of the acceleration environment encountered in aircraft accidents.

Head Impact

The subject of head injury due to impact is one of the most thoroughly investigated areas of biomechanical research. The large amount of published literature over the past thirty years discussed the mechanisms of injury to the brain and spinal cord, levels of acceleration necessary to cause concussion and skull fracture, establishment of tolerance levels, mechanical properties of various parts of the head and, more recently mathematical models describing its biodynamic response to impact. The continuing investigation of head injury is indicative of the complexity of the problem and of the lack of good design information for head protection.

In the area of tolerance limits, a curve was proposed by Lissner, et al. [18] in 1960 and subsequently modified by Patrick et al. [19]. This curve became known as the Wayne Tolerance Curve which is now widely used in automotive safety research. Figure 4 shows the Wayne Curve plotted on log paper. The ordinate is effective acceleration, which is an 'average' acceleration of the skull measured at the occipital bone for impacts of the forehead against plane, unyielding surfaces. The abscissa is the duration of the pulse. The history of the development of this curve is somewhat difficult to trace since it involves data accumulated over a period of more than 30 years. Briefly, the following steps were involved in the evolution of this curve:

(a) It was observed clinically that linear skull fracture is usually associated with unconsciousness or a mild concussion [20].

(b) The acceleration level and pulse duration necessary to cause skull fracture in a human cadaver head are approximately the human tolerance limit to impact, since loss of consciousness is in the ragne of a minor reversible injury. The fracture data provided points for the curve up to 0.006 sec.

(c) Experimental animals were concussed by pressure pulses of varying magnitudes and durations [21].

(d) The pressure pulses measured in the parietal and temporal regions of the human cadaver head [18, 22] were compared with the animal data and the corresponding acceleration measurements were used to provide data points for the curve between 0.006 and 0.01 sec.

(e) Thylong duration end of the curve or the asymptotic value of 42g was obtained from human volunteer data by Stapp [42] and others. Patrick et al. [19] considered this to be too low, since human volunteers have survived $-G_x$ accelerations of over 45g (without a blow to the head). They recommended that the value of the asymptote be raised to 80g. This curve provided the original Federal Motor Vehicle Safety Standard (FMVSS) for head impact. The head injury criterior was set at 80g for a cumulative period not exceeding 3 milliseconds [23].

The Wayne Tolerance Curve was also the basis of several indices of injury severity. Among the most popular is the Severity Index (S.I.) due to Gadd [24, 25]. The idea of a single number to represent tolerance for various regions of the body is attractive from the

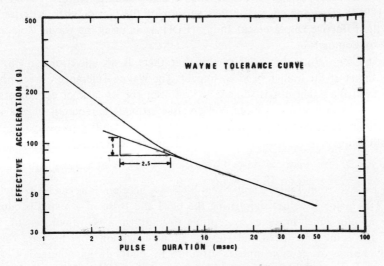

Fig. 4. Wayne Tolerance Curve for Head Impact (19)

point of view of the safety engineer who must satisfy a host of safety requirements. Gadd's severity index for head injury is given by the equation

$$SI = \int_0^T a^n \, dt \leqslant 1000$$

where   a   =   acceleration in g's
            n   =   weighting factor, 2.5 for head impacts
           dt   =   time interval
           T   =   pulse duration, $0.0025 \leqslant T \leqslant 0.050$

The weighting factor of 2.5 is primarily based on a straight line approximation of the Wayne Tolerance Curve, as shown in Figure 4. The negative slope of the line is 2.5 between 2.5 and 50 milliseconds. A value of 1000 was taken as the threshold of serious internal head injury for frontal impact. Other justifications for the choice of these numbers are given by Gadd [25] but the proof of their validity lies with further experimental verification. Recent work by Hodgson et al. [26] on the correlation of fracture with the severity index revealed that linear fracture of 9 cadaver skulls occurred between 540 and 1760 with a median value of 910. More work is required to overcome objections by other investigations on the use of the severity index as a reliable tolerance limit. Nevertheless, the most recent version of FMVSS No. 208 requires that the resultant acceleration of the head shall not exceed a severity index of 1000, calculated by the method described in SAE Information Report J885 a [27].

Several other indices of injury have been proposed. Slattenschek [28] introduced a criterion of injury using a simple spring-mass model for the head. The displacement of the mass representing the brain with respect to that representing the skull was limited to 0.092 in. If the relative displacement exceeded this value, injury is assumed to have occurred. The basis for this criterion is still the Wayne Tolerance Curve. Brinn and Staffeld [29] have proposed a modified version of the Slattenschek criterion and have called it the Effective Displacement Index (EDI), again using the Wayne Tolerance Curve as a basis for comparison.

It can be seen from this brief survey that there is no universal agreement on the tolerance limits of the human head to impact. The Wayne Tolerance Curve appears to be the only available guide but is subject to criticism for its lack of documentation. For aircraft crash environments, a more conservative estimate is required since a temporary loss of consciousness and of orientation may mean the inability to escape the holocaust that usually follows. Much work in this area remains to be done.

Vertebral Injury

It can be seen from Tables 2 and 3 that high $+G_z$ accelerations can be expected during a crash. The major structure supporting the head and torso of a seated occupant is the vertebral column which becomes susceptible to injury in a $+G_z$ environment. Tolerance limits of the spine have been the subject of investigation for the past thirty years, principally becase of ejection injuries sustained by pilots during emergency egress from disabled aircraft. The discussion on tolerance of the spine to impact can be better

TABLE 2

Summary of Design Pulses Corresponding to the 95th
Percentile Accident of Rotary- and Light Fixed-Wing Aircraft (13)

Impact Direction	Velocity Change (fps)	Peak G	Average G	Pulse Duration "T" Second
Longitudinal (Cockpit)	50	30	15	0.104
Longitudinal Passenger Compartment	50	24	12	0.130
Vertical	42	48	24	0.054
Lateral	25	16	8	0.097

TABLE 3

Summary of Design Pulses Corresponding to the 95th
Percentile Accident of Fixed-Wing Transport Aircraft (13)

Impact Direction	Velocity Change (fps)	Peak G	Average G	Pulse Duration "T" Second
Longitudinal (Cockpit)	64	26	13	0.153
Longitudinal (Cabin)	64	20	10	0.200
Vertical	35	36	18	0.060
Lateral (Cockpit)	30	20	10	0.093
Lateral (Cabin)	30	16	8	0.116

understood with the simultaneous consideration of the mechanism of injury a subject which is far less controversial than that for the head. It is somewhat unfortunate that spinal impact injury was associated with whole-body tolerance to $+G_z$ acceleration. Early work on vertebral fracture due to ejection was carried out by Watts et al. [30] and others [31, 32] on fully-restrained human volunteers and animals. The tolerance curve given by Eiband [11] was based on these data (Figure 5). Because the torso was highly restrained this curve is generally taken to represent whole-body tolerance to $+G_z$ acceleration. It was found that vertebral fracture would occur above a level of approximately 20g and that this was a limit for minor injury. The mechanism of spinal injury was assumed to be due to pure compression, since the torso was not allowed to flex forward. Studies on the strength of isolated vertebral segments were generally limited to uni-axial compression when the load was applied along the longitudinal axis of the spine [33, 34, 35]. A mathematical model proposed by Payne [36] reinforced the concept of fracture due to compression alone. His model was simply the base-excitation

EIBAND'S TOLERANCE CURVE

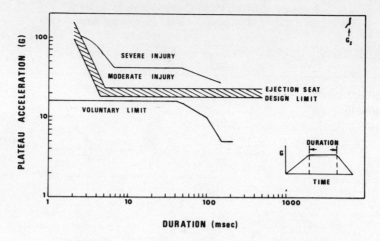

Fig. 5. Eiband's Tolerance Curve for +G_z Acceleration (11)

problem of a single degree-of-freedom spring-mass system. Subsequent models by Liu and Murray [37] and Toth [38] also considered axial compression of the spine. All of the references quoted thus far were published before 1967. The prevailing concept during the sixties was that restraint systems should be used but that not much could be done about

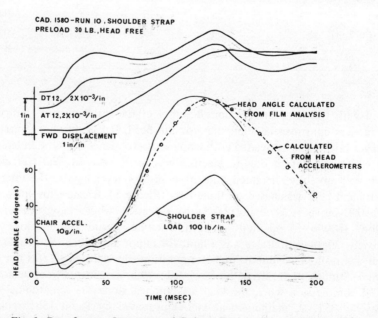

Fig. 6. Data from an Instrumented Cadaver During +G_z Acceleration (42)

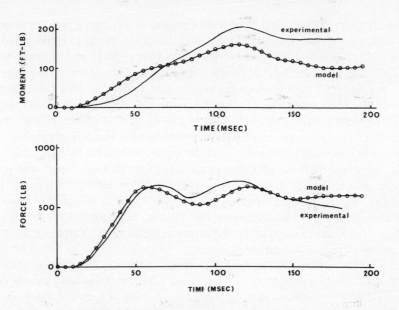

Fig. 7. Forces and Moments at TII Measured During a Simulated Ejection with a Human Cadaver as Test Subject (42)

reducing the compression on the vertebrae in the presence of a given level of $+G_z$ acceleration.

In 1968, King and Vulcan [30] reported that the vertebral column was subject to significant bending as well as compression during $+G_z$ acceleration. From anatomical considerations, it can be seen that the weight of the chest wall, the thoracic contents, and part of the arm and shoulder complex exert a bending moment on the spine via the rib cage. The eccentricity of this load is magnified by $+G_z$ acceleration as forward flexion of the head and torso takes place. It was postulated that fracture resulted from the combined action of bending and axial compression [40]. In tests on intact cadavers, it was found that compressive strain along the anterior aspects of the vertebral bodies were higher than that along the lateral aspects. Furthermore, the strain-time history had two peaks, the first one corresponding to the dynamic overshoot of an elastic system and the second due to an increase in bending strain as the torso reached its maximum forward position in flexion. A typical example of the data is shown in Figure 6 [41] and the forces and moments sustained by a vertebra are shown in Figure 7 [41].

In view of this recent information, work is now in progress to utilize the restraint system to raise the level of fracture. By moderately hyperextending the lumbar spine, the average fracture level for cadavers was raised from 10.4 ± 3.79 g to 17.75 ± 5.55 g, with a confidence level of better than 95% [42]. The age of the cadavers used averaged about 60 years and since the strength of vertebrae in the second decade of life is twice that in the sixth [43] the projected fracture level can be raised from roughly 20 g to 36 g.

Thus, tolerance limits for the spine are highly dependent on the restraint system used and Eiband's curve [11] is a simplified representation of a complex problem. The use of a severity index to denote tolerance is also unsatisfactory. Stech [44] proposed such an index in 1963. It is called the Dynamic Response Index (DRI), and is based on Payne's model [36] described earlier. A linear single-degree-of-freedom system can have a dynamic overshoot as high as twice the steady-state value. The DRI takes into account this overshoot but ignores the bending phenomenon which plays a dominant role in causing vertebral fracture. Physically, DRI is the ratio of the force in the spine to the body weight. For a simple base-excitation model, this force is proportional to the acceleration of the mass, which can exceed the input acceleration by a factor equal to the dynamic overshoot. Thus the DRI is an acceleration tolerance value in g's which accounts for the overshoot. The recommended peak DRI for 5% probability of injury is 18. If the overshoot is 50% the allowable input acceleration is 12g. For low rates of onset, there is minimal overshoot and the limit becomes 18g. That is, for a given g-level the DRI is solely a function of the rate of onset of the input acceleration pulse. Vulcan's results [42] are not in agreement with this conclusion.

In an aircraft crash it is not sufficient to consider the $+G_z$ component of acceleration alone. The $-G_x$ component aggravates the forward rotation of the torso and the combined effect may lower the spinal tolerance limit considerably. Further research in this area should be directed towards a combined experimental and analytical study of spinal response to simultaneous input of $+G_z$ and $-G_x$ acceleration.

Injury to the Extremities

To maintain mobility after a crash, it is imperative that there be no fracture of the bones of the lower extremities. For a seated occupant, the long bones, the knee cap and the ankle are most vulnerable. Fractures of the ankle and the lower third of the tibia-fibula complex most commonly occur in ground impact. The lower leg pivots about the knee joint and is capable of violent motion, resulting in a second collision with instrument panels, seat backs, control pedals or structural intrusions through the floor [15]. Unfortunately, published literature on impact tolerance of the lower leg is non-existence. Snyder [2] quotes results from four unpublished research studies on cadaver legs. Three of these provided tibial fracture loads that varied from 1000 to 3000 lb. Impact durations were not given. The fourth study by Mather was the most extensive. A total of 318 fresh cadaver legs were impacted midway between the angle and the knee by dropping a 9.2-lb. weight. the energy required for fracture was analyzed statistically, yielding the following results:

Fracture Energy (ft-lb)	% of Population Fractured
25.4	5
53.3	50
85.1	95

Forces and impact durations are not available.

Tolerance data for the femur have been obtained by several investigators. For longitudinal impact at the knee along the axis of the bone, Patrick et al. [45] reported data on 10 embalmed cadavers. There were 9 fractures out of 20 legs from 950 to 2400 lb. The other femurs, however, sustained 1500 to 3850 lb. without fracture. The pulse duration was of the order of 40 milliseconds. Human volunteers tolerated 800 to 1000 lb. with only minor pain in the knee. Mather [46] obtained energy levels necessary for the fracture of 44 pairs of fresh femurs. The bones were loaded statically and impacted by dropping a weight on the anterior surface, midway between the condyles and the head. The mean static energy absorbing capacity was 21.15 ft-lb. while the dynamic value averaged 31.33 ft-lb. The difference between the means was very significant. A t-test showed that $P < 0.05\%$ for the observed difference. Impact forces and durations were again not available. Recent work by Cooke and Nagel [47] on cadaver knees showed that no fractures were produced at peak forces below 1600-2000 lb. for the range of energies examined (225-600 ft-lb.). The duration varied from 5 to 21 milliseconds.

Patrick et al. [45] recommended that the tolerance limit for the knee-thigh-hip complex be set at 1400 lb. per leg for knee impact forces. This value has been adopted as an injury criterion in FMVSS No. 208.

SCALING OF ANIMAL INJURY DATA

The injury patterns and mechanisms observed in animals are usually applicable to humans, particularly if sub-human primates or large mammals are used. The major problem is the establishment of human tolerance limits based on animal tolerance data. To attain a similar pattern of injury, the input acceleration or force is usually higher for the animal, and the duration shorter. The work of Kazarian et al. [48] is a typical example. They subjected rhesus monkeys to $+G_z$ acceleration and observed vertebral and soft tissue injuries. The accelerations ranged from 25-900g and the total time duration was from 2-22 milliseconds. The equivalent acceleration magnitude and duration for the human is unknown since no satisfactory scaling laws are available. The same problem is encountered in head injury research as evidence by the work of Ommaya et al. [49] who applied a scaling postulate originally due to Holbourn [50] to obtain concussive levels for man, using data from sub-human primates. There is no firm experimental evidence that their scaling is correct.

A general systematic approach would be to use the principles of dimensional analysis and to obtain a set of dimensionless parameters which govern a given impact situation. The most significant of these can be used to obtain the various ratios such as time, acceleration, and force, assuming of course, that man and animal are geometrically similar.

The major objection to this approach is that we have neglected to consider the physiological differences between species. Such variables may not be quantifiable and thus cannot be accounted for in any dimensional analysis. Relatively little is known about biochemical stress, for example [2].

DISCUSSION & SUMMARY

A survey was made of existing information regarding human tolerance limits to impact. Particular emphasis was placed on the head, the spine and the lower extremities because of their importance in aircraft crash environments. The data are for uni-directional accelerations and are inadequate for multi-directional crash environménts. However, off-axis impact tolerance data are not available. Several attempts have been made to obtain voluntary limits, for space vehicle applications [51, 52], but very little is known about the injuries resulting from the cumulative effect of a combined acceleratio input. As a matter of fact, the injury limits for lateral acceleration are also not available [2].

The tolerance limits for the thoracic and abdominal regions were not discussed in this paper, but a considerable amount of information is available with regard to injury pattern and mechanisms. Snyder's reviews[2, 15] contain most of the pertinent information for these areas.

From the point of view of design for aircraft crashworthiness, recent work by Haley et al [53] and Schwartz [54] on load-limiting devices to cushion vehicle impact and by Reed and Avery [55] on the design of crashworthy structures for attenuating longitudinal impacts represents significant steps towards increased survivability of aircraft crashes. The fact that some aircraft seats are presently designed to withstand a forward deceleration ($-a_x$) of 9g while human volunteers have survived impacts of over 40g is contrary to the principle of passenger protection. Swearingen [56] stated recently that the fatality rate in general aviation accidents is at least seven times as high as that for automotive transportation. He attributes this to improper design and was able to document that engineering design changes can sharply reduce the death and injury rate in general aviation accidents.

Many of the safety standards for automotive occupants can be used termporarily as a guide for aircraft design until more accurate information becomes available. The need for tolerance limits for multi-directional impacts is quite apparent. Existing tolerance values also require adjustment or confirmation. In particular, the re-evaluation of head tolerance limits is a top priority item. Severity indices can only be regarded as a temporary measure and not as a permanent solution to the problem of tolerance. It is hoped that future research will produce more realistic limits for the prevention of fatalities and the reduction of injuries.

REFERENCES

1. A. H. Hasbrook, Severity of injury in light plane accidents: a study of injury rate, aircraft damage, accident severity, impact angle, and impact speed involving 1596 persons in 913 light plane accidents. AV-CIR-6-SS-105, Sept. 1959.
2. R. G. Snyder, State-of-the art—human impact tolerance. SAE Paper No. 700398, Revised Aug. 1970.
3. L. M. Patrick and G. Grime, Applications of human tolerance data to protective systems: Requirements for soft tissue, bone, and organ protective devices. *Impact Injury and Crash Protection,* (E. S. Gurdjian, W. A. Lange, L. M. Patrick, and L. M. Thomas, Eds.), Charles C. Thomas, Springfield, Ill., 1970, Chap. XX, pp. 444-473.

4. G. A. Holcomb and M. Huheey, A minimal Compression fracture of T-3 as a result of impact. *Impact Acceleration Stress,* NAS-NRC, Publ. 977:191, 1962.
5. J. H. Henzel, N. P. Clarke, G. C. Mohr and E. B. Weis, Jr., *Compression fractures of thoracic vertebrae apparently resulting from experimental impact, a case report.* Aerospace Medical Laboratories, Wright-Patterson AFB, Ohio, Rept. AMRL-TR-65-134, Aug. 1965.
6. H. DeHaven, Mechanical analysis of survival in falls from heights of fifty to one hundred and fifty feet. *War Med.* 2:586, 1942.
7. R. G. Snyder, Terminal velocity impacts into snow. *Military Med.* 131(10):1290-1298, Oct., 1966.
8. F. W. Kiel, Hazards of military parachuting. *Milit. Med.* 130:512-521, 1965.
9. J. P. Stapp, Human tolerance to severe, abrupt deceleration. *Gravitational Stress in Aerospace Medicine* (O. H. Gauer and G. D. Zuidema, eds.). Little, Brown, Boston, 1961, pp. 165-188.
10. J. P. Stapp, Voluntary human tolerance levels. *Impact Injury and Crash Protection* (E. S. Gurdjian, W. A. Lange, L. M. Patrick, and L. M. Thomas, eds.). Charles C. Thomas, Springfield, Ill., 1970, pp. 308-349.
11. A. M. Eiband, Human tolerance to rapidly applied acceleration: A survey of the literature. National Aeronautics and Space Admin., Washington, D.C., NASA Memo No. 5-19-59E, June, 1959.
12. C. F. Gell, Table of Equivalents for acceleration terminology: Recommended for general international use by the Acceleration Committee of the Aerospace Medical Panel, AGARD. *Aerospace Med.* 23(12):1109-1111, Dec., 1961.
13. J. W. Turnbow, D. F. Carroll, J. L. Haley, Jr. and S. H. Robertson, *Crash survival design guide.* USA AVLABS Tech. Rept. 67-22, U.S. Army Aviation Material Labs., Ft. Eustis, Va., July 1957. (Revised Aug., 1969).
14. Dept. of Transportation, Fed. Aviation Admin. *FAA statistical handbook of aviation.* 1968, p. 240.
15. R. G. Snyder, Occupant impact injury tolerances for aircraft crashworthiness design, SAE Paper No. 710406, 1971.
16. D. J. Marrow, Analysis of injuries of 1942 persons in 1442 light place accidents. CAA Medical Service Records, Washington, D.C. Unpublished data, 1949.
17. H. DeHaven, The site, frequency, and dangerousness of injury sustained by 800 survivors of light plane accidents. Dept. of Public Health and Preventive Medicine, Cornell Univ. Medical College, New York, July 1952.
18. H. R. Lissner, M. Lebow and F. G. Evans, Experimental studies on the relation between acceleration and intracranial pressure changes in man. *Surg., Gynec., and Obstet.* 111:329-338, 1960.
19. L. M. Patrick, H. R. Lissner and E. S. Gurdjian: Survival by design-Head protection. *Proc. of the 7th Stapp Car Crash Conf.* (D. M. Severy ed.) Charles C. Thomas, Publ. Springfield, Ill. 1965.
20. E. S. Gurdjian, H. R. Lissner, V. R. Hodgson and L. M. Patrick, Mechanism of head injury, Chap. 8, *Clinical Neurosurgery* 12, pp. 112-128, Congress of Neurological Surgeons, 1965.
21. E. S. Gurdjian, J. E. Webster and H. R. Lissner: Observations on the mechanism of brain concussion, contusion and laceration. *Surg., Gynec. and Obstet.* 101:688-890, 1955.
22. E. S. Gurdjian, H. R. Lissner, F. G. Evans, L. M. Patrick and W. G. Hardy, Intracranial pressure and acceleration accompanying head impacts in human cadavers, *Surg., Gynec. and Obstet.* 113:185-190, 1961.
23. Fed. Motor Vehicle Safety Standards, Motor Vehicle Safety Standard No. 201. Occupant Protection in Interior Impact—Passenger Cars. U.S. Dept. of Transportation, Fed. Hwy. Admin., Nat. Hwy. Safety Bureau, Effective Jan. 1, 1968.

24. C. W. Gadd, Criteria for injury potential, *Impact Acceleration Stress,* Publication 977, NAS-NRC, pp. 141-145, 1962.
25. C. W. Gadd, Use of a weighted-impulse criterion for estimating injury hazard, *Proc. of the 10th Stapp Car Crash Conf.,* SAE, New York, 1966, pp. 164-174.
26. V. R. Hodgson, L. M. Thomas and P. Prasad, Testing the validity and limitations of the severity index. *Proc. of the 14th Stapp Car Crash Conf.* Publ. by SAE pp. 169-187, 1970.
27. SAE Information Report, Human tolerance to impact conditions as related to motor vehicle design – SAE J885a, *SAE Handbook,* 1971.
28. A. Slattenschek, Behavior of motor vehicle windscreens in impact tests with a phantom head. Technische Hochschule, Vienna, Austria. *Automobiltechnische Zeitschrift,* 70(7):233-241, July, 1968.
29. J. S. Brinn and S. E. Staffeld, Evaluation of impact test accelerations: A damage index for the head and torso. Paper 700902 publ. in *Proc. of the 14th Stapp Car Crash Conf.,* P-33. New York: SAE, 1970.
30. D. T. Watts, E. S. Mendelson, H. N. Hunter, A. T. Kornfield and J. R. Popper, Tolerance to vertical acceleration required for seat ejection. *Aviation Med.* 18(6):554-564, 1947.
31. H. E. Savely, W. H. Ames and H. M. Sweeney, Laboratory tests of catapult ejection seat using human subjects. Memo Rep. No. TSEAL 695-66C, AAF, 1946.
32. J. P. Stapp, Tolerance to abrupt deceleration. AGARDograph No. 6, *Collected papers on Aviation Medicine,* Butterworths Sci. Publ. pp. 122-169, 1955.
33. O. Perey, Fracture of the vertebral end plate in the lumbar spine. *Acta Ortho. Scand.,* Suppl., 25, 1957.
34. F. H. Evans and H. R. Lissner, Biomechanical studies on the lumbar spine and pelvis, *J. Bone & Joint Surg.,* 41-A:278-290, 1959.
35. T. Brown, R. J. Hansen and A. J. Yorra, Some mechanical tests on the lumbosacral spine with particular reference to the intervertebral discs. *J. Bone & Joint Surg.,* 39A(5):1135-1164, Oct., 1957.
36. P. R. Payne, The dynamics of human restraint systems, *Impact Acceleration Stress,* Publ. 977, NAS-NRC, pp. 195-257, 1962.
37. Y. K. Liu and J. D. Murray, A theoretical study of the effect of impulse on the human torso, *Biomechanics,* pp. 167-186, Ed. by Y. C. Fung, ASME, N. Y., N. Y., 1966.
38. R. Toth, Multiple degree-of-freedom, non-linear spinal model. *Proc. of the 19th Annual Conf. on Eng. in Med. and Biol.,* p. 102, 1966.
39. A. I. King, A. P. Vulcan and L. K. Cheng, Effects of bending on the vertebral column of the seated human during caudocephalad acceleration. *Proc. of the 21st Annual Conf. on Engineering in Med. and Biol.,* Nov., 1968.
40. A. I. King and A. P. Vulcan, Elastic Deformation Characteristics of the Spine, *J. of Biomechanics,* 1971.
41. A. P. Vulcan and A. I. King, Forces and moments sustained by the lower vertebral column of a seated human during seat-to-head acceleration. *Dynamic Response of Biomechanical Systems* ASME, N. Y., N. Y., 1970, pp. 84-100.
42. C. L. Ewing, A. I. King and P. Prasad, Structural considerations of the human vertebral column under $+G_z$ impact acceleration, Paper No. 71-144 at the AIAA 9th Aerospace Sciences Meeting, N. Y., N. Y., Jan. 25-27, 1971.
43. J. McElhaney and V. Roberts, Mechanical properties of cancellous bone. Preprint, AIAA Paper No. 71-111. Presented at AIAA 9th Aerospace Sciences Meeting, Jan. 26, 1971.
44. E. L. Stech, *The variability of human response to acceleration in the spinal direction,* Report No. 122-109, Frost Engineering Dev. Corp., Contract AF 33(657)-9514, 1963.

45. L. M. Patrick, C. K. Kroell and H. J. Mertz, Jr., Force on the human body in simulated crashes. *Proc. of the 9th Stapp Car Crash Conf.,* Univ. of Minn. Press, Minneapolis, pp. 237-259, 1966.
46. B. S. Mather, Impact tolerance of the human leg, unpublished test data. Personal communication to R. G. Snyder, 1967.
47. F. W. Cooke and D. A. Nagel, Biomechanical analysis of knee impact. Paper 690800 published in *Proc. of the 13th Stapp Car Crash Conf.,* P-28, N. Y., SAE, 1969.
48. L. E. Kazarian, J. W. Hahn and H. E. von Gierke, Biomechanics of the vertebral column and internal organ response to seated spinal impact in the rhesus monkey (macaca mulatta) *Proc. of the 14th Stapp Car Crash Conf.,* SAE, N. Y., N. Y., pp. 121-143, 1970.
49. A. K. Ommaya, P. Yarnell, A. E. Hirsch and E. H. Harris, Scaling of experimental data on cerebral concussion in sub-human primates to concussion threshold for man. *Proc. of the 11th Stapp Car Crash Conf.,* Publ. by SAE, N. Y., N. Y., pp. 73-80, 1967.
50. A. H. S. Holbourn, Private Communication, Oct. 13, 1956 to Dr. Sabina Strich.
51. W. K. Brown, J. D. Rothstein and P. Foster, Human response to predicted Apollo landing impacts in selected body orientations. *Aerospace Med.* 37:394-398, 1966.
52. E. B. Weis, N. P. Clarke and J. W. Brinkley, Human response to several impact acceleration orientations and patterns. *Aerospace Med.* 34:1122-1129, 1963.
53. J. L. Haley, Jr., J. W. Turnbow and R. E. Klemme, Evaluation of 1000-4000 pound load-limiting devices for use in aircraft seats and cargo restraint systems AvSER Memo Rpt. M69-2, Contract No. DAAJ02-67-C-0004, 1969.
54. M. Schwartz, *Dynamic testing of energy attenuating devices,* Phase report, Rpt. No. NADC-AC-6905, 1969.
55. W. H. Reed and J. P. Avery, *Principles for improving structural crash wortiness for STOL and CTOL aircraft,* USAAVLABS Tech. Rept. 66-39, Contract No. DA44-177-AMC-254(T), 1966.
56. J. J. Swearingen, General aviation structures directly responsible for trauma in crash decelerations. Civil Aeromedical Inst., Office of Aviation Med., Fed. Aviation Admin., Oklahoma City, March, 1971.

RESPONSE OF A SEAT-PASSENGER SYSTEM TO IMPULSIVE LOADING

JACK A. COLLINS, PhD and JAMES W. TURNBOW, PhD

Departments of Mechanical Engineering and Engineering Mechanics
Arizona State University

Abstract—This paper presents a summary of a study of the dynamic response of an aircraft seat-passenger system to impulsive loading typical of aircraft crash situations.

A brief description of the computer model "SIMULA" is presented and selected data from 305 separate cases which have been studied are discussed. Maximum system forces, displacements, velocities, and accelerations are presented as functions of velocity change, aircraft deceleration, crash pulse shape, passenger weight, and seat belt slack. Data from both single and coupled parameter studies are included. A correlation of "SIMULA" results with experimentally obtained data is made.

INTRODUCTION

With the rapid growth of air transportation subsequent to World War II, increasing attention has been focused on the causes of aircraft passenger injuries in severe, but potentially survivable crashes. In many accidents the probable cause of injuries has been traced to failure of the occupant's seat restraint system. These failures have permitted seated occupants to be thrown against objects inside or, in some cases, outside the aircraft. For some accidents in which the occupant's restraint system did not fail, injuries have been caused by the flailing of body extremities into adjacent seats.

Unfortunately, it appears that some aircraft accidents will continue to occur regardless of efforts expended to prevent them. With this premise in mind, the Aviation Crash Injury Steering Committee* of the Cornell-Guggenheim Aviation Safety Center met in 1964 to specify certain tasks which should be conducted in the crash injury field. The Committee divided the tasks into the following four broad areas.

1. An investigation of crash-induced floor accelerations for transport aircraft.

2. An investigation of the dynamic response of seat-passenger systems to crash-induced floor accelerations through an experimentally verified computer aided analysis of the problem.

3. An investigation of the influence of strength and load-deformation characteristics of the seat and restraint system.

---

*This Committee is composed of representatives from the following agencies: United States Air Force, United States Navy, United States Army, Civil Aeronautics Board, Federal Aviation Agency, Air Force Institute of Pathology, National Aeronautics and Space Administration, and Aircraft Industries Association/Air Transport Association.

4. An investigation of new concepts in seat-system design.

This paper presents a summary of results from studies of the second broad area listed above, including the response characteristics of both seat and passenger to crash-induced acceleration pulses at the aircraft floor.

COMPUTER MODEL

The results presented in this paper culminate a three year effort to develop an analysis and associated computer program to predict the forces and deflections of a forward facing aircraft seat due to a crash impulse, as well as the displacements, velocities, and accelerations of the passenger's body. This analysis and associated computer program will henceforth be referred to by the program name "SIMULA". The computer program SIMULA is designed to yield a digital and a graphical display of important forces, displacements, positions, velocities, accelerations, and jerks (time derivatives of acceleration) for the seat and the passenger, all as a function of time, for any desired combination of the physical parameters of the system and the configuration of the acceleration-time impulse applied to the system.

A simplified schematic of the seat-passenger system, as used in the analysis, is shown in Figure 1. The system is assumed to be two-dimensional with a plane of symmetry in which all masses are located. The passenger's body is lumped into eight discrete masses, M1 through M8, connected with seven inextensible members representing the neck, arms, legs, and torso. The resistances within the body due to muscular action are simulated by a system of nonlinear springs with damping, placed at the joints. The seat is assumed to be rigid except in the legs and cushion and to have assignable mass and moment of inertia properties. Also, the seat legs and cushion can be assigned a wide variety of different force-deflection characteristics. The nonlinear springs (with damping) representing the seat belt, shoulder harness, and seat cushion are shown in Figure 1. The deformability of the seat legs is not illustrated, but is provided for in the program.

In Figure 1 the seat cushion is illustrated as being attached to the pelvic mass. However, to account for the fact that the seat cushion reaction is actually distributed over the passenger's thigh with a changing center of pressure, a variable moment was introduced on the thigh at the pelvis. Also, it should be noted that the deformability of the passenger must be considered in arriving at reasonable load-deflection characteristics for the shoulder harness (if employed), the seat belt, and the seat cushion. Several experiments were conducted during the study to aid in arriving at reasonable body deformability characteristics for use in comparing computer simulations with actual experimental seat-passenger system behavior.

In program SIMULA the motion of the seat and passenger are given in terms of the non-Newtonian X-Y coordinate system attached to the aircraft floor as shown in Figure 1. Throughout this paper the terms "Horizontal" and "Vertical" refer respectively to directions *parallel to* and *normal to* the aircraft floor. Twelve differential equations were obtained, corresponding to twelve independent degrees of freedom for the system, in twelve generalized coordinates. These are the "X" and "Y" displacements of the seat

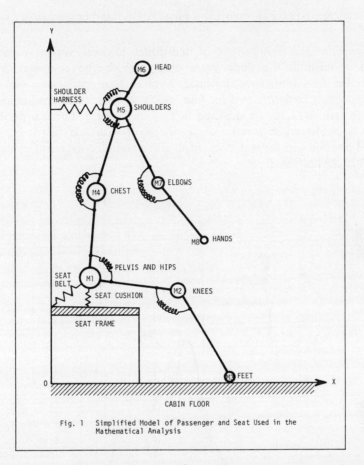

Y

M6 HEAD

SHOULDER
HARNESS

M5 SHOULDERS

M7 ELBOWS

M4 CHEST

M8 HANDS

PELVIS AND HIPS

SEAT
BELT   M1

M2 KNEES

SEAT CUSHION

SEAT FRAME

M3 FEET

0   X

CABIN FLOOR

Fig. 1   Simplified Model of Passenger and Seat Used in the
Mathematical Analysis

belt attachment point at the seat, the seat rotation angle, the seat belt angle, the seat belt deflection and seven angles fixing the angular positions of the seven body segments. These twelve equations were solved for the coordinates and their derivatives using a modified Runge-Kutta method. Appropriate subroutines were used to calculate the X and Y displacements, velocities and accelerations of the body masses M1 through M8 and the forces of interest.

Finally it should be noted that a horizontal "load-limiter" is incorporated in program SIMULA between the seat and the floor. Physically this feature is the equivalent of a mechanical device which will accept any imposed force up to a preselected level without deflection, and when the preselected limiting load is reached will deform without bound, never accepting a load greater than the limiting load. For example a spring-loaded slip-clutch is one form of load-limiter often used as a safety device in rotating industrial machinery. In the presentation of results it will be noted that the use of a load-limiting device between the seat and floor of the aircraft has important implications in terms of the maximum forces and accelerations imposed on the passenger and the seat under crash conditions.

EXPERIMENTAL VERIFICATION OF COMPUTER PROGRAM

Many experimental verifications of individual physical system parameters were performed throughout the three years during which the computer program was developed. The final verification, however, involved four acceleration sled experiments with corresponding computer runs for four widely varying sets of conditions. The results of one such test, MTT-19, are depicted in Figures 2 and 3. For this particular test, a dummy was positioned in a test seat on the sled, with seat belt only, in a slightly jack-knifed position and the seat oriented to give a vertical component of acceleration of 27 percent of the horizontal pulse.

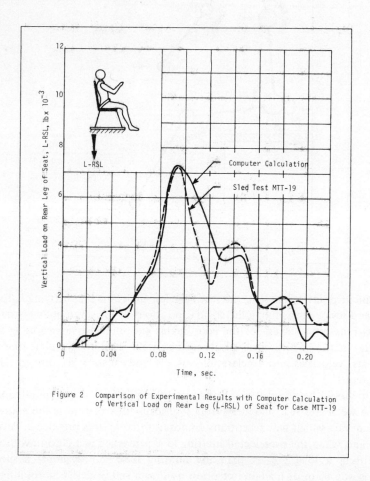

Figure 2   Comparison of Experimental Results with Computer Calculation of Vertical Load on Rear Leg (L-RSL) of Seat for Case MTT-19

Figures 2 and 3 indicate the comparison between computer calculations and experimental measurements for the vertical load in the rear leg, and the horizontal shear load at the floor. Other tests yielded results similar to these with discrepancies in the

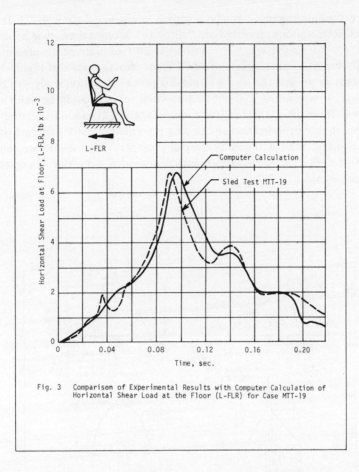

Fig. 3   Comparison of Experimental Results with Computer Calculation of
Horizontal Shear Load at the Floor (L-FLR) for Case MTT-19

worst case ranging from 20-30 percent, but for most cases less than 20 percent. Comparisons of other system responses, such as seat belt extension and passenger kinematics also showed very acceptable agreement between computer predictions and experimental measurements.

COMPUTER ANALYSIS

Following the experimental verification phase, a sequence of 305 carefully selected computer runs were planned and executed to study the effects in response of the seat-passenger system to a variety of crash pulses and physical characteristics of the system. It is possible to present only a small sample of the interesting information gathered from these computer runs due to constraints of space, clarity, and time. Only the *maximum values* of selected forces, displacements, velocities, and accelerations during the time of the whole impulse have been shown in Figures 5 through 22. It should be noted that the accelerations presented are "absolute" accelerations, that is, accelerations with respect to the earth; while the velocities and displacements presented are all measured with respect to the airframe or the X-Y coordinate system of Figure 1.

In an attempt to present the results of the computer simulation study in a concise but meaningful way, it was noted that the responses of interest could be naturally divided into two types. These are (1) seat-related responses, and (2) passenger-related responses. For this reason, it may be noted that all the data presented in the Figures are presented in mated pairs of plots. The first figure of the pair presents certain seat-related data and the second figure of the pair presents certain passenger-related data. For example, Figures 5 and 6 form such a pair.

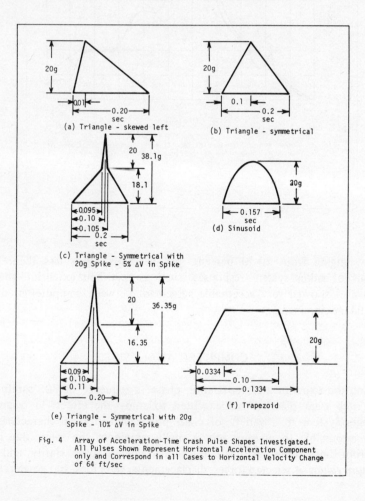

Fig. 4   Array of Acceleration-Time Crash Pulse Shapes Investigated.
All Pulses Shown Represent Horizontal Acceleration Component
only and Correspond in all Cases to Horizontal Velocity Change
of 64 ft/sec

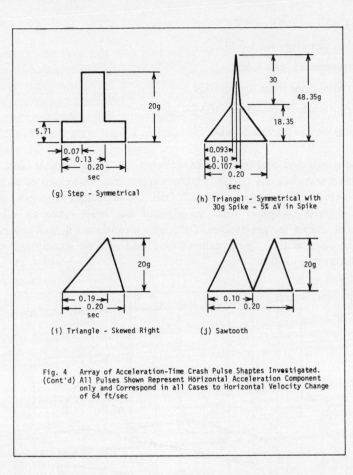

(g) Step - Symmetrical

(h) Triangel - Symmetrical with
    30g Spike - 5% ΔV in Spike

(i) Triangle - Skewed Right   (j) Sawtooth

Fig. 4   Array of Acceleration-Time Crash Pulse Shaptes Investigated.
(Cont'd)   All Pulses Shown Represent Horizontal Acceleration Component
           only and Correspond in all Cases to Horizontal Velocity Change
           of 64 ft/sec

Eight seat-related responses were selected as being of primary importance to a seat designer. These responses appear on the first of each mated pair of plots presented in Figures 5 through 22. Only the maximum value of each response is plotted for a given computer run, no attempt being made to show the time correspondence among the peak values. Since the peak values do not all occur at the same instant of time, the reader is cautioned that force equilibrium conditions are not met by assuming that peak values of all forces act simultaneously. One should therefore not expect force equilibrium among the peak values presented.

The eight seat-related responses selected for this presentation are as follows. Note that each response is accompanied by a symbol and a plotting code which will help in the interpretation of the data presented.

(1) Total seat belt load L-SBLT   _____

(2) Shoulder harness load L-SHAR . . . .

(3) Seat Cushion load L-SCSN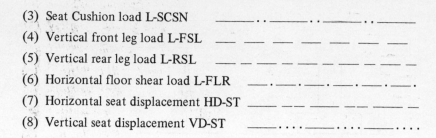

(4) Vertical front leg load L-FSL

(5) Vertical rear leg load L-RSL

(6) Horizontal floor shear load L-FLR

(7) Horizontal seat displacement HD-ST

(8) Vertical seat displacement VD-ST

Six passenger-related responses were selected as being typical and of primary importance to a seat designer. These responses appear on the second of each mated pair of plots presented in Figures 5 through 22. Only the maximum value of each response is plotted for a given computer run. No attempt has been made to show the time correspondence among the peak values. The reader is cautioned that in general these peak values do not occur at the same instant of time, and such an assumption might lead to wholly erroneous conclusions.

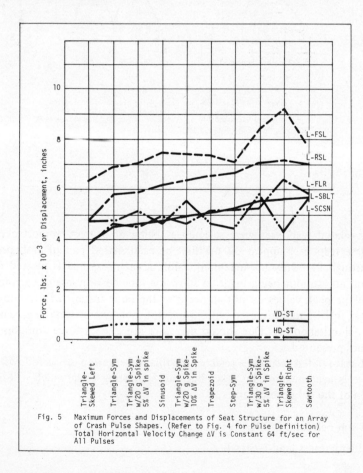

Fig. 5   Maximum Forces and Displacements of Seat Structure for an Array of Crash Pulse Shapes. (Refer to Fig. 4 for Pulse Definition) Total Horizontal Velocity Change ΔV is Constant 64 ft/sec for All Pulses

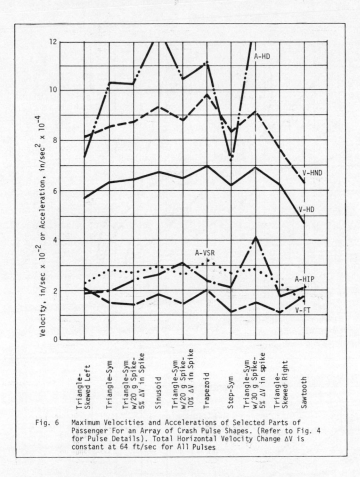

Fig. 6   Maximum Velocities and Accelerations of Selected Parts of Passenger For an Array of Crash Pulse Shapes. (Refer to Fig. 4 for Pulse Details). Total Horizontal Velocity Change ΔV is constant at 64 ft/sec for All Pulses

The six passenger-related responses selected for this presentation are as follows. Note that each response is accompanied by a symbol and a plotting code which will help in the interpretation of the data presented in figures 5 through 22.

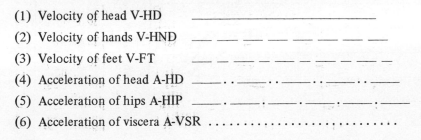

(1) Velocity of head V-HD

(2) Velocity of hands V-HND

(3) Velocity of feet V-FT

(4) Acceleration of head A-HD

(5) Acceleration of hips A-HIP

(6) Acceleration of viscera A-VSR

For the purpose of the following discussion, it will be expedient to define two broad categories into which the parametric investigations may be divided. The first category will

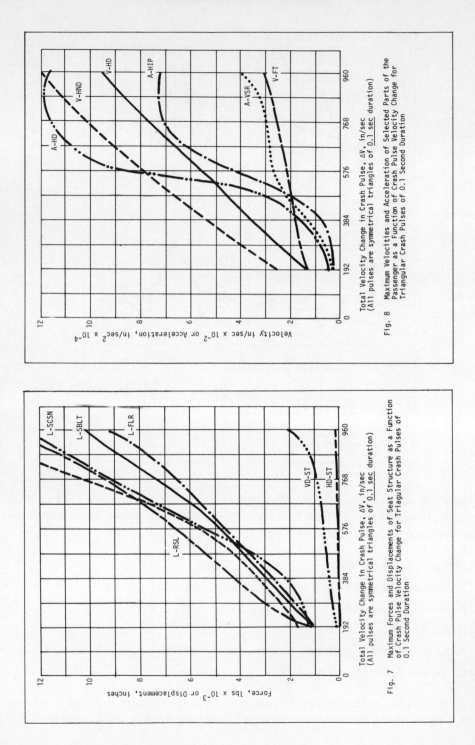

Velocity in/sec x 10^{-2} or Acceleration, in/sec^2 x 10^{-4}

Total Velocity Change in Crash Pulse, ΔV, in/sec
(All pulses are symmetrical triangles of 0.1 sec duration)

Fig. 8   Maximum Velocities and Acceleration of Selected Parts of the
Passenger as a Function of Crash Pulse Velocity Change for
Triangular Crash Pulses of 0.1 Second Duration

Force, lbs x 10^{-3} or Displacement, inches

Total Velocity Change in Crash Pulse, ΔV, in/sec
(All pulses are symmetrical triangles of 0.1 sec duration)

Fig. 7   Maximum Forces and Displacements of Seat Structure as a Function
of Crash Pulse Velocity Change for Triangular Crash Pulses of
0.1 Second Duration

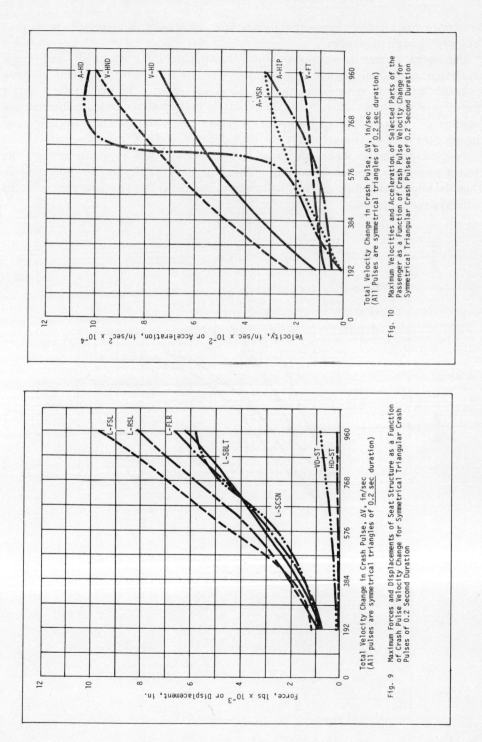

Total Velocity Change in Crash Pulse, ΔV, in/sec
(All Pulses are symmetrical triangles of 0.2 sec duration)

Fig. 10   Maximum Velocities and Acceleration of Selected Parts of the
Passenger as a Function of Crash Pulse Velocity Change for
Symmetrical Triangular Crash Pulses of 0.2 Second Duration

Total Velocity Change in Crash Pulse, ΔV, in/sec
(All pulses are symmetrical triangles of 0.2 sec duration)

Fig. 9   Maximum Forces and Displacements of Seat Structure as a Function
of Crash Pulse Velocity Change for Symmetrical Triangular Crash
Pulses of 0.2 Second Duration

be termed "uncoupled" parametric studies and the second category will be termed
"coupled" parametric studies.

Uncoupled parametric studies are those investigations for which all crash pulse
characteristics and all physical parameters of the system are held constant at some
specified "standard" value except for one parameter which is allowed to range over a
spectrum of reasonable values. Thus, the effects on the system response are observed as a
function of one variable parameter with all others arbitrarily fixed and any effects of
changing the fixed parameters are thereby excluded or uncoupled from the results.

Coupled parametric studies are those investigations in which two or more of the
system parameters are sumultaneously changed in many or all possible combinations.
Thus the interrelationships and cross influences of one parameter on another are reflected
in the system responses.

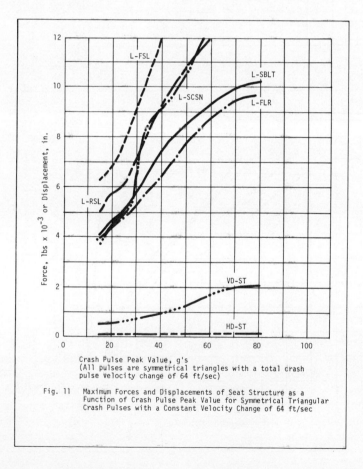

Fig. 11   Maximum Forces and Displacements of Seat Structure as a
          Function of Crash Pulse Peak Value for Symmetrical Triangular
          Crash Pulses with a Constant Velocity Change of 64 ft/sec

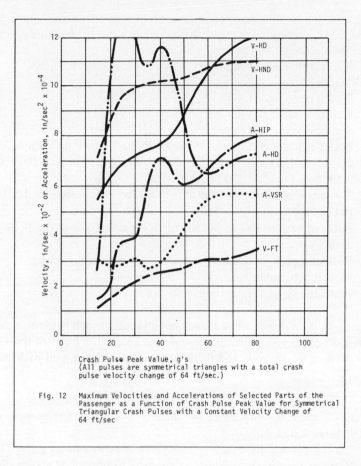

Crash Pulse Peak Value, g's
(All pulses are symmetrical triangles with a total crash
pulse velocity change of 64 ft/sec.)

Fig. 12   Maximum Velocities and Accelerations of Selected Parts of the
Passenger as a Function of Crash Pulse Peak Value for Symmetrical
Triangular Crash Pulses with a Constant Velocity Change of
64 ft/sec

UNCOUPLED PARAMETRIC STUDIES

Much can be learned regarding trends and sensitivity of a system to a given parameter
by performing uncoupled parametric studies. Figures 5 through 22 present a portion of
the results of 17 different uncoupled parametric studies encompassing a total of 177
different sets of conditions and requiring 177 computer runs with the SIMULA program.
To accomplish this study a "standard run" was established on the basis of experience and
judgment, for which reasonable and typical numerical values were assigned to each input
parameter for the system.

While it is beyond the scope of this paper to list all of the approximately 200 constants required as input to the SIMULA program, some of the more significant values used in the "standard run" are:

(1) Shoulder harness — none used
(2) Seat belt slack — none used
(3) Weight of occupant — 170 lb.
(4) Weight of seat — 25 lb.
(5) Seat belt — curve D Figure 23
(6) Buckling load front seat leg — 15,000 lb. at 1.0 inch deflection.
(7) Elastic limit rear seat leg — 10,000 lb. at 1.0 inch deflection. (slope of secondary branch = 3000 lb/in.)
(8) Symmetrical triangular pulse in X-direction — 0.20 seconds duration and 20 g peak as shown in Figure 4 (b).
(9) Ratio of vertical (Y) to horizontal (X) acceleration at the floor — 0.50 (30° angle).
(10) Resultant peak acceleration at floor — approximately 22.4 g.
(11) Velocity change in X-direction — 64 ft/sec.
(12) Front-to-rear seat leg spacing — 16 inches

The technique used was to allow a range of variation of each parameter, one at a time, while holding all other parameters at the standard run value. Eight of the 17 parameters which were studied in this way, and figure numbers associated with each parameter, are listed below.

(1) Shape of acceleration-time crash pulse, Figures 4, 5, 6.
(2) Crash pulse severity based on velocity change associated with crash pulse, Figures 7, 8, 9, 10.
(3) Crash pulse severity based on peak value of crash pulse, Figures 11, 12.
(4) Ratio of vertical acceleration component to horizontal acceleration component during crash pulse, Figures 13, 14.
(5) Passenger weight, Figures 15, 16.
(6) Horizontal load limiter setting, Figures 17, 18.
(7) Force-deflection characteristics of seat belt and passenger abdomen in series, Figures 19, 20.
(8) Seat belt slack, Figures 21, 22.

Figure 24 shows a typical response of a passenger without shoulder harness to a trapezoidal pulse of 0.20 seconds duration with a ratio of Y to X acceleration at the floor of $a_V/a_H = 0.20$.

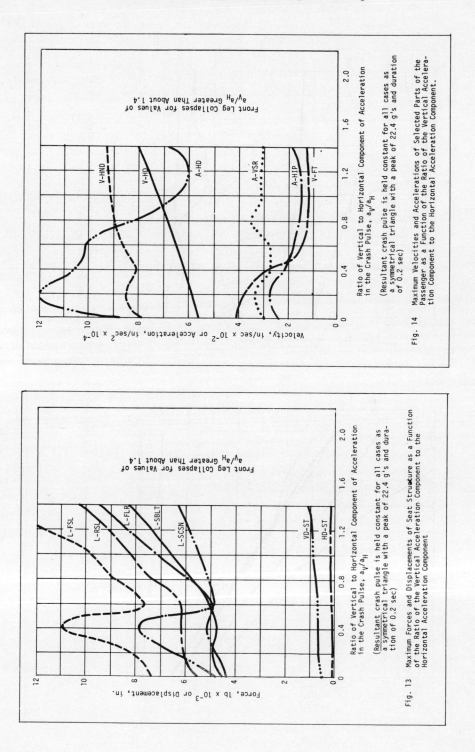

Fig. 14 Maximum Velocities and Accelerations of Selected Parts of the Passenger as a Function of the Ratio of the Vertical Acceleration Component to the Horizontal Acceleration Component.

Fig. 13 Maximum Forces and Displacements of Seat Structure as a Function of the Ratio of the Vertical Acceleration Component to the Horizontal Acceleration Component

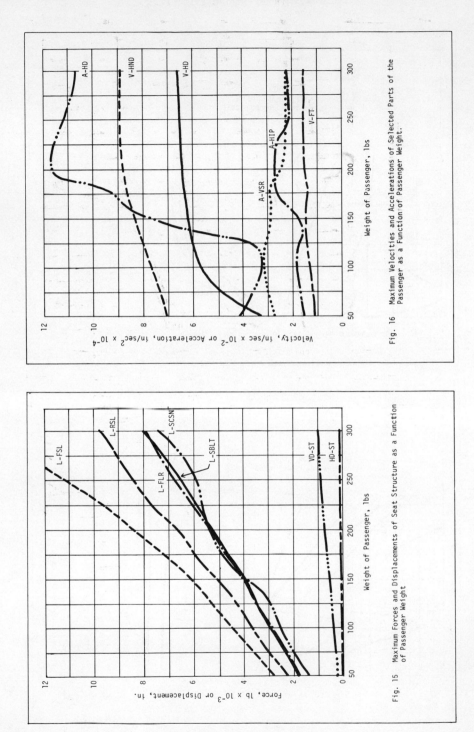

Fig. 16   Maximum Velocities and Accelerations of Selected Parts of the Passenger as a Function of Passenger Weight.

Fig. 15   Maximum Forces and Displacements of Seat Structure as a Function of Passenger Weight

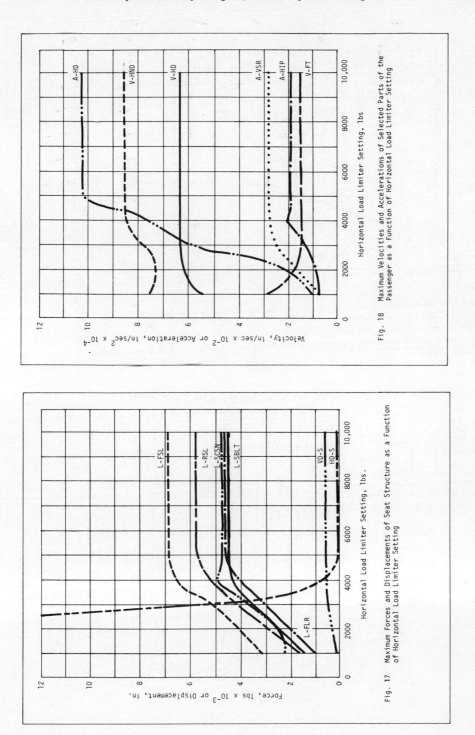

Fig. 18 Maximum Velocities and Accelerations of Selected Parts of the Passenger as a Function of Horizontal Load Limiter Setting

Fig. 17 Maximum Forces and Displacements of Seat Structure as a Function of Horizontal Load Limiter Setting

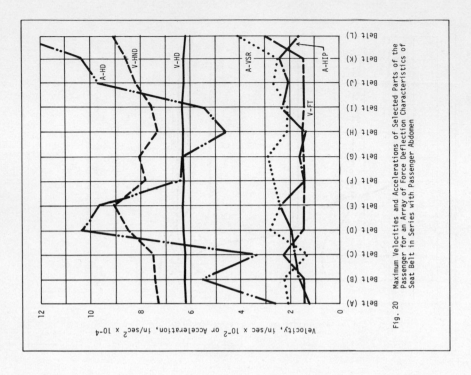

Fig. 20   Maximum Velocities and Accelerations of Selected Parts of the
Passenger for an Array of Force Deflection Characteristics of
Seat Belt in Series with Passenger Abdomen

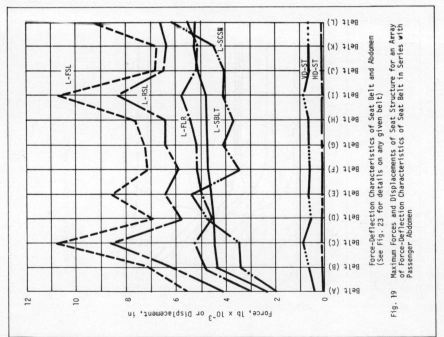

Fig. 19   Maximum Forces and Displacements of Seat Structure for an Array
of Force-Deflection Characteristics of Seat Belt in Series with
Passenger Abdomen

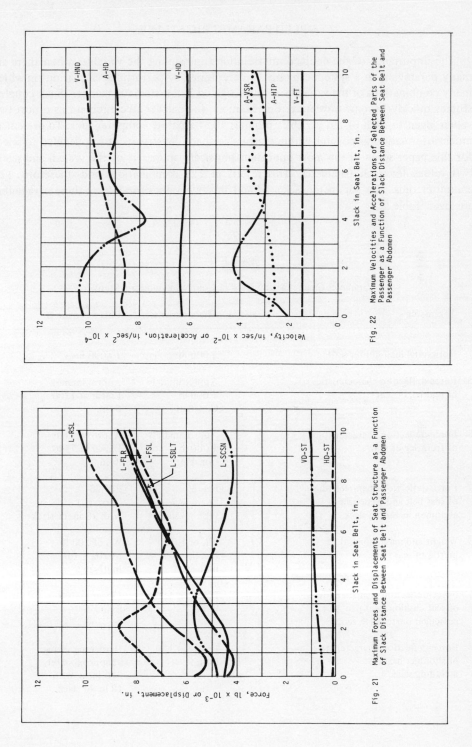

Fig. 22 Maximum Velocities and Accelerations of Selected Parts of the Passenger as a Function of Slack Distance Between Seat Belt and Passenger Abdomen

Fig. 21 Maximum Forces and Displacements of Seat Structure as a Function of Slack Distance Between Seat Belt and Passenger Abdomen

COUPLED PARAMETRIC STUDIES

The importance of the coupled interrelationships among the variables when there are many parameters in a system can only be determined by performing calcuations in which many combinations of the variables are investigated in a systematic manner. Such coupled studies quickly become unwieldly. For example, to consider all combinations of just two values given to each of ten variables requires 1024 separate computer runs. To present as much information as possible within reasonable space constraints, it was decided to select for this paper seven of the most important physical parameters of the system and select two values for each parameter. This results in 128 combinations and, therefore, 128 computer runs. The parameters selected and the two values for each of these parameters are given Table I.

TABLE I.

Parameter Characteristics Used in Coupled Parametric Study

Parameter	Characteristic A	Characteristic B
1. Horizontal load limiter setting	3,000 lb	10,000 lb
2. Force-deflection characteristics of rear leg of seat	Approximately 4,000 lb Constant	Approximately Linear at 1200 lb/in.
3. Force-deflection characteristics of front leg of seat	Approximately 4,000 lb Constant	Approximately Linear at 7500 lb/in.
4. Force-deflection characteristics of seat belt and passenger abdomen including slack	Fig. 23 Curve (A) With no slack	Fig. 23 Curve (L) With 10 in. of slack
5. Weight and mass moment of inertia of seat	Wt. of 1.0 lb Inertia = 0.8 in-lb-sec^2	Wt. of 200 lb Inertia = 160 in-lb-sec^2
6. Force-deflection characteristics of seat cushion, seat pan, and passenger buttocks in series	Linear at 1500 lb/in.	Linear at 500 lb/in.
7. Force-deflection characteristics of shoulder harness including slack	Linear at 120 lb/in.	Hardening spring at approximately 300 lb/in. with 10 in. of slack

The parameter values under the column heading "Characteristic A" of Table I were those values which yielded generally minimal values of all responses (forces, velocities and accelerations) for the 177 uncoupled cases studied. The parameter values under the column heading "Characteristic B" of Table I were those values which yielded generally maximal values of all responses for the 177 uncoupled cases, except that any combinations of parameters which resulted in a computer calculated collapse of the seat were excluded from consideration when selecting the "Characteristic B" parameter values.

All combinations of "Characteristic A" and "Characteristic B" for the seven parameters of Table I were computer calculated. This resulted in a 128 case net of responses which are summarized in Figure 25.

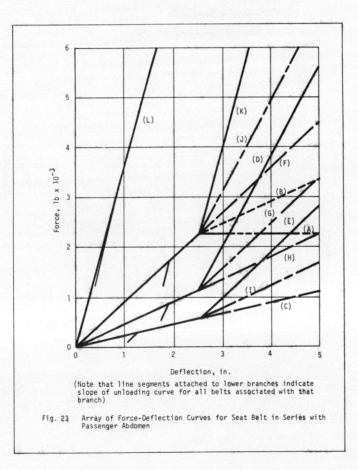

(Note that line segments attached to lower branches indicate slope of unloading curve for all belts associated with that branch)

Fig. 23   Array of Force-Deflection Curves for Seat Belt in Series with Passenger Abdomen

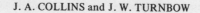

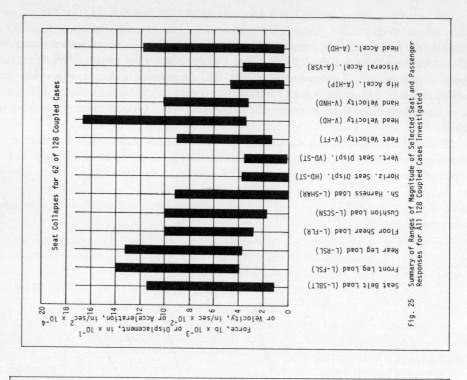

Fig. 25 Summary of Ranges of Magnitude of Selected Seat and Passenger Responses for All 128 Coupled Cases Investigated

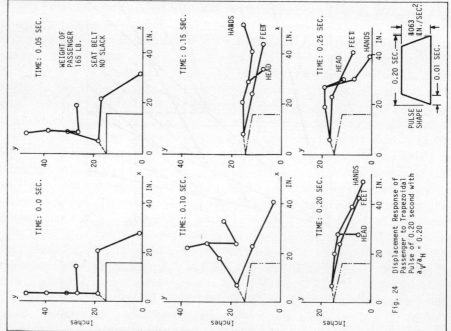

Fig. 24 Displacement Response of Passenger to Trapezoidal Pulse of 0.20 second with $a_V/a_H = 0.20$

SUMMARY OF RESULTS

While no attempt will be made to make sweeping generalizations based on the results of the calculations summarized in Figures 1 through 25, the following observations were made from the results obtained in the study. The observations deal only with peak values and do not in any way reflect the time-response of the system.

(1) Maximum responses of seat and passenger are relatively insensitive to shape of the acceleration-time crash pulse if the associated velocity change is constant.

(2) Short duration spikes during the acceleration-time pulse have little effect on maximum response of the seat or passenger.

(3) There seem to be critical values of the ratio of the vertical component of acceleration to the horizontal component of acceleration during the crash pulse which produce much higher passenger and seat maximum responses than adjacent higher and lower values of a_V/a_H.

(4) Maximum responses of the seat are very sensitive to increasing passenger weight, but maximum responses of the passenger are only moderately sensitive to passenger weight up to about 200 lb., and very insensitive above 200 lb.

(5) Load limiting is very effective in reducing the maximum responses of both seat and passenger. Some stroking of the load limiter must of course be accepted.

(6) Maximum response of the seat is only moderately sensitive to seat weight and mass moment of inertia, individually or in combination, and maximum response of the passenger is insensitive to these variables.

(7) Integrally load-limited seat belts significantly reduce the maximum responses of seat and passenger. Some associated belt extension must be accepted with such belts.

(8) If load limited seat belts are excluded, the response sensitivity of seat and passenger to a wide variety of force-deflection seat belt characteristics is moderate.

(9) Seat response is moderately sensitive to seat belt slack, but passenger response is relatively insensitive to belt slack.

(10) Seat and passenger maximum responses are relatively insensitive to shoulder harness force-deflection characteristics, but the *existence* of a shoulder harness significantly decreases the passenger responses while significantly increasing certain seat responses such as the forces in front and rear legs. This observation could probably be extrapolated to aft facing seats as well.

(11) Seat and passenger maximum responses are relatively insensitive to force-deflection characteristics of the seat cushion, seat pan, and passenger buttocks.

(12) Forces and accelerations of seat and passenger are significantly reduced by increasing the coefficient of friction between seat and passenger up to a value of about 0.4, but very insensitive to further increase in the friction coefficient.

(13) In computing the 128 coupled cases presented in Figure 25, no individual set of input parameter values cited in Table I resulted in seat collapse. Yet when running the 128 *coupled* cases, 62 cases, or nearly half the cases, resulted in collapse of the seat. The seat designer must therefore be clearly advised that coupling effects in this system are very significant and using the uncoupled data of Figures 1 through 25 without due regard for adverse coupling effects may result in unexpected seat failures.

CLOSURE

The results presented are miniscule when compared with the information recorded in the digital and graphical records of the 305 computer runs made prior to the preparation of this paper. The uncoupled data presented in Figures 5 through 22 do, however, give a good indication of the sensitivity of maximum values of seat and passenger response to a wide variety of physical system parameters and crash pulse characteristics. The coupled-case study depicted in Figure 25 illustrates the dangers associated with neglecting coupling effects.

In performing any investigation of a particular seat, particular crash configuration or particular set of physical system parameters, it would be far superior to utilize the computer program SIMULA for the specific system being investigated rather than try to extrapolate, interpolate, or otherwise combine the limited results presented in this paper.

ACKNOWLEDGMENTS

The results presented in this paper are taken primarily from work conducted by the School of Engineering at Arizona State University for Dynamic Science, a division of Marshall Industries, under National Aeronautics and Space Administration Contract NSR 33-026-003. The work was based on earlier studies conducted jointly by the Aviation Safety Engineering and Research Division of the Flight Safety Foundation and the School of Engineering at Arizona State University under U.S. Army Aviation Materiel Laboratories Contracts DA 44-177-AMC-254(T) and DA 44-177-AMC-360(T). Funds for these studies were provided in equal amounts by the National Aeronautics and Space Administration, the United States Army, the United States Navy, and the United States Air Force. The authors gratefully acknowledge the many contributions made in the effort by Nils O. Mykelstad, I. Irving Pinkel, John H. Enders, and Joseph L. Haley, Jr.

REFERENCES

1. U.S. Army Aviation Materiel Laboratories, *Aircraft passenger-seat-system response to impulse load,* USAAVLABS TR 67-17, Ft. Eustis, Va., August 1967, 291 pp.
2. U.S. Army Aviation Materiel Laboratories, *Floor accelerations and passenger injuries in aircraft accidents,* USAAVLABS TR 67-16, Ft. Eustis, Va., May 1967, 46 pp.
3. U.S. Army Aviation Materiel Laboratories, *Crash survival design guide,* USAAVLABS TR 67-22, Ft. Eustis, Va., July 1967, 253 pp.
4. J.A. Collins, W.B. Bickford, J.P. Avery, J.W. Turnbow, M.T. Wilkinson and S.E. Larsen, *Crash worthiness study for passenger seat design,* Engineering Research Center, Arizona State Univ., June 1968, 171 pp. (unpublished).

THE ANALYSIS OF UNDERDETERMINED AND OVERDETERMINED SYSTEMS

L. KNOPOFF and D. D. JACKSON

Institute of Geophysics, University of California, Los Angeles

Abstract—The "structure" of the earth is defined to be a set of parameters describing the density, rigidity, and compressibility of the earth as a function of depth. The observed velocities of seismic surface waves, together with the mass and moment of inertia of the earth, form a set of data which are related in a quasi-linear way to the unknown "structures". The first part of this paper presents a tutorial exposé of the methods which have been heretofore available for the solution of problems with degrees of freedom in the data which are different from the degrees of freedom in the model parameters.

The relationship between the data and the structures is fundamentally underdetermined but, with even slight errors in the data, may be overconstrained as well. We present a method of inverting the data, which is stable for such a system, and which indeed takes advantage of this underdeterminacy to test hypotheses directly related to physical interpretation of the structure. The inversion scheme consists of "building" an inverse matrix from the best determined eigenvectors of the Jacobian matrix. A functional definition of "best-determined" is provided which not only produces a stable inverse, but also helps to elucidate the real freedom one has in selecting structures consistent with the data. This inversion scheme is applicable to any linearizable system of equations relating θ data to N unknowns, independent of θ and N.

INTRODUCTION

Suppose that we have observations of certain properties of the earth, such as travel-times of body waves, phase velocities of surface waves or periods of free oscillations. We construct a model to try to fit these observations. Let the observations be f_α and let the model parameters be ℓ_n. In general, if we try to generate the theoretical values of the quantities f as a consequence of the postulated model ℓ_n, we find that they do not match the observations. Let the differences between the observed and the computed quantities be called "data" and be given the symbols g_α; let the changes in the model required to obtain an exact fit to the data be $\Delta\ell_n$. Then, by a simple Taylor expansion, we see

$$g_\alpha = \frac{\partial f_\alpha}{\partial \ell_n} \Delta\ell_n + \frac{1}{2} \frac{\partial^2 f_\alpha}{\partial \ell_n \partial \ell_m} \Delta\ell_n \Delta\ell_m + \ldots$$

Assuming that second and higher order terms may be neglected,

$$g_\alpha = G_{\alpha n} m_n \tag{1}$$

where we have let $m_n = \Delta\ell_n$, $G_{\alpha n} = \partial f_\alpha / \partial \ell_n$.

We can now state our problem in the following form. Let g_α represent "data" such as travel-times of body waves, periods of free oscillation, phase velocities of surface waves, mass and moment of inertia, etc. These are the result of observations on the surface of the earth and are suitably represented in discrete form. We may assume without loss of generality that the data have been written so that they have the same dimensionality and are statistically independent. Let m_n be the parameters of the "real earth" which we wish to find, again suitably represented in discrete form, and let μ_n be some estimate of m_n, which we shall call the "model". There exists some theoretical operator G which generates the data from the model. Let this operator be linear in the sense that it does not change significantly as we move from model to model in our search for the properties of a solution. We solve the equation

$$g_\alpha = G_{\alpha n} m_n \tag{1}$$

where these are simultaneous linear equations in the unknowns m_n.

In this discussion, Greek indices represent the observations and Latin indices represent the model parameters. The summation convention applies except where explicitly indicated. Let the number of degrees of freedom of the observations be θ. Let the number of degrees of freedom of the model parameters be N:

$$\alpha, \beta, \gamma, \ldots \text{etc} = 1, 2, \ldots \theta$$
$$\ell, j, m, n, \ldots \text{etc} = 1, 2, \ldots N$$

We discuss several cases, including those for which N is less than, equal to or greater than θ.

II. $N < \theta$, THE OVERCONSTRAINED SYSTEM

In this case, we assume that the data are not noise-free. In general, the theoretical values derived from the model will not be consistent with all θ values of the data since the calculated values of the data will have too few degrees of freedom. Thus, we assume that the observations of the g_α differ from the theoretical values because of noise, i.e., statistical uncertainty in the observations. We must make some postulate, according to a theory of the errors in the observations, of the residuals in the observations. Often, lacking a more sophisticated theory, the least-squares procedure is invoked; this method leads to simple mathematics. In the least-squares procedure, we assume *a priori* that the residuals $(G_{\alpha n} m_n - g_\alpha)$ are normally distributed. We construct

$$\text{Min}_m \, (G_{\alpha n} m_n - g_\alpha)(G_{\alpha k} m_k - g_\alpha)$$

and get

$$G_{\alpha k}\,(G_{\alpha n}m_n - g_\alpha) = 0 \qquad\qquad (2)$$

If the matrix product $G_{\alpha k}G_{\alpha n}$ is non-singular, the solution to (2) is

$$m_n = g_\alpha G_{\alpha k}\,(G_{\beta k}G_{\beta n})^{-1} \qquad\qquad (3)$$

If weights are introduced into (2), we get other solutions than (3).

It should be recognized that we have constrained the solution by specifying the depths at which the N unknowns are located. For example, the g_α may be short period surface wave data, but we may have arbitrarily chosen a *mathematically valid* set of unknowns, m_n such as variables representing the properties of the earth's core. Variations in these variables do not have a large influence on the values of g_α. One can always get a solution from (2), but it may be physically unreasonable (e.g., densities of 10^3 cgs or negative moduli, etc.), since no constraints are placed upon the model m_n. Appropriate constraints are called by Keilis-Borok "inequalities of state". The imposition of inequalities of state turns the linear problem (2) into a non-linear one.

On the other hand, we might have taken the m_n as representing near-surface parameters that would be more appropriate, namely to choose m_n in those places where $G_{\alpha n}$ are large, but no definite rule can be written about this since almost any assumption about the location of the parameters m_n will lead to a solution, no matter how implausible.

III. N = θ, THE WELL-POSED PROBLEM

In this case, we have as many unknowns, m_n as we have data. Assuming that the data represent independent measurements (i.e., that the rows of the matrix $G_{\alpha n}$ are linearly independent), the solution to (1) becomes

$$m_n = g_\alpha\,G_{\alpha n}^{-1}$$

Cases II and III are both complicated for the reasons cited above. To repeat, we get a different solution for different choices of the places in the earth where the unknown m_n are to be located; no rule can be state;in an unbiased way about determining where these sites are to be chosen, although we may wish to place these where the $G_{\alpha n}$ are large.

IV. $\theta < $ N, THE UNDETERMINED SYSTEM

This is a case of major geophysical interest. Our demands for information concerning the earth's interior will always exceed in detail the availability of the quantity of independent data to determine the structure. We assume, in this section, that the data are noise-free; that is, the repetitions of the geophysical measurements will lead to the same g_α. We also assume once again that the data represent independent measurements, so that the rows of $G_{\alpha n}$ are linearly independent.

We note the following problem. If we guess perfectly, i.e., if ℓ_n is the real earth, then $g = 0$ and $m = 0$, which is consistent with (1). But, in the cases $N > \theta$, there are many models which satisfy the data equation. If we start with any one of these models, we also get $g = 0$. But, in this case, we cannot assume that ℓ_n is the real earth. Thus, some property of the matrix G must be singular. This property of the matrix G is the subject of this section.

If $g \neq 0$, our problem is evidently to take the rectangular matrix $G_{\alpha n}$ and increase its dimensions from $\theta \times N$ to $N \times N$ by "inventing" some new data. Since we can always invent new numbers easily, e.g., by some Monte Carlo procedure, the problem is evidently to invent the new data artfully and in a controlled way, so that we know the limitations on the process. When this is done, we will understand the limitations on the validity of the solution.

It is evident by inspection that any choice of fictitious "data" which will permit one to expand and "square up" the matrix G will lead to a solution by the methods of Part III. Thus, there exists a continuum of solutions to the data equation and it is assumed that the real earth is one of these solutions.

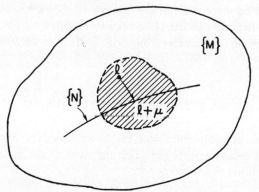

Fig. 1. Space of earth models $\{M\}$ and of solutions to the data equation $\{N\}$. The shaded region represents the subspace in which G, and hence A, does not change significantly from model to model. If we invent new data appropriately, we may generate models which lie outside the range of slowly varying G.

Consider the space of earth models $\{M\}$. The solutions to equation (1) are represented by a hyperspace in $\{M\}$ which is called $\{N\}$ (see Figure 1). We select an initial model ℓ_n in $\{M\}$ and, by "inventing" some data arbitrarily, we proceed to a solution in $\{N\}$ which is $\ell_n + \mu_n$. The real earth also lies in $\{N\}$ in this noise-free case. Different methods of inventing "new data" generate different members of $\{N\}$.

The following procedure illustrates one method of generating models. Indeed, it generates some of the possible μ_n in one step from ℓ_n onto $\{N\}$, but not all of the models μ that can be generated in a one-step process. This special case is

$$\mu_n = g_\alpha f_{nm} \, G_{\beta m} \, (f_{pq} G_{\alpha p} G_{\beta q})^{-1} \tag{4}$$

where f_{nm} is an arbitrary, symmetric non-singular matrix of weighting factors. The necessary and sufficient condition for the existence of the inverse matrix in equation (4) is the independence of the θ measurements, which we have taken as an assumption. It may be seen by direct substitution that the model satisfies the data, i.e.,

$$G_{\alpha n}\mu_n = g_\alpha \tag{5}$$

The special case $f_{nm} = \delta_{nm}$ was considered by Smith and Franklin (1969).

It is possible to show that the result (4) is the consequence of finding a solution to the problem

$$\text{Min } f_{kn}^{-1} (m_k - \mu_k)(m_n - \mu_n) \tag{6}$$

subject to the condition (5). The proof follows by substituting $\mu_n = b_\alpha G_{\alpha m} f_{mn}$ and minimizing (6) with respect to a choice of the vector b_α. The physical significance of the criterion (6) is that we have constructed in (6) a weighted autocorrelation function. Minimizing (6) is equivalent to minimizing the weighted power spectral differences between the model and the real earth[1]. Thus, these models have a maximum smoothness in the sense that jumps in differences $(m_n - \mu_n)$ would have a broad spectrum and this would be minimized. In this postulate, we have assumed that the models generated to fit (1) have some stochastic property that permits the application of this smoothing criterion.

V. DELTANESS

An important concept for understanding undetermined systems is that of the "resolving kernel" (Backus and Gilbert, 1968, 1970). Let it be supposed that the model μ_i can be found by operating on equation (1) with a $N \times \theta$ matrix $a_{i\alpha}$, that is

$$\mu_i \equiv a_{i\alpha} g_\alpha. \tag{7}$$

---

[1] The reader may be amused to consider the similar problem

$$\text{Min } f_{kp}^{-1} (m_k - f_{km}\mu_m)(m_p - f_{pn}\mu_n)$$

where we let $\mu_m = b_\alpha G_{\alpha m}$ and the minimization is taken with respect to a choice of the vector b_α. The solution to this problem is

$$\mu_p = g_\beta G_{\alpha p} (f_{km} G_{\alpha m} G_{\beta k})^{-1}$$

and does not satisfy the data equation (5). The physical significance of this problem, if there is any, is not obvious at present, although the formal minimization would appear to be similar.

Then, by equation (1),

$$\mu_i = a_{i\alpha}G_{\alpha j}m_j \tag{8}$$

We now define a matrix

$$A_{ij} \equiv a_{i\alpha}G_{\alpha j} \tag{9}$$

and we have

$$\mu_i \equiv A_{ij}m_j \tag{10}$$

Though there may be infinitely many solutions m_j which satisfy (1), μ_i is uniquely defined in terms of the data by equation (7) for a given matrix $a_{i\alpha}$. Thus, the matrix A_{ij} maps each member of the set $\{N\}$, into the same single model μ_i. Although μ_i will not in general satisfy (1), it has a useful physical significance. By virtue of (9), each element of μ_i is a weighted average of the values of m_j, the weighting function being the ith row of A_{ij}. If the ith row is "deltalike" near $i = j$, that is $A_{ij} \approx \delta_{ij}$, then μ_i is a reasonable estimate of the physical property m_j, averaged over the ith layer. For this reason, each row of the matrix A_{ij} is referred to as a "resolving kernel". It would be pleasant if $A_{ij} = \delta_{ij}$, i.e., the solution to (1), would be unique. This is easily seen from equation (10): suppose there exist two solutions m_j' and m_j'', and $A_{ij} = \delta_{ij}$. Then, by (10),

$$m_i = m_j' = m_j''$$

that is, the solution must be unique. Because $N > \theta$ we know that the solution is not unique, therefore $A_{ij} \neq \delta_{ij}$. Because there will now be many solutions m_j for which (10) holds, we must have

$$\det A_{nk} = 0 \tag{11}$$

Backus and Gilbert (1968) propose to make A_{ij} as deltalike as possible by selecting the generating matrix $a_{i\alpha}$ according to the following prescription: Let us design a "penalty matrix", J_{ij}, and minimize the quantity

$$w_i = \sum_{j=1}^{N} J_{ij} A_{ij}^2 \tag{12}$$

for each value of i, subject to the constraint that

$$\sum_j A_{ij} = 1 \tag{13}$$

The quantities J_{ij} are arbitrary, but clearly they play the role of weighting factors. It is relatively simple to show that the ith row can be made most delta like if J_{ij} is chosen so that it is least at the diagonal element i=j and greatest far from the diagonal, and increases monotonically as one goes away from the diagonal. Backus and Gilbert suggest

$$J_{ij} = (i-j)^2 \tag{14}$$

Thus, the prescription involves making each row separately as deltalike as possible. Because of this, we have temporarily dropped the summation convention and index summation is indicated explicitly for equations (12) through (18).

We introduce Lagrangian multipliers $2\gamma_i$ and minimize, from (13) and (12),

$$\Psi_i = \sum_{j=1}^{N} J_{ij} A_{ij}^2 + 2\gamma_i \sum_{j=1}^{N} A_{ij} \tag{15}$$

Making use of equation (9),

$$\Psi_i = \sum_j^N J_{ij} \left(\sum_\alpha^\theta a_{i\alpha} G_{\alpha j}\right)^2 + 2\gamma_i \sum_\alpha^\theta \sum_j^N a_{i\alpha} G_{\alpha i} \tag{16}$$

We differentiate (16) with respect to $a_{i\alpha}$ and thus we solve

$$\sum_\alpha^\theta \sum_j^N J_{ij} (a_{i\alpha} G_{\alpha j}) (G_{\beta j}) + \sum_j^N \gamma_i G_{\beta i} = 0 \tag{17}$$

together with the constraint condition of normalization (13),

$$\sum_\alpha^\theta \sum_j^N a_{i\alpha} G_{\alpha j} = 1 \tag{18}$$

Equations (17) and (18) are $(\theta + 1)$ simultaneous equations in the unknowns $a_{i\alpha}$ (i is fixed) and γ_i (i is fixed). Solving these simultaneous equations for $a_{i\alpha}$, we repeat the process for each row by changing the index i. From the complete matrix $a_{i\alpha}$, we can substitute in (9) and generate A_{nm}. If we wish, we can substitute in (7) and generate μ_i, the model, although, as will be shown below, this is not crucial to the discussion.

Although each row of the matrix A_{ij} is made as deltalike as possible, the deltaness of any given row does not depend on the properties of any other row explicitly. Thus, the matrix A_{ij} generated by this procedure is in general non-symmetric.

We can further ask, does the model μ_i generated in $(12) - (18)$ satisfy the data equations in a way that the models generated in IV did? If the model does satisfy (5), namely

$$g_\alpha = G_{\alpha k}\,\mu_k \tag{5}$$

then from (7)

$$g_\alpha = G_{\alpha k}\,a_{k\beta}g_\beta$$

will be an identity. This will require that

$$G_{\alpha k}\,a_{k\beta} = \delta_{\alpha\beta} \tag{19}$$

Because of the non-interactive properties of the rows of $a_{i\alpha}$, there is no reason for (19) to be satisfied. Hence, the A_{ij} matrix obtained for the deltaness criterion does not generate values μ_i that satisfy the data equation (5) in general.

The result that the model μ_i obtained from (7) does not satisfy (5) is not distressing. We have found two important features which are perhaps more valuable than a "solution" *per se*. First, we have received a good deal of guidance in formulating intelligent questions about the earth. We know that, where the resolving kernel is "deltalike", we may ask for detailed information about the earth with some hope of reliability in the result of the inquiry. Second, we have learned something about the physical properties of the regions of the earth, even if they are only "average" properties. This information may be sufficient to discriminate between competing theories, and furthermore, this information is unique, rather than model-dependent.

For those who cannot be satisfied unless equation (1) is satisfied, an exact solution may be found by equation (4). For this case, by analogy with equation (7), we have the generating matrix

$$a_{i\alpha} = f_{im}G_{\beta m}\,(f_{pq}\,G_{\alpha p}\,G_{\beta q})^{-1}$$

and the resolution matrix

$$A_{ij} = a_{i\alpha}G_{\alpha j} = f_{im}\,G_{\beta m}\,G_{\alpha j}\,(f_{pq}\,G_{\alpha p}\,G_{\beta q})^{-1} \tag{20}$$

Unfortunately, this does not equal the identity matrix, since we have assumed that the matrix $G_{\alpha j}$ is underdetermined. This matrix of resolving kernels can be pretty good, in that it can be shown that it minimizes

$$\sum_{i=1}^{N}\left(\sum_{j=1}^{N}\sum_{m=1}^{N} f_{kj}^{-\frac{1}{2}}\,A_{jm}\,f_{mi}^{\frac{1}{2}}-\delta_{ki}\right)^2$$

for each k. This is a somewhat different criterion from that of Backus and Gilbert, but is an equally valid one. Clearly, the resolving kernels which are optimum under one criterion will be less than optimum under some other criterion, so that the two methods will lead to somewhat different results.

VI. OTHER SOLUTIONS: $\theta < N$

At the outset of section IV, it was stated that it was easy to find solutions μ_n to (5) merely by "inventing" data. We are evidently surfeited with solutions. In the solutions of equation (4), a family of such solutions has been generated through the choice of the weighting functions f_{ij}. These solutions have the property of providing maximum smoothing in a spectral sense.

Perhaps the simplest of solutions, if solutions are all that are sought, is to interpolate among the data to obtain $N-\theta$ new "data" points, to assume these are statistically independent (which they are not) and to solve the problem by the method of section III. But perhaps we might wish to generate more solutions by less obvious methods.

We have seen that the Backus-Gilbert approach leads to considerable insight which, as it turns out, is more important than the fact that it does not lead to a model which satisfies the data equations. Let us try to see if it is possible to use this technique repeatedly to get a solution to (1) by iteration. Since the model given by (11) does not satisfy (1), we can compute the discrepancy in the data as

$$\Delta g_\alpha = g_\alpha - G_{\alpha n}\mu_n$$

and now solve for the "correction" to the model $\Delta\mu$ as

$$\Delta\mu_n = a_{n\alpha}\Delta g_\alpha$$

By direct substitution, we get

$$\Delta\mu_n = a_{n\alpha}(g_\alpha - G_{\alpha m}\mu_m)$$

or

$$\Delta\mu_n = \mu_m(\delta_{mn}A_{mn})$$

Thus, we can set up the iteration as the matrix operation

$$<\mu> = \left\{ \sum_{n=0}^{\infty} (I-A)^n \right\} \mu$$

with $<\mu>$ the final result. The question now is whether the operator in braces converges, in the sense that $<\mu>$ now satisfies (1). But, if $<\mu>$ satisfies (1), and μ given by (11) does not, then we have found a matrix A' in the equation

$$m = <\mu> = A'\mu$$

which is evidently the inverse to (10). But we have indicated that the matrix A does not have any inverse. Hence we conclude that the iteration does not converge to a solution to the data equation.

A modification of the deltaness method can be obtained which proceeds to a solution μ on N in one step, while still retaining the essence of the deltaness criterion (12). Let us append to the problem (16) the additional constraint that the data equations (5) are to be satisfied. This constraint is written as a condition on the matrix elements a in (20), since (20) now holds. In (20), the matrix a is a dual of G ; a is not G^{-1} because $\theta \neq N$. The constraints are explicitly stated as (19).

We write, using our summation signs explicitly,

$$\text{Min } \sum_{j}^{N} J_{ij} \left(\sum_{\alpha}^{\theta} a_{i\alpha} G_{\alpha j}\right)^2 + 2\gamma_i \sum_{\alpha}^{\theta} \sum_{j}^{N} a_{i\alpha} G_{\alpha j}$$

$$+ 2 \sum_{\beta=1}^{\theta} \sum_{j=1}^{N} \sum_{\alpha=1}^{\theta} \pi_\beta g_{\alpha} a_{j\alpha} G_{\beta j}$$

for our minimization problem, with γ and π the Lagrangian multipliers. Differentiating with respect to $a_{i\alpha}$, we get

$$\sum_{\beta}^{\theta} \sum_{j}^{N} J_{ij} a_{i\beta} G_{\alpha j} G_{\beta j} + \gamma_i \sum_{j}^{N} G_{\alpha j} + \sum_{\beta}^{\theta} \pi_\beta g_{\alpha} G_{\beta i} = 0 \tag{21}$$

These are now $N\theta$ simultaneous equations in the N unknowns $a_{i\beta}$, which together with θ constraints of type (19) and the N constraints of type (18) form a complete set, since there are N multipliers γ and θ multipliers π . We solve (21), (19) and (18) for the $a_{i\beta}$. The solution to this problem, that of maximum deltaness with the constraint that (5) be satisfied, cannot lead to an averaging kernel that is as deltalike as that without the latter constraint. It is clear, in view of (20) and (9), that A is the product of G with its dual a , the dual being defined in (19).

VII. WHICH SOLUTION IS "BEST"?

Hints have been given above in the discussion of the deltaness criterion that perhaps the obtaining of a solution μ_n to the data equation

$$g_\alpha = G_{\alpha n} \mu_n \tag{5}$$

may not be of the greatest importance. Let us investigate why.

Glutted with solutions, we approach the problem established at the outset of this section, namely, which one is the "best"? The response to this problem must be whichever solution has the most aesthetic appeal. The question is meaningless since, without additional insight, there is no way to select from among the solutions the appropriate one, m_n.

However, we inquire what properties the solutions have in common. From (1) and (5), we see

$$G_{\alpha n}\mu_n = G_{\alpha n}m_n \tag{22}$$

Let us take the scalar product of (22) with the matrix $a_{m\alpha}$ defined in (10) or with $f_{mp}G_{\beta p} (f_{ij}G_{\beta j}G_{i\alpha})^{-1}$ as in (8). From (8) and (10), we have

$$A_{mn}\mu_n = A_{mn}m_n \tag{23}$$

Thus, the weighted average $A_{mn}\mu_n$ of any solution is the same as that for any other solution, including the real earth. Hence, the averaging kernels with maximum deltaness, corresponding to the problem (12)-(18), give the best resolution possible for the determination of the properties of the interior where these properties are now expressed as weighted averages. Where A_{mn} is most like δ_{mn}, all solutions, including those generated by the methods outlined above, will have properties close to one another and close to that for the earth. If A_{mn} in its optimum circumstance is not very deltalike, then this remains the best we can do: namely, the local average of all solutions μ_n will be an average over a broad range of indices m; we cannot hope to learn more, for the given data set g_α.

Thus, one need not be concerned about the failure of the deltaness procedure to provide a solution. Any solution that satisfies (5) will have the same local average properties as the real earth, no matter how the solution was derived. There are many averaging kernels A_{mn}, as for example those given by (8). These averaging kernels provide correct local averages for all solutions, whether the solution was derived by the method which yielded the corresponding A_{mn} or not. The most deltalike kernels A_{mn} of (17)-(18) provide the "best" local average estimates, i.e., involve averaging over as small a range of model parameters as possible and apply to any solution. Thus, the best that can be said for any of the solutions is that they can be obtained by the explicit methods listed above.

A note of caution must be introduced. The concept of a resolving kernel as developed here depends on the linearity of the problem (1). The resolving kernels calculated by any of the above techniques will be meaningful only within the range of linearity about the starting model. In actual fact, "structures" may exist which satisfy the data equations and yet are outside this range of linearity. When making physical interpretations based on the "resolvability" of certain features, care must be taken to ensure that all solutions of physical interest are indeed within the linear range of the starting model.

VIII. THE SIMULTANEOUSLY UNDETERMINED AND OVERDETERMINED CASE

Suppose that $\theta < N$ as before, but that the data are not noise-free. Indeed, in the treatment of this section and of Section IX, we may have $\theta < N$ or $N < \theta$, although the former case is the more likely to occur. In this case, from equation (5),

$$\delta g_\alpha = G_{\alpha n} (\delta \mu_n)$$

where δg_α and $\delta \mu_n$ are the uncertainties in the data and the corresponding ones in the model. Backus and Gilbert (1970) propose that the problem be studied by computing the debilitation of the deltaness, i.e., the broadening of the A_{nm} matrices, and hence the loss of resolution in the determination of the average properties of the earth structure.

We estimate the variance in the model

$$\epsilon_{ii} = \delta_i^2 \text{ (summation convention does not apply)} \tag{24}$$

From (7), this quantity is

$$\sum_{\alpha,\beta} a_{i\alpha} \delta g_\alpha a_{i\beta} \delta g_\beta = \epsilon_{ii} \tag{25}$$

It is evidently not possible to minimize both (24) and (16) at the same time, but it is possible to minimize a linear combination of these two expressions, namely

$$\text{Min} \left\{ \sum_{j}^{N} J_{ij} \left(\sum_{\alpha}^{\theta} a_{i\alpha} G_{\alpha j} \right)^2 + 2\gamma_i \sum_{\alpha}^{\theta} \sum_{j}^{N} a_{i\alpha} G_{\alpha j} \right.$$

$$\left. + f \sum_{\alpha,\beta}^{\theta} a_{i\alpha} a_{i\beta} \delta_\alpha \delta g_\beta \right\} \tag{26}$$

with respect to a choice of the matrix $a_{\alpha i}$ for a given weighting factor f. It should be noted that the quantity being minimized in (25) is not the same as that minimized in (6) with the factor

$$f_{kn} = \delta_{kn},$$

since in (6) we have summed over the repeated indices. Berry and Knopoff (1967) have given one example of the calculation of ϵ_{ii} from surface wave data.

The problems raised earlier still apply to this case, namely where one chooses the unknown model parameters. In addition, especially in the presence of noise, it is not clear what θ, the number of degrees of freedom in the data, may be. We address ourselves to these questions in the next section.

IX. A SECOND SOLUTION TO THE SIMULTANEOUSLY OVER- AND UNDERDETERMINED CASES

We now present an alternative solution to (1) in the case in which we have more degrees of freedom in the model than in the data, but the data are also encumbered with experimental uncertainty. The details are given in a paper by Jackson (1972). To start the operation, we premultiply both sides of (1) by a matrix $H_{k\alpha}$ which is a "pseudo-inverse". The result is a model μ such that

$$\mu_j = H_{j\alpha}\, G_{\alpha k}m_k \equiv A_{jk}m_k \qquad (27)$$

The inverse matrix H (a θ by N matrix) will have been chosen successfully if three criteria are met, of which two are a composite of the criteria already set forth in II and V. These three criteria are

i) The model fits the data as well as possible, that is $c_1 = (G_{\alpha k}\mu_k - g_\alpha)\,(G_{\alpha j}\mu_j - g_\alpha)$ is small.

ii) The matrix A_{kj} is "close" to a delta function δ_{kj}, i.e., the quantity $c_2 = (A_{kj} - \delta_{kj})\,(A_{kj} - \delta_{kj})$ is small. We have already seen that μ may be interpreted as a weighted average of the values for *any solution* to the underdetermined problem (which has infinitely many solutions). If A is strongly peaked near $k = j$, then the jth component is resolvable using the given data. Even if the problem is not underdetermined in the sense of a system without the constraint of (i) above (i.e., as in V), this discussion for the matrix A still holds. If A is not sharply peaked, we must settle for an average of the values of m_k over a wider range.

iii) Small errors in the data will not cause large errors in the model. If the errors in the data are uncorrelated, we have

$$\text{var } \mu_k = L_{k\alpha} \text{ var } g_\alpha \qquad (28)$$

where $L_{k\alpha} = H_{k\alpha}^2$ i.e., L is the matrix whose elements are the squares of the as yet unknown matrix H. Here "var" is the variance operator and we would like var μ_k to be small for each value of k.

In the case of problem III ($\theta = N$), the optimum operator H is simply

$$H_{k\alpha} = G_{k\alpha}^{-1}$$

since criteria (i) and (ii) are satisfied exactly. However, the expected error in the model var μ_k may be large if $G_{\alpha k}$ is nearly singular.

We digress for the purpose of discussion below and describe the well-known construction of the inverse of a square, symmetric, non-singular matrix $G_{\alpha k}$ ($\theta = N$). This inverse is expressed in terms of its normalized eigenvectors $V_{\alpha Я}$ and its eigenvalues $\lambda_Я$, $Я = 1, \ldots, Ю; Ю = N = \theta$. $V_{\alpha Я}$ is the αth component of the $Я^{th}$ eigenvector of G. Since G is symmetric, $V_{\alpha Я}$ is symmetric. The inverse is

$$G_{k\alpha}^{-1} = \sum_{Я=1}^{Ю} V_{kЯ} \lambda_Я^{-1} V_{Я\alpha} \qquad Я = N = \theta \qquad (29)$$

For the purposes of the later discussion, we have introduced a separate symbol $Ю$ for the number of degrees of freedom of the index $Я$, although in this case of symmetry $Ю = N = \theta$, as noted. We further assume that the eigenvalues λ can be placed in order of decreasing magnitude without loss of generality. Non-singularity of $G_{\alpha k}$ implies that none of the eigenvalues $\lambda_Я$ vanishes and near singularity means that at least one of the $\lambda_Я$ is small in some sense. In terms of criterion (iii), (28) becomes

$$\mathrm{var}\,\mu_k = \sum_{Я=1}^{Ю} (V_{\alpha Я}\lambda_Я^{-1}V_{Яk})^2 \,\mathrm{var}\,g_\alpha \qquad (30)$$

If G is nearly singular, the presence of a small value of λ^{-1} in (30) causes the variance of the model to be large.

We can now write the "optimum" inverse matrix $H_{k\alpha}$ for the general matrix $G_{\alpha k}$. This inverse is

$$H_{m\beta} = \sum_{Я=1}^{Ю} V_{mЯ}\lambda_Я^{-1} U_{\beta Я} \qquad (31)$$

where $V_{mЯ}$ is the mth component of the $Я^{th}$ eigenvector of the matrix $G'_{nm} = G_{n\alpha}G_{\alpha m}$ and $U_{\beta Я}$ is the βth component of the $Я^{th}$ eigenvector of the matrix $G''_{\alpha\beta} = G_{\alpha n}G_{n\beta}$. G'_{nm} and $G''_{\alpha\beta}$ will both have positive eigenvalues $\lambda_Я^2$, of which the first $Ю$ will be common to both, and all remaining eigenvalues must be zero. Without loss of generality, we may use the positive square root of λ^2 to obtain the values used in (31). $Ю$ must be less than or equal to the smaller of N (the numbers of degrees of freedom for G') and θ (the number of degrees of freedom for G''). The result (30) is obtained by "squaring up" G by adding enough zeroes so that it is now a square singular matrix of the order of the larger of θ or N.

The matrix (31) has the following properties:

a) If $G_{\alpha k}$ is non-singular, $Ю = N = \theta$ then $H_{m\beta} = G_{m\beta}^{-1}$, the usual matrix inverse. If G is non-singular and symmetric, H is obviously the same as given in (28).

b) If $G_{\alpha k}$ is underdetermined only, $Ю = \theta < N$, the model μ_k given by (27) satisfies the data equation (1) exactly and $c_1 = 0$. In this case, H will be the optimum in the sense that c_2 is minimized. We note further that the model which is obtained minimizes (6) in the case $f_{kn} = \delta_{kn}$.

c) If $G_{\alpha k}$ is overdetermined only, $Ю = N < \theta$, then H is such that $c_2 = 0$ and c_1 is minimized, thereby making this result the usual "least-squares" inverse.

d) If $G_{\alpha k}$ is both underdetermined and overdetermined, H is such that both c_1 and c_2 are minimized, although neither is now zero. $Ю$ is the effective number of degrees of freedom for both the data and the model.

It is evident that the inverse $H_{m\beta}$ defined by (31) optimizes criteria (i) and (ii) without regard to (iii). The variance of the model is given by

$$\text{var } \mu_m = \sum_{\beta=1}^{\theta} \left\{ \sum_{я=1}^{ю} V_{mя} \lambda_я^{-1} U_{\beta я}{}^2 \right\} \text{var } g_\beta \tag{32}$$

and may be large if $\lambda_{ю}$ is very small. A modification of equation (31) allows us to compromise very slightly on criteria (i) and (ii) in order to control the variance of the model. We may truncate the summation over я in equation (31) at some smaller value than ю, say, и. As we choose и smaller and smaller, we degrade the resolution and the fit to the data.

The trade-off between resolution and variance (Fig. 2) yields a rational basis for selecting и. In addition, we can now provide a procedure whereby we can respond to the

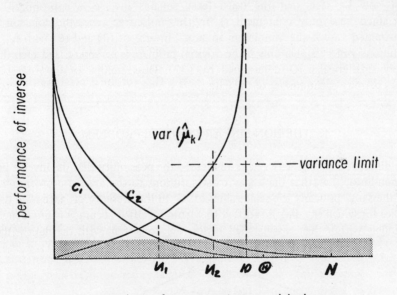

number of eigenvectors used in inverse

Fig. 2. Schematic tradeoff curve showing rms residual c_1, resolving parameter c_2, and the variance of the model var ($\hat{\mu}_k$), as a function of the assumed value of и used in constructing the inverse matrix H. Each of these quantities is considered "good" if it is small. The case shown is both underdetermined and overconstrained. Possible choices of и which would have merit for specific cases include $и_1$, the smallest value for which the residual c_1 is consistent with the standard deviation of the data (shown by shaded area); $и_2$, that choice which gives the best resolution and smallest residual without violating *a priori* limits on the variance of the model; and ю , that which gives the best resolution and smallest residual regardless of variance. Most standard "least-squares" programs would implicitly assume $и = \theta$, which would lead to attempted inversion of a singular matrix, and hence to failure.

problem alluded to in sections II and III, namely to avoid the pitfall of selecting as unknown parameters, quantities which are not particularly sensitive to the data available. In addition, we have the thorny problem of determining the number of degrees of freedom θ in the data; this problem is quite sensitive since the number of data points do not yield θ simply, in view of the fact that the fit to data, in the presence of noise, by a criterion such as (i) implies smoothing of some sort.

The resolution to this problem involves the setting of an upper limit for the allowable variance in the model var μ_m. We now increase $И$ by going to smaller and smaller values of λ, implying the use of more and more eigenvectors, until the variance calculated by (32) exceeds the allowable limit. Replacing $Ю$ by $И$ in (d), we have now determined the number of degrees of freedom in the data. We conclude that the number of degrees of freedom in the data does not depend on the data themselves; $И$ depends only on the eigenvectors and eigenvalues of the matrix G, (i.e., on those model parameters which are sensitive to the data) and the variance of the data.

It can now be seen that the usual least squares inverse is appropriate to the overconstrained case only (statement i). In this noise-free case, the solution to the underconstrained case is the "minimum squares" inverse of (4) and (6) with $f_{kn} = \delta_{kn}$. These solutions will be stable only if the original problem is parameterized such that θ or N is safely smaller than $Ю$ determined by (32). Thus, neither of these least-squares approaches permits optimum use of the data.

X. THE NON-LINEAR INVERSE PROBLEM

In the discussion of Fig. 1, we have alluded to the possibility that the inverse problem may be non-linear, i.e., that there may exist solutions to the inverse problem which do not have the same or nearly the same values of G. In the presence of inaccurate data, the model space for which the data is satisfied to within specified accuracy is broadened, as in Fig. 3. Models within the cross-hatched region are consistent with the data within pre-

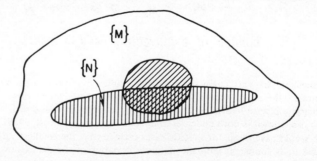

Fig. 3. Space of earth models $\left\{ M \right\}$ and of solutions $\left\{ N \right\}$ to the data equation including possible uncertainties in the data. The diagonally shaded region represents the subspace of linearity, e.g., G varies only slightly in this region.

assigned uncertainties in the data. Models within the diagonally shaded region may have G matrices which are reasonably close to one another.

In the case of the non-linear problem, the elegant procedures of linear matrix theory are inappropriate. Here the most convenient approach is through more direct techniques. We may initiate a search in the model space $\{M\}$ until we find a solution in the hyperspace $\{N\}$. This can be done either by Monte Carlo methods or more systematically. Having found one solution in $\{N\}$, others can be found as well in $\{N\}$ by searching in the neighborhood of the first solution along some lattice in parameter space. The search is continued by moving through parameter space, exploring the neighborhood of each successful solution in turn, until the outlines of a singly connected hyperspace $\{N\}$ are determined. The boundary is, of course, determined by the errors of fit to the data. In the case of multiply connected $\{N\}$, starting points in the remaining lobes of $\{N\}$ must be found by random search or other methods. A computer search procedure which avoids re-investingating lattice points already studied and which explores the neighborhood of successful solutions in turn has been constructed; this is called the "hedgehog method".

One difficulty with the hedgehog procedure is immediately apparent. If the lattice spacing in $\{N\}$ in Fig. 2 is too fine, much computer time is spent in a neighborhood for which G is slowly changing. If the lattice spacing in $\{N\}$ is too course, perhaps only one model point, or even no points, in $\{N\}$ will be determined. Present research is concerned with the coupling of the techniques of section IX with those of X to reduce this uncertainty, namely to determine the appropriate lattice dimension of search in the model space.

REFERENCES

1. G. Backus and J. F. Gilbert, The resolving power of gross earth data, *Geophys. J. R. Astron. Soc.*, **16**, (1968) 169-205.
2. G. Backus and J. F. Gilbert, Uniqueness in the inversion of inaccurate gross earth data, *Phil. Trans. R. Soc.*, London, **266**, (1970) 123-192.
3. M. J. Berry and L. Knopoff, Structure of the upper mantle under the western Mediterranean basin, *J. Geophys. Res.*, **72**, (1967), 3613-3626.
4. D. D. Jackson, Interpretation of inaccurate, insufficient or inconsistent data, *Geophys. J., R. Astron. Soc.*, (1972) in press.
5. M. L. Smith and J. R. Franklin, Geophysical application of generalized inverse theory, *J. Geophys. Res.*, **74**, (1969) 2783-2785.

APPLICATIONS OF PHOTOELASTICITY TO ELASTODYNAMICS

JAMES W. DALLY

Abstract—The elastodynamic problem is classified into two different areas including periodic or vibrating fields and pulse or stress-wave fields. A brief review of the application of dynamic photoelasticity to each of these areas is covered. In particular, the use of photoelasticity to study stress waves propagating at high velocities is examined in detail. The influence of the high rates of loading on the material fringe value is considered. It is shown that the material fringe value f_σ can be treated essentially as a constant in a dynamic photoelastic analysis using the conventional stress optic law.

A theory of dynamic recording of isochromatic fringe patterns propagating at stress wave velocities is given. The concept of fidelity of recording systems is introduced and then illustrated for records produced from a spark gap and a pulsed ruby laser. Two recording systems in common use today, based on a multiple spark array and a pulsed ruby laser, are described and compared.

Applications of photoelasticity to elastodynamic problems arising in the fields of mechanics, geophysics, mining and non-destructive testing are briefly introduced.

NOMENCLATURE

A   transmission and reflection coefficients

c   wave velocity

C   contrast on negative

d   flaw depth

E^* normalized exposure of film

E   modulus of elasticity

f   material fringe value

F   body force intensity

h   model thickness

I   intensity of light

J   first strain invarient

N   fringe order

P   dilatational wave

R   Rayleigh wave

S   distortional wave

t   time

t_e exposure time

T   optical transmission coefficient

u   displacements

x,y,z   coordinates

λ,μ Lamé constants

λ_1 wave length

υ Poisson's ratio

τ $(\sigma_1 - \sigma_2)$ difference in principal stresses

ρ mass density

INTRODUCTION

In the most general terms, the area of elastodynamics includes solutions to boundary value problems for the field quantities, stress, strain and displacement which vary with time. The dynamical equations in terms of the displacements u_i are:

$$\mu \Delta^2 u_i + (\lambda + \mu) \partial J/\partial x_i + F_i = \rho \, \partial^2 u_i/\partial t^2 \qquad (1)$$

Solutions of these equations must satisfy the initial and boundary conditions which are prescribed as functions of space co-ordinates x_i and time t for each specific boundary value problem of dynamical elasticity. Significant progress has been made in developing theoretical procedures in elastodynamics as indicated in texts by Love [1], Kolsky [2], Goldsmith [3], Fung [4], and Ewing, Jardetsky and Press [5]. Recent research achievements are summarized in a thorough review by Miklowitz [6]. In spite of these developments, theoretical results are difficult to achieve and for bodies with complex geometries the analytical approach becomes extremely involved. It is in these applications that dynamic photoelasticity can be used most effectively to provide a full field visualization of the dynamic state of stress.

The dynamical area can be subdivided into two different classes of problems namely vibrations and stress wave propagation. With vibration problems, the loading period and the observation period are usually of about the same duration and relatively long. As the problems are often three dimensional and involve signficant effects of structural dampening both internal and external, scaling is difficult if not impossible to accomplish with photoelastic models. For this reason, only the elastodynamical problems dealing with wave propagation will be treated here.

Wave propagation problems usually involve very short loading times with pulses of only few μ sec in duration and observation periods ranging from 50 to 500 μ sec. Dilatational (P), shear (s) and Rayleigh (R) waves are formed which propagate into the model with respective velocities given by*:

---

* For a two-dimensional generalized plane stress problems where the elastodynamic field quantities are averaged across the thickness of the model. Rayleigh wave velocity corresponds to $\upsilon = 1/3$.

$$c_1 = \kappa c_2$$
$$c_2 = (\mu/\rho)^{1/2}$$
$$\kappa^2 = 2/(1-v) \tag{2}$$
$$c_R = 0.92\, c_2$$

TABLE I

Properties of Photoelastic Model Materials

		CR-39	Homalite 100	Epoxy
Modulus of Elasticity	E_d	535,000	700,000	690,000 psi
Poisson's Ratio	v_d	0.32	0.31	0.30
Wave velocities in	c_1	70,000	83,000	83,000 in/sec
	c_2	41,000	49,000	49,000 in/sec
	c_R	38,000	45,000	44,000 in/sec
Material fringe value**	f	110	130	100 psi-in fringe

**at 5000 A°

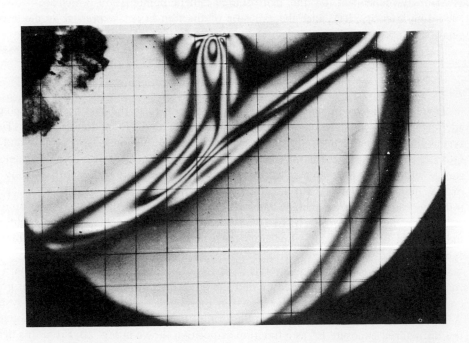

Fig. 1. Photoelastic Fringe Pattern Representing the P, PS, S and R Waves Propagating in a Half-plane (t = 190 μsec).

Wave speeds for three different high-modulus photoelastic materials shown in Table I indicate that the velocities range from about 40,000 to 83,000 in/sec which is sufficiently high to require ultra-high speed recording techniques. A typical example of a dynamic isochromatic fringe pattern illustrating the photoelastic representation of stress wave propagation in a half-plane is presented in Fig. 1. This photograph indicates the full-field visualization of the dynamic state of stress which can be accomplished by using photoelasticity if adequate recording methods are used.

FIDELITY OF DYNAMIC RECORDING

The theory for the dynamic exposure of a fringe pattern recorded in a polariscope was established by Dally, Henzi and Lewis [7] for a plane stress wave traveling with the velocity c in the x direction. The difference in the principal stresses in this plane stress wave was denoted as $\tau(x,t)$ and the instantaneous intensity of light emerging from the polariscope was identified as the transmission coefficient $T(x,t)$ given by

$$T(x\text{-}ct) = \frac{T_o}{2} \left\{ 1 + \cos \left[\frac{2\pi h}{f} \tau(x\text{-}ct) \right] \right\} \tag{3}$$

The record of the dynamic fringe pattern characterized by the transmission coefficient is strongly dependent on the light source-camera combination which controls the recording intensity, $I(t)$, since the exposure of the film $E(x)$ is expressed by the integral relation:

$$E(x) = \int_o^\infty I(t) \, T(x\text{-}ct) \, dt \tag{4}$$

Three different recording systems have been developed in the past decade which give adequate results in dynamic photoelastic studies with high modulus models. These systems include the high-speed framing camera, the multiple-spark-gap assembly and the Q-switched laser system. The high-speed framing camera operates with constant-intensity source with an exposure time t_e which varies with framing rate (t_e is approximately 1/3 of the interframe interval). The spark gap light source provides an expoentially decaying intensity independent of framing rate with t_e established at the 1/3 − peak-intensity points at typically $0.5 \, \mu$ sec. Q-switched lasers provide a triangular pulse of light also independent of framing rate; however, the value of t_e measured at the 1/3 points is typically $0.10 \, \mu$ sec.

The ability of each of these systems to accurately record a dynamic fringe pattern is a function of the stress wave velocity c and the stress distribution. To illustrate this point, consider the triangular stress pulse shown in Fig. 2A recorded with a spark-gap-system. Analysis of this pulse using Eq. (4) and a form of $I(t)$ associated with the spark gap gives the normalized exposure E* as a function of position shown in Fig. 2B. From this figure, it is clear that the normalized exposure varies between maximum and minimum values with the range of variation inversely proportional to the values of the slopes of the

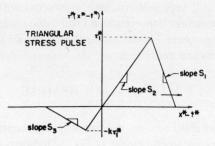

Fig. 2A. Triangular Stress Pulse $S_1 = 1.0$, $S_2 = 0.5$, $S_3 - 0.2$.

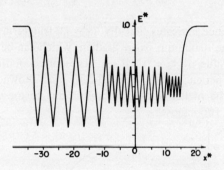

Fig. 2B. Normalized Exposure as a Function of x^*.

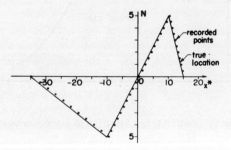

Fig. 2C. True and Recorded Fringe as a Function of x^*.

triangular stress pulse. The positions of the maximum and minimum values of the normalized exposure correspond to positions of the light and the dark fringes on the photographic record and hence permit the determination of the recorded fringe order position illustrated in Fig. 2C. The true distribution of the triangular pulse is also shown in Fig. 2C and the comparison indicates that the maximum fringe order is recorded precisely; pulse distortion is negligible and time or position shift is quite small.

The contrast is of particular importance in establishing a dynamic fringe pattern with adequate resolution. The contrast is defined as the difference in the normalized exposure between adjacent maxima and minima: thus

$$C = E^*_{[N/2]} - E^*_{[(N+1)/2]} \tag{5}$$

The contrast was established for the spark gap system recording a ramp stress distribution. The results obtained, shown in Fig. 3, indicate that the contrast decreases monotonically with the slope S of the dimensionless ramp distribution. The fringe gradient is related to the slope S since at some fixed time:

$$S = \frac{d\tau^*}{dx^*} = (\frac{c\,t_e\,h}{f})\,(\frac{d\tau}{dx}) = c\,t_e\,\frac{dN}{dx} \tag{6}$$

Laboratory experience indicates that $dN/dx \approx 20$ fringes/in., for a P wave propagating at 75,000 in/sec, is the maximum fringe gradient which can be recorded with a spark gap system exhibiting a value of $t_e = 0.5\,\mu$ sec. The corresponding minimum value of C is about 0.2. Similar experience with a pulsed-ruby-laser showed that the resolution limit was 120 fringes/in. The higher resolution limit is due to the very short duration of the laser pulse where $t_e \approx 100$ nanosec.

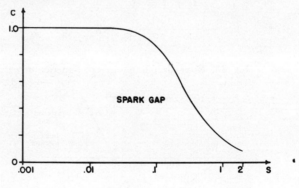

Fig. 3. Contrast C as a Function of Stress Gradient.

It is evident that high resolution photoelastic fringe patterns can be recorded with the spark gap system and that the fidelity of these patterns is excellent. Marked improvements are possible with the pulsed laser; however, this particular recording system has not yet been adapted for multiple frame applications.

DYNAMIC RECORDING SYSTEMS

While many different optical or electro-optical systems can be effectively employed to record dynamic fringe patterns, it appears that the multiple spark gap assembly and the pulsed-ruby-laser system each offer individual capabilities which make them adaptable to many different elastodynamic problems.

The multiple-spark-gap camera, developed originally by Cranz and Schardin [8] in 1929 has been used extensively in Europe in fluid dynamics, ballistics and fracture investigations; however, its application in dynamic photoelasticity has been much more limited. Christie [9] in England described the first application of the camera to photoelasticity in 1955 and Wells and Post [10] introduced the camera in the United States in 1957. The multiple spark gap camera contains a pulse circuit for the sequential firing of spark gaps and an optical system for separation of the multiple recordings to obtain individual images. The pulse circuit which is shown in Fig. 4A is a series connected

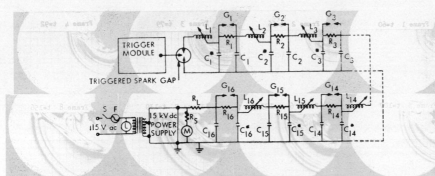

Fig. 4A. Series Connected L-C Line Energizing an Array of Spark Gaps.

line of L-C loops. When switched to ground through the triggered spark gap the first L-C loop oscillates from its operating voltage of +15KV to a negative value of about -10KV. This oscillation produces a voltage difference of 25KV across gap G_1 and it ionizes and the discharge of capacitor C_1 produces a short but intense flash of light across gap G_1. The sequence continues down the line and all gaps are fired. The inter-frame interval (or the framing rate is controlled by adjusting the inductance in each of the L-C loops. The number of gaps is arbitrary; however, experience indicates that a 4 by 4 array of 16 gaps is sufficient to cover most dynamic events.

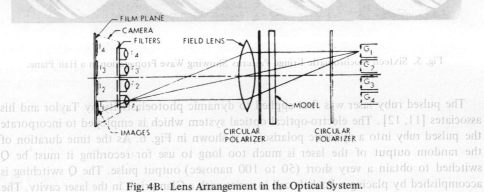

Fig. 4B. Lens Arrangement in the Optical System.

314   J. W. DALLY

The light from spark gap G_1 is collected by the field lens, in the optical system shown
in Fig. 4B, and focused on the camera lens ℓ_1. In a similar manner, the light from the
other spark gaps $G_2, G_3, G_4 \ldots$ is focused on camera lenses $\ell_2, \ell_3, \ell_4 \ldots$. With this optical
arrangement it is possible to form one image for each spark gap with the image formed by
the light from one and only one spark gap. With this simple image separation method, a
time-sequenced series of 16 photographs (for the authors camera) is obtained on a fixed
sheet of film in a commercial view camera. A typical set of dynamic fringe patterns
obtained for the half-plane loaded with small explosive charge on the boundary is shown
in Fig. 5. Fringe patterns characteristic of the dilatational, shear, von-Schmidt and
Rayleigh waves are clearly defined over the entire dynamic event.

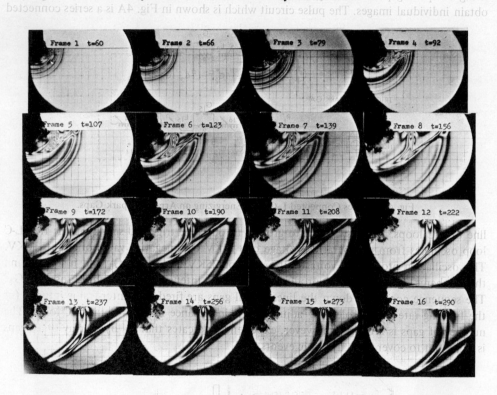

Fig. 5. Sixteen Isochromatic Fringe Patterns Showing Wave Propagation in a Half Plane.

The pulsed ruby laser was first applied to dynamic photoelasticity by Taylor and his
associates [11, 12]. The electro-optical optical system which is employed to incorporate
the pulsed ruby into a dynamic polariscope is shown in Fig. 6. As the time duration of
the random output of the laser is much too long to use for recording it must be Q
switched to obtain a very short (50 to 100 nanosec) output pulse. The Q switching is
accomplished by placing a Pockels cell in series with a Glan prism in the laser cavity. The

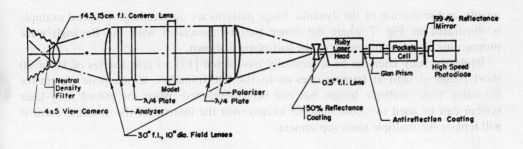

Fig. 6. Pulsed-Ruby Laser Dynamic Polariscope.

Pockels cell is an electrically-actuated wave plate which is maintained at the quarter-wave plate voltage as the laser rod is optically pumped. Maintenance of the Pockels cell as a quarter-wave plate inhibits reinforcement until the voltage on the cell is suddenly reduced to zero and the wave plate is optically removed from the cavity. At this instance, the laser issues a very intense but short pulse of light which is used to record the dynamic fringe pattern in a commercial view camera.

As the exposure time is very short and the light issued from the ruby laser is highly monochromatic and sufficiently intense for proper exposure of the infrared film, the

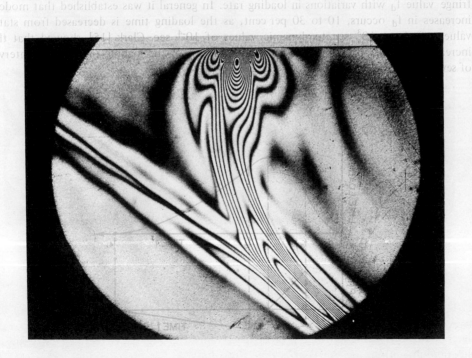

Fig. 7. Fringe Pattern of a Rayleigh Wave Recorded with the Pulsed-Ruby Laser.

quality and resolution of the dynamic fringe patterns are outstanding. A typical example is illustrated in Fig. 7 where the fringe pattern associated with the Rayleigh wave propagating along the boundary of the half-plane is shown.

While the ruby laser can be sequentially modulated [12] to give a series of 10 or 20 short pulses of light at framing rates up to 100,000 frame/sec, an adequate system for recording these multiple images has not been developed. When the pulsed ruby laser system can be used to record multiple images over the interval of the dynamic event it will replace the multiple spark gap camera.

THE STRESS OPTIC LAW

The order of the fringes in each dynamic fringe pattern is established and the fringe order is converted to the difference in the principal stresses using the conventional stress optic law.

$$\tau = N f_d / h \tag{7}$$

During the 50's several investigators [13, 14, 15] examined the applicability of this static form of the stress optic law and in particular studied the changes in the dynamic material fringe value f_d with variations in loading rate. In general it was established that modest increases in f_d occurs, 10 to 30 per cent, as the loading time is decreased from static values of about 10^3 sec to dynamic values of 10^{-4} sec. Clark [15] showed that the increase in f_d was approximately linear with the log of the loading time over an interval of seven time decades.

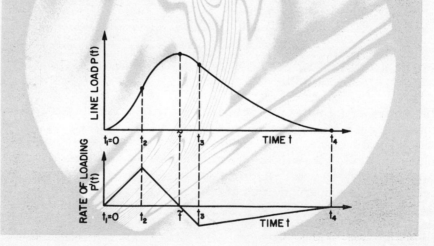

Fig. 8A. Theoretical Loading Function.

The fact that the material fringe value f_d is not a constant does not produce significant errors in wave propagation experiments. A Fourier analysis of a Rayleigh pulse generated by an explosive charge in an epoxy model indicated that more than 90 per cent of the energy in the pulse was contained in the wave length decade between 1 and 10 in. This concentration of the energy of stress wave pulses over a very narrow region of the wave length spectrum limits the variation of f_d to small changes. While the dynamic material fringe value may be somewhat higher than the static material fringe value, the former can, from a practical viewpoint, be treated as a constant without introducing serious errors.

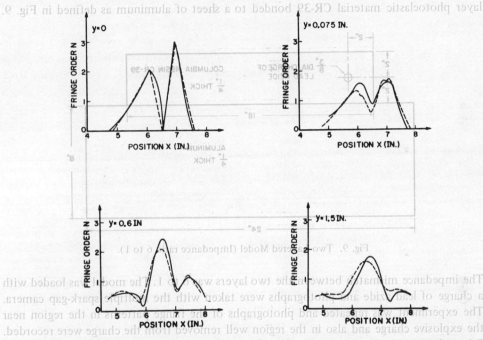

Fig. 8B. Comparison of Theoretical and Experimental Results for the Rayleigh Wave. —Experimental ----Theoretical

The validity of Eq. (7) was verified by Thau and Dally [16] by considering the subsurface characteristics of the Rayleigh wave propagating in a half-space. A theoretical loading function for a line load P(t) with a load-time profile is shown in Fig. 8A. This profile is a series of three parabolas with the end points of connecting parabolas having the same value of P(t) and P'(t). The parameters defining the three parabolas were adjusted to give a solution for the stresses due to the Rayleigh wave on the free boundary of the half-plane which were in close agreement with the results obtained from a dynamic photoelastic experiment. Having matched theory and experiment on the boundary where y = 0 as indicated in Fig. 8B, comparisons were made along lines parallel to the boundary at positions y = 0.075, 0.6 and 1.5. The results show a close correspondence between theory and experiment and serve to verify the extension of static form of the stress optic law to the dynamic case.

APPLICATIONS

Dynamic photoelasticity is an extremely effective method to employ in the investigation of a wide variety of applications in many fields. In an effort to show the versatility of the method four different applications will be briefly introduced and results summarized.

The first application deals with the geophysics problem of stress wave propagation in a two-layered model [17]. Thyexperiment consisted of the fabrication of a model with a layer photoelastic material CR-39 bonded to a sheet of aluminum as defined in Fig. 9.

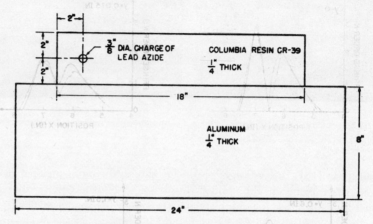

Fig. 9. Two Layered Model (Impedance ratio 6 to 1).

The impedance mismatch between the two layers was 6 to 1. The model was loaded with a charge of lead azide and photographs were taken with the multiple spark-gap camera. The experiment was repeated and photographs of the fringe patterns in the region near the explosive charge and also in the region well removed from the charge were recorded. Select photographs from the series of 32 frames obtained are presented in Fig. 10.

Wave propagation in the two layered model is extremely complex and nine different waves are predicted to occur in the top layer. These include the incident dilatational wave P_1, the reflected dilatational waves $P_1 P_1$ at both the free boundary and the interface, the reflected shear waves $P_1 S_1$ at the boundary and the interface, and the four head waves generated by the refracted waves propagating in the lower layer $P_1 P_2 P_1, P_1 P_2 S_1, P_1 S_2 P_1, P_1 S_2 S_1$.

Six of these nine admissible wave fronts are identified in the fringe patterns shown in Fig. 10. Early in the dynamic event, the incident P_1 wave and the reflected $P_1 S_1$ waves exhibited the largest amplitude. However, late in the event, the three head waves $P_1 P_2 S_1$, $P_1 S_2 P_1$ and $P_1 S_2 S_1$ were the dominant waves as they completely eroded the cylindrical wave front associated with the incident P_1 wave. The $P_1 P_2 P_1$ and the $P_1 P_2$ waves were not of sufficient magnitude to produce a photoelastic response and as such they are of minor importance.

The second application treats the three-dimensional problem of stress wave propaga-

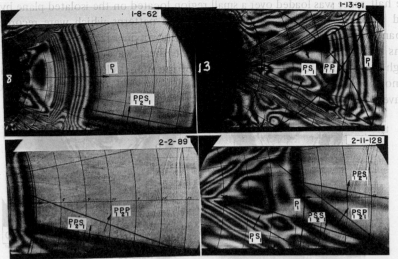

Fig. 10. Select Photoelastic Fringe Patterns Compared with Predicted Positions of Wave Fronts.

tion in a half-space [18]. Utilization of photoelasticity in three-dimensional cases requires the isolation of a plane in the model. In dynamic photoelasticity the isolation is accomplished by embedding the complete polariscope in the model as illustrated in Fig. 11. As the circular polariods which are cemented in the joints of the model are very thin (0.002 in.) and made of a polymer closely matching the properties of the model materials, the influence of the polariods on the dynamic stress field is negligible.

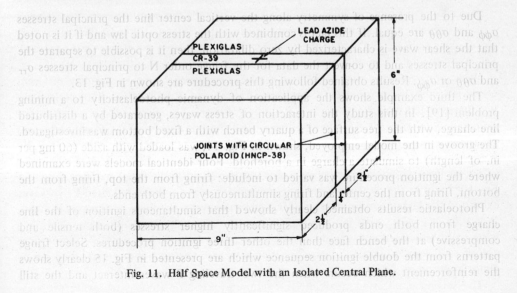

Fig. 11. Half Space Model with an Isolated Central Plane.

The half space was loaded over a small region located on the isolated plane by a charge of lead azide. Select frames showing the fringe patterns which were recorded as the disturbance propagated through the body are shown in Fig. 12. Examination of these patterns show the leading dilatation wave, the shear wave, and the von-Schmidt. The Rayleigh wave which occurs at the surface of the model closely trailing the shear wave could not be studied since the recording time was too short to permit separation of the two waves.

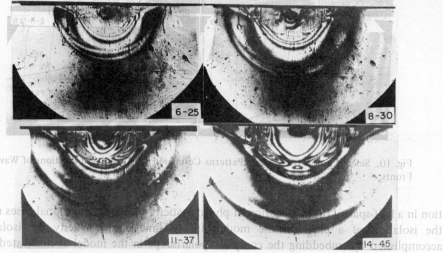

Fig. 12. Dynamic Fringe Patterns for Half Space Model.

Due to the presence of symmetry along the vertical center line the principal stresses $\sigma_{\phi\phi}$ and $\sigma_{\theta\theta}$ are equal. If this fact is combined with the stress optic law and if it is noted that the shear wave is characterized by zero dilatation, then it is possible to separate the principal stresses and to convert the data for the fringe order N to principal stresses σ_{rr} and $\sigma_{\theta\theta}$ or $\sigma_{\phi\phi}$. Results obtained following this procedure are shown in Fig. 13.

The third example shows the application of dynamic photoelasticity to a mining problem [19]. In this study the interaction of stress waves, generated by a distributed line charge, with the free surface of a quarry bench with a fixed bottom was investigated. The groove in the model employed, shown in Fig. 14, was loaded with azide (60 mg per in. of length) to simulate a charge in a borehole. Four identical models were examined where the ignition procedure was varied to include: firing from the top, firing from the bottom, firing from the center and firing simultaneously from both ends.

Photoelastic results obtained clearly showed that simultaneous ignition of the line charge from both ends produced significantly higher stresses (both tensile and compressive) at the bench face than the other three ignition procedures. Select fringe patterns from the double ignition sequence which are presented in Fig. 15 clearly shows the reinforcement which occurs when the two traveling P waves interact and the still

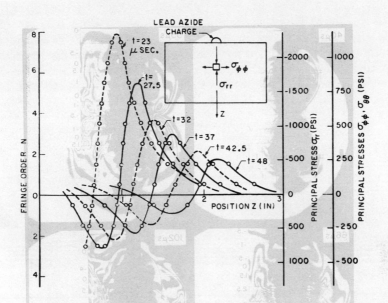

Fig. 13. Fringe Orders and Principal Stresses Associated with the S Wave on the Axis of Symmetry.

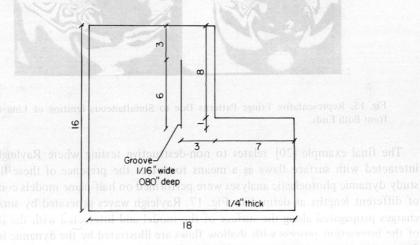

Fig. 14. Bench Face Model.

more significant interaction which produces extremely high boundary stresses when the two S waves interact. The distribution of the stresses as a function of position along the bench face showing the tensile stresses which are so important in producing rock fragmentation are given in Fig. 16.

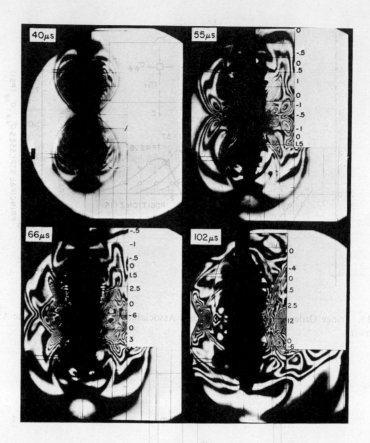

Fig. 15. Representative Fringe Patterns Due to Simultaneous Ignition of Line Charge from Both Ends.

The final example [20] relates to non-destructive testing where Rayleigh waves are interacted with surface flaws as a means to detect the presence of these flaws. In this study dynamic photoelastic analyses were performed on half-plane models containing slits of different lengths as defined in Fig. 17. Rayleigh waves generated by small explosive charges propagated along the surface of the model and interacted with the slit. Features of the interaction process with shallow flaws are illustrated by the dynamic isochromatic fringe patterns given in Fig. 18. For these shallow flaws it is evident that a significant portion of the R wave is transmitted in all three instances. There is a marked range in the reflected and scattered waves as the ratio of the flaw depth to wave length d/λ is increased. For $d/\lambda = 1/32$ there is no visible reflected Rayleigh wave RR and the scattered RS is just discernable. As d/λ increases to 1/16 and 1/8 the RR wave becomes clearly evident and significant amounts of energy are scattered by means of RP (Rayleigh induced dilatational wave), RS (Rayleigh generated shear wave) and RPS (shear wave induced by the RP wave).

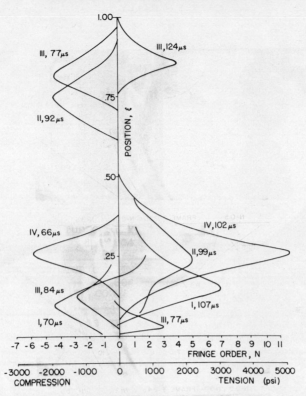

Fig. 16. Tangential Stresses as a Function of Position and Time.

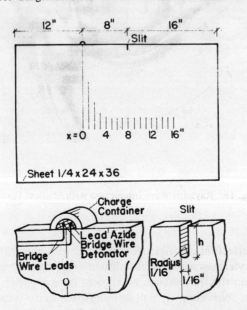

Fig. 17. Model with Slit Representing a Surface Flaw.

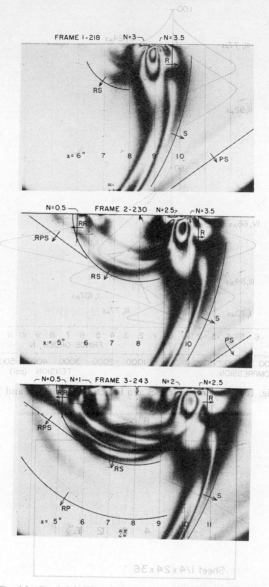

Fig. 18. Rayleigh Wave Interaction with Shallow Flaws.

Transmission and reflection coefficients associated with the transmitted and reflected components of the Rayleigh wave were established from the fringe patterns as shown in Fig. 19. It is evident that the transmission coefficient A_t approaches zero as d/λ increases; however, the reflection coefficient A_r will not approach unity because of the energy lost due to the generation of body waves such as the RP, RS, and RPS waves.

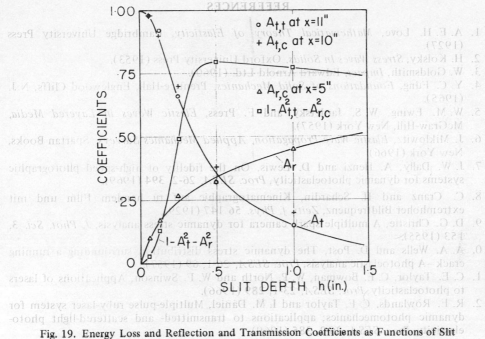

Fig. 19. Energy Loss and Reflection and Transmission Coefficients as Functions of Slit Depth d.

CONCLUSIONS

Dynamic photoelasticity has been developed to a point where it can be applied to a wide variety of problems arising in the fields of structures, mechanics, geophysics, mining technology and non-destructive testing. High speed recording methods suitable for dynamic photoelasticity are currently available and relatively easy to employ. Future developments with pulsed ruby lasers will provide still further improvements in the recording procedures.

More experimental effort is necessary in the development of controlled loading methods, model materials and scaling laws. This future experimental activity coupled with new theoretical developments should enable us to make signficant progress in problems of elastodynamics.

ACKNOWLEDGMENTS

The authors work in dynamic photoelasticity over the past six years has been sponsored by the National Science Foundation. The author wishes to express his appreciation to Drs. John Ide and Michael Gaus for their support and encouragement.

REFERENCES

1. A. E. H. Love, *Mathematical Theory of Elasticity*, Cambridge University Press (1927).
2. H. Kolsky, *Stress Waves In Solids*, Oxford University Press (1953).
3. W. Goldsmith, *Impact*, Edward Arnold Ltd. (1960).
4. Y. C. Fung, *Foundations of Solid Mechanics*, Prentice-Hall, Englewood Cliffs, N.J. (1965).
5. W. M. Ewing, W. S. Jardetsky and F. Press, *Elastic Waves in Layered Media*, McGraw-Hill, New York (1957).
6. J. Miklowitz, *Elastic Wave Propagation, Applied Mechanics Surveys*, Spartan Books, New York (1966).
7. J. W. Dally, A. Henzi and D. Lewis, On the fidelity of high-speed photographic systems for dynamic photoelasticity, *Proc. SESA*, 26-2, **394** (1969).
8. C. Cranz and H. Schardin, Kinematographic auf ru hendem Film und mit extremhoher Bildfrequenz, *Zeits. f. Phys.* **56** 147 (1929).
9. D. G. Christie, A multiple spark camera for dynamic stress analysis, *J. Phot. Sci.* **3**, 153 (1955).
10. A. A. Wells and D. Post, The dynamic stress distribution surrounding a running crack—A photoelastic analysis, *Proc. SESA*, 26-1, **69** (1957).
11. C. E. Taylor, C. E. Bowman, W. P. North and W. F. Swinson, Applications of lasers to photoelasticity, *Proc. SESA*, 23-1, **289** (1966).
12. R. E. Rowlands, C. E. Taylor and I. M. Daniel, Multiple-pulse ruby-laser system for dynamic photomechanics; applications to transmitted- and scattered-light photoelasticity, *Proc. SESA*, 26-2, **385** (1969).
13. M. M. Frocht, Studies in dynamic photoelasticity with special emphasis on the stress optic law, *International Symposium on Stress Wave Propagation in Materials*, Interscience Publishers, 91 (1960).
14. J. W. Dally, W. F. Riley and A. J. Durelli, A photoelastic approach to transient stress problems emphasizing law modulus material, *J. Appl. Mech.* **26**, 4, (1959).
15. A. B. J. Clark, Static and dynamic calibration of a photoelastic model material -CR-39, *Proc. SESA*, **14**, 1, 195 (1956).
16. S. A. Thau and J. W. Dally, Subsurface characteristics of the Rayleigh Wave. *Int. J. Engrg. Sci.* **7**, 37 (1969).
17. W. F. Riley and J. W. Dally, A photoelastic analysis of stress wave propagation in a layered model, *Geophysics*, **31**, 5, 881 (1966).
18. J. W. Dally and W. F. Riley, Initial studies in three-dimensional dynamic photoelasticity, *J. Appl. Mech*, **34**, 2, 405 (1967).
19. H. W. Reinhardt and J. W. Dally, Dynamic photoelastic investigation of stress wave interaction with a bench face, *Trans SME/AIME*, 250, 1, 35 (1971).
20. H. W. Reinhardt and J. W. Dally, Some characteristics of Rayleigh Wave interaction with surface flaws, *Materials Evaluation*, 28, 10, 213 (1970).

SOME RECENT EXPERIMENTAL INVESTIGATIONS IN STRESS-WAVE PROPAGATION AND FRACTURE

H. KOLSKY

Division of Applied Mathematics
Brown University, Providence, Rhode Island 02912

Abstract—This paper describes preliminary results of three distinct experimental programs at present being carried out by the author and his colleagues at Brown University. These are:

(a) Rubber is found to have a stress-strain curve which is concave upwards for tensile strains, i.e., the tangent modulus increases with increasing strain and also the attentuation becomes small at high strains. As a result of this, tensile pulses of reasonably large amplitudes are found to develop shock fronts when they are propagated along rubber filaments which are already subjected to large tensile pre-strains. Experiments illustrating this effect are described, and a discussion of the conditions for such shock-wave formation is given.

(b) When a fracture occurs in a brittle solid, there is a sudden local change in stress and a compressive stress pulse is propagated away from the crack. Experiments are described concerning the generation of such pulses for Hertzian fractures, simple tensile fractures, and flexural fractures. The theoretically predicted pulse shapes are compared with those observed experimentally.

(c) When a longitudinal stress pulse is reflected at normal incidence at the boundary between two solids of different properties, two pulses are, in general, produced. A reflected pulse is sent back, and a transmitted pulse is propagated in the second medium. When one or both of the solids are viscoelastic, both the transmitted and reflected pulses have shapes which differ from that of the incident pulse. In particular, if the values of the acoustic impedances cross at some frequency ω, called the *cross-over frequency*, dc pulses of duration comparable with $1/\omega$ are reflected as s-shaped ones. Experiments which show this phenomenon are described.

INTRODUCTION

This paper is an outline of some of the experimental work at present being carried out in Brown University by the author and his colleagues, and three distinct problems in the field of stress wave propagation are considered. The progress so far made on each problem is described, and the plans for the extension of the work are discussed. The three problems are:

(a) A study of the formation of shock fronts in tensile pulses propagated along filaments of natural rubber which have been pre-stretched quasi-statically. A property of rubber, and of many other polymers is that the value of the "tangent elastic Young's modulus," $(d\sigma/de)$, increases with increasing longitudinal tensile strain so that one might expect a large tensile pulse to develop a shock front as it travels along a filament or a thin rod of the material. This effect is, however, normally masked by the very high attenuation of these materials to the high frequency components of the pulse. Fortunately in rubber, this attenuation falls off very markedly when the material is put in a state of high tensile pre-strain. Thus, if a rubber specimen is pre-strained in a quasi-static manner, the

327

attenuation of incremental tensile pulses is found to be very much smaller than it is with an unstrained specimen. The large pre-strain thus achieves two separate functions, it brings the material well into the non-linear region of its stress-strain curve, so that the magnitude of an incremental pulse which might be expected to develop a shock front within a reasonable distance of travel can be comparatively small, and it at the same time, brings the rubber into a region where attenuation is low, so that the tendency to shock wave formation is not inhibited.

(b) The second investigation is concerned with the stress pulses generated when a brittle solid fractures. When such a fracture occurs, the stress around the crack suddenly changes from the large tensile value which it had just before fracture commenced, to zero. Consequently, a compressive pulse is generated, and travels through the medium. Such pulses can be produced by Hertzian fractures in glass blocks, and also by the failure of glass rods in simple tension or in flexure. The shapes of the pulses generated for all these cases have been measured and are compared with those predicted theoretically. In order to make such predictions, a model of the fracture process has to be hypothesised. For the problem of the simple tensile fracture of a circular rod this was found to be fairly straightforward. For Hertzian impacts the elastic wave patterns set up are quite complicated even in the absence of fractures, and the analysis of the fracture waves produced leads to even more complicated analysis. For the fractures produced in the flexural loading of bars, the problem is complicated by the fact that after the wave reaches the neutral axis it would, in the absence of unloading waves, reach a region of compression where it might be expected to stop propagating. Its propagation into the initially compressed half of the bar results entirely from the effect of tensile waves of unloading, and the analysis has so far had to be of only a qualitative nature.

(c) When a stress pulse approaches a boundary between two media normally, the boundary conditions require that in general, two pulses be generated. One of these pulses, (the reflected pulse) is propagated back into the first medium, while the other pulse (the transmitted pulse) is propagated forward into the second medium. When both media are elastic the transmitted pulse is of the same shape and the same sign as the incident pulse, while the reflected pulse is always of the same shape, but is of the same sign only when the second medium has a higher acoustic impedance than the first. If it has a lower acoustic impedance there is a change of sign on reflection, and an incident tensile pulse produces a reflected compressional pulse and vice versa. The amplitudes of the transmitted and reflected pulses depend on the relative values of the acoustic impedances of the two media, and if these have the same value, the reflected pulse is of zero amplitude, i.e., no pulse is reflected.

When one or both of the media are viscoelastic the situation is no longer so simple, the shapes of both the reflected and transmitted pulses differ from each other, and also from that of the incident pulse. In order to determine these shapes, the incident pulse must be expressed as a Fourier Integral, and the reflection of each individual Fourier component must be considered separately. (In practice the Fourier Integral is approximated to by means of a Fourier sum with a finite number of terms.) The acoustic impedances of the viscoelastic media are complex and depend on frequency, and thus the relative amplitudes of the reflected and transmitted components also vary with frequency, further the reflected and transmitted components are no longer in phase with each other or with the phase of the incident component. The result of all this is that the shapes of the two pulses can be calculated only by what, without the aid of a digital computer, would be a laborious numerical Fourier Synthesis.

It is found, in practice, that when the viscoelastic losses are reasonably small and the two media have acoustic impedances which are of widely different magnitudes, the three pulses have shapes which are, in fact, closely similar. When, however, the acoustic impedances have values which are close to each other, so that the difference between them changes markedly over the frequency range comprising the relevant Fourier spectrum of the pulse, the difference between the shape of the reflected pulse and the shape of the incident one can be very large indeed. This effect is most marked when there is a *cross-over frequency*, i.e., a frequency at which the two media have acoustic impedances of equal magnitude, and this frequency lies in the middle of the relevant Fourier spectrum. Under these conditions the components with frequencies on one side of this cross-over frequency are reflected as if the second medium had a higher acoustic impedance than the first, while the components with frequencies on the other side are reflected as if they were entering a medium of lower acoustic impedance. As shown by S. S. Lee and the author, the result of such a situation is that a d.c. pulse is reflected as an s-shaped one.

The point on the pulse where the 's' cuts the time axis depends on the value of the cross-over frequency and very recently M. Moffett working here at Brown has shown experimentally how a d-c pulse can change shape so that its reflection turns from a d-c pulse of compression to a d-c pulse of tension through a set of intermediate s-shaped intermediate profiles. Moffett demonstrated this effect by propagating longitudinal stress pulses along a system of two rods, (polystyrene and polyvinyl chloride) in end to end contact, and varying the temperature so that the cross-over frequency traversed the relevant Fourier spectrum of the pulse.

TENSILE SHOCK WAVES IN RUBBER

In a non-dispersive medium each section of a plane stress pulse is propagated at a velocity c+v. c is here the value of $[S/\rho^{1/2}]$ for the particular amplitude of the section, and S is the value of the *tangent modulus* $d\sigma/de$, for that section of the pulse. v is the local particle velocity produced by the disturbance, and in fluids it is generally the fact that v is comparable with the velocity c, which results in shock fronts being produced in compression.

In contrast, for solids it is the non-linearity of the stress-strain relation, which results in S increasing with increasing pressure, and this produces such shock fronts. Thus the value of the quantity c becomes greater at high compressive stresses, and this results in the velocity of high amplitude waves exceeding those of low amplitude, so that shock fronts develop. This effect has been utilized extensively by a number of workers in the study of (pvT) relations of solids at high pressures, and a review of the field is given by

Duvall [1]. No further discussion of this type of shock wave is therefore given here.

There appears hitherto to be no record of experimental work on the production of shock waves in tension, although some theoretical work on plastic waves by Lee and Tupper [2] predicts the possibility of the formation of such shock fronts in steel specimens, as a result of the increase of $d\sigma/de$ with strain in the plastic region. This increase would here be a result of the strain hardening of the steel specimen.

Some years ago K. W. Hillier and the author [3] developed an experimental technique for measuring the dynamic tangent moduli of solids by propagating sound waves along filaments of rubbers and polymers, while these filaments were being stretched quasi-statically. The basis of this technique is shown schematically in Fig. 1. An infinite train of sound waves, of sonic or ultrasonic frequency, is propagated along a filament of

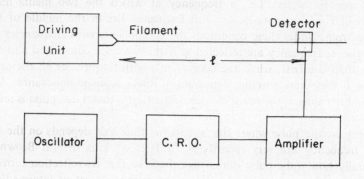

Fig. 1. Apparatus for wave propagation measurements in filaments.

the material, and the phase and amplitude of the vibration is measured as a function of the distance from the source [the sound source can be either a loudspeaker, a magnetostrictive rod or a quartz crystal]. From these measurements the velocity and the attentuation for sinusoidal waves at that frequency can be determined, and while the measurements are being made, the filament can be slowly stretched. Thus the phase velocity and the attenuation can be determined as functions of the longitudinal strain in the specimen. Hillier [4] later used this method for studying a number of natural and synthetic rubbers, and Fig. 2 shows a plot of the results he observed for natural rubber at a temperature of 10°C.

It may be seen that as the tensile strain increases, two phenomena occur. First, the velocity of propagation increases very markedly, e.g., at a strain of 300% it has increased to a value more than six times the value for unstretched rubber, and is still continuing to increase rapidly with increasing pre-strain. Secondly, the attenuation of the 4 Kc sound waves, which is very high for unstretched rubber 0.43 nepers/cm, has fallen to a very tiny fraction of this value, 0.01 nepers/cm, when the pre-strain has reached 300%. These two facts make it highly probable that when a tensile pulse of moderate amplitude is propagated along a rubber specimen which has already been pre-stretched by a large amount, the pulse will tend to develop a shock front. Thus, we have first a situation where the dynamic tangent modulus $d\sigma/d\epsilon$ is increasing rapidly with increasing tensile strain, and secondly a medium which no longer smooths sharp pulses by attenuating the high frequency components present in them.

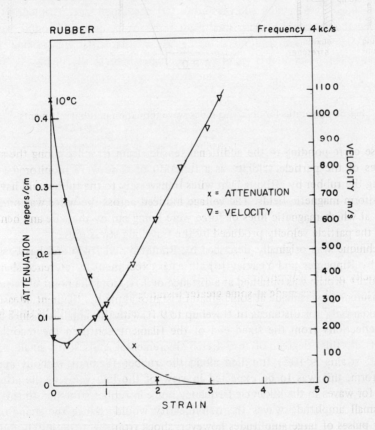

Fig. 2. Velocity of propagation and attenuation of waves for natural rubber as a function of longitudinal strain.

In order to investigate if the development of such shock fronts could be observed experimentally, the arrangement shown schematically in Fig. 3 was set up by the author. The rubber used was a length of vulcanized gum stock which was of square cross-section (½″ x ½″). Thus in a typical set of experiments the specimen was prestretched by means of an electrically driven winch to five times its initial length, so that the pre-strain was 400%, the total length of the specimen was then 13 ft. A length of 10 inches of this prestretched rubber was then stretched even further, by means of a small hand winch, and the additional stretch was maintained by means of a piece of piano wire. The incremental tensile pulse was generated by volatilising the wire electrically. When this occurred a

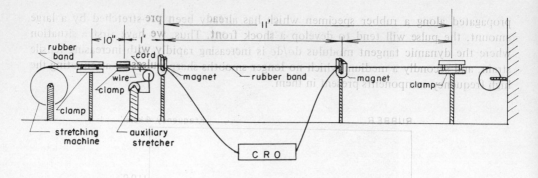

Fig. 3. Apparatus for observing shock wave formation in rubber filaments.

tensile pulse corresponding to the additional tensile strain travelled along the specimen. The profiles of the particle velocity as a function of time were monitored at various points along the rubber by affixing light wires transversely to the rubber and having these wires in uniform magnetic fields. The voltage induced across the wires was proportional to the rate at which magnetic lines of force were being cut by the wire and hence gave a measure of the particle velocity produced by the travelling stress pulse.

This technique was originally described by Ramberg and Irwin and has been further developed by Ripperger and Yeakley [6]. In each experiment, a reference shape of the particle velocity profile was obtained at a distance of 1 ft. from the point of release, and a second measurement was made at some greater distance along the filament. Measurements could be taken only for distances of travel up to 9 ft. with this apparatus since at greater distances reflections from the fixed end of the filament confused the records. It was found that at this degree of pre-stretch incremental pulses of small amplitude (incremental strains < 10%) travelled along the rubber filament without appreciable change in form, this was to be expected in view of the low value of the attentuation coefficient for waves in the kilocycle frequency range in rubber stretched to this amount. For such small amplitude waves the non-linearity would change the shape only very slowly. For pulses of large amplitudes however, shock fronts were definitely found to be produced. Thus where the 10″ section of the prestretched rubber was further extended to a length of 11″, so that the strain in this section was now 450%, and the experiment was carried out by volatilising the retaining wire, an incremental pulse of 50% strain travelled along the filament. The photographic records of the observed particle velocity profiles for this pulse at distances of travel of 1, 3, 5, 7 and 9 ft. along the filament are shown in Fig. 4, and it may be seen how a shock front is steadily built up as the tensile pulse travels along the filament.

The technique affords an extremely convenient method of observing the formation of shock fronts in materials when different degrees of attentuation of stress waves are present, since on the one hand specimens of one type of rubber can be pre-stretched to different extents before performing experiments with them (see Fig. 2), and on the other, different materials with widely different attenuation coefficients can be employed. The

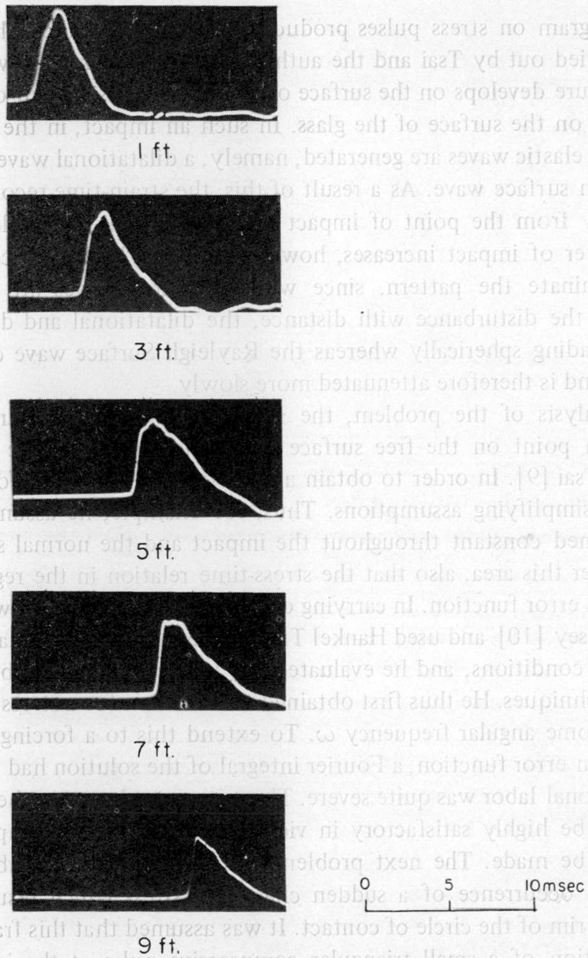

1 ft.

3 ft.

5 ft.

7 ft.

9 ft.

Fig. 4. Oscilloscope records of pulses in stretched rubber.

analytic treatment of shock wave formation in the presence of viscoelastic attenuation is an extremely difficult one, and experimental data of the observed effects would thus be particularly valuable. Another line of work which may prove interesting is the study of the propagation of compressive pulses in pre-stretched filaments. Here one would expect to find 'shock tails' rather than shock fronts and the apparatus as it stands could be used for this purpose. Very preliminary results of the work on shock fronts has been reported by the author [7], and work is at present continuing at Brown, where it is hoped to follow the lines described above, and to obtain more exactly quantitative data on the phenomena.

STRESS PULSES PRODUCED BY BRITTLE FRACTURES

In the program on stress pulses produced by brittle fractures, the first experimental work was carried out by Tsai and the author [8] on the elastic waves generated when a Hertzian fracture develops on the surface of a glass block as a result of the normal impact of a steel ball on the surface of the glass. In such an impact, in the absence of fracture, three types of elastic waves are generated, namely, a dilatational wave, a distortional wave and a Rayleigh surface wave. As a result of this, the strain-time records on the surface of the glass away from the point of impact are very complicated indeed. As the distance from the center of impact increases, however, the Rayleigh surface wave contribution tends to dominate the pattern, since while all three waves are attentuated by the divergence of the disturbance with distance, the dilatational and distortional contributions are spreading spherically whereas the Rayleigh Surface wave contribution spreads cylindrically and is therefore attenuated more slowly.

In the analysis of the problem, the Lamb solution for a transient force applied normally to a point on the free surface of a semi-infinite elastic half-space has been extended by Tsai [9]. In order to obtain a numerical solution Tsai found he had to make a number of simplifying assumptions. Thus, for example, he assumed that the area of contact remained constant throughout the impact and the normal stress was uniformly distributed over this area, also that the stress-time relation in the region of contact is in the form of an error function. In carrying out the analysis, Tsai followed the treatment of Miller and Pursey [10] and used Hankel Transforms of both the equations of motion and the boundary conditions, and he evaluated the inverse transforms by means of contour integration techniques. He thus first obtained the solution for a stress varying sinusoidally with time at some angular frequency ω. To extend this to a forcing function which had the shape of an error function, a Fourier integral of the solution had to be evaluated, and the computational labor was quite severe. The agreement between theory and experiment was found to be highly satisfactory in view of the number of simplifying assumptions which had to be made. The next problem was to treat the probable distortion of the record by the occurrence of a sudden change in stress which resulted from a tensile fracture at the rim of the circle of contact. It was assumed that this fracture corresponded to the generation of a small triangular compressive pulse at the instant of maximum compressive stress and Fig. 5 shows how such an additional pulse will propagate over the surface of the glass block. The figure shows the pulse shape as it is assumed to be at the point of generation and at two distances remote from the center of the area of contact. The observed oscillograph records when fractures did occur were found to conform closely to the shapes shown here.

The next study of waves produced by brittle fractures was carried out by Phillips [11] who first investigated the very much simpler problem of a glass rod subjected to simple tension. A theoretical treatment of this problem was given some years ago by Miklowitz [12]. Under these conditions a tensile break will eventually occur, so that the stress over the fracture surface will decrease from its value at the commencement of fracture to zero in a very few microseconds and generate compressive pulses which will travel away from the fracture surfaces in both directions at the velocity of elastic waves.

Were the fracture to originate at the center of the rod and spread outwards radially this would be the entire effect. Since, however, the fracture starts at the surface of the rod and spreads cylindrically from the point of initiation, the stress distribution has a net moment about the axis of the rod and consequently also sets up flexural waves in the rod during the fracture process.

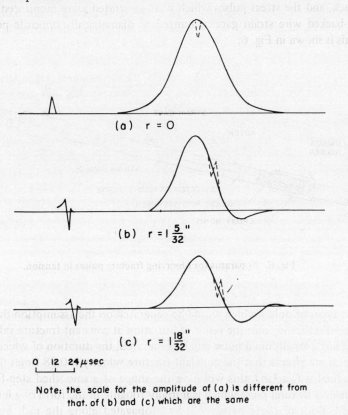

(a) r = 0

(b) r = $1\frac{5}{32}$"

(c) r = $1\frac{18}{32}$"

0 12 24 μsec

Note: the scale for the amplitude of (a) is different from
that of (b) and (c) which are the same

Fig. 5. Pulses produced by Hertzian Impact and Fracture.

The stress on the fractured part of the cross-section is clearly zero, but some assumption must be made about the stress distribution over the unfractured sections of the cross-section. Here two extreme assumptions are possible.

The first of these is that, since the fracture is travelling extremely rapidly across the section, the stress does not change during the fracture process. The other extreme assumption is that the stress completely readjusts itself to the value one would expect in quasi-static loading. The second assumption leads to rather unrealistic results, so that Phillips, like Miklowitz, worked on the basis of a stress distribution which was either zero or retained the value it had just before fracture commenced.

Since experimentally it was important to know where the fracture was going to occur, a small notch was put on the glass rod with a file. The rods which were ½″ (nominal) diameter were stretched in an Instron testing machine which extended them slowly until fracture occurred. The cathode-ray oscilloscope used for recording, was triggered by the breaking of a thin coat of silver paint which was painted on the glass in the expected path of the crack, and the stress pulses which were generated were monitored by two B-L-H C-8 paper-backed wire strain gages mounted at diametrically opposite positions on the rod. All this is shown in Fig. 6.

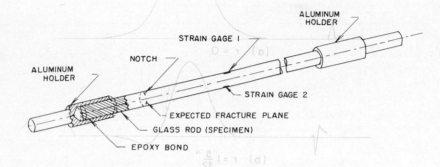

Fig. 6. Apparatus for observing fracture pulses in tension.

The two types of pulses which would be generated on the assumption that the fracture propagates cylindrically from the point of initiation at constant fracture velocity is shown in Fig. 7. Thus a longitudinal pulse will be generated, the duration of which will be about 4 microsec. if we assume that the constant fracture velocity is 0.38 times the rod velocity for longitudinal waves, and this will have the shape of a smoothed step-function, while simultaneously a flexural pulse which is initially roughly in the form of a letter U will also be produced. Both of these pulses will be propagated along the rod, but whereas the longitudinal one will change in form only very slightly, as a result of radial inertia effects, the flexural pulse will change in shape rapidly because of the inherent dispersive nature of flexural wave propagation.

The oscillograph records from the two strain gages will consequently both show records with shapes very similar to that of the longitudinal pulse in Fig. 7, which will arrive first, these will be followed by the disturbances caused by the flexural pulse and the latter will change rapidly in shape with increasing distances of travel. In predicting the wave shapes, it would have been desirable to use the exact three-dimensional Pochhammer-Chree treatment for flexural waves. This, however, led to impossibly difficult computational problems, so that Phillips used instead the more simple Timoshenko treatment. Although this treatment does not give more than two modes of propagation (when in fact an infinite number can be propagated), the predictions for the first mode are extremely close to those given by the exact theory (see Davies [13]).

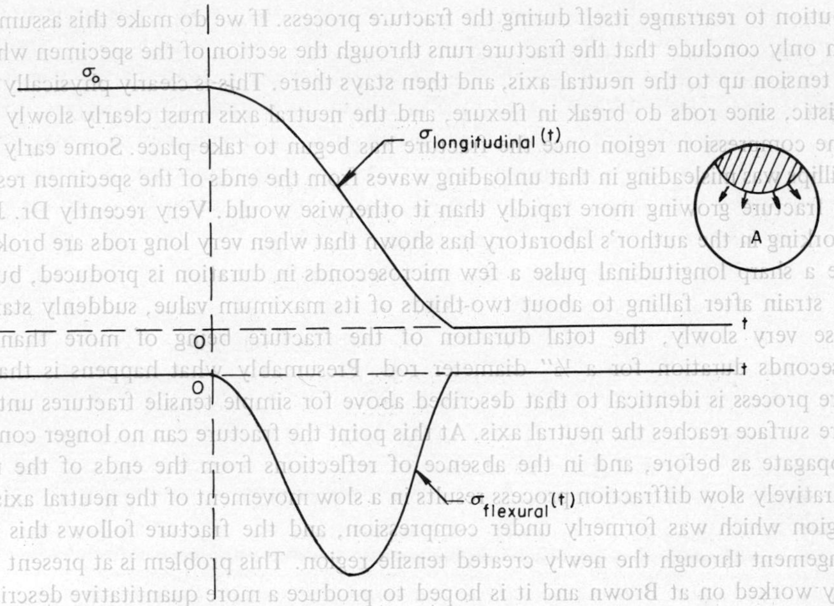

Fig. 7. Expected pulse shapes from tensile fracture.

Figure 8 shows a comparison between the strain-time record observed experimentally on one of the strain gages after a distance of travel of 6.2 in. from the fracture surface, and a number of points calculated on the basis of the theory outlined above. It may be seen that the agreement is highly satisfactory particularly in view of the many simplifying assumptions which have had to be made.

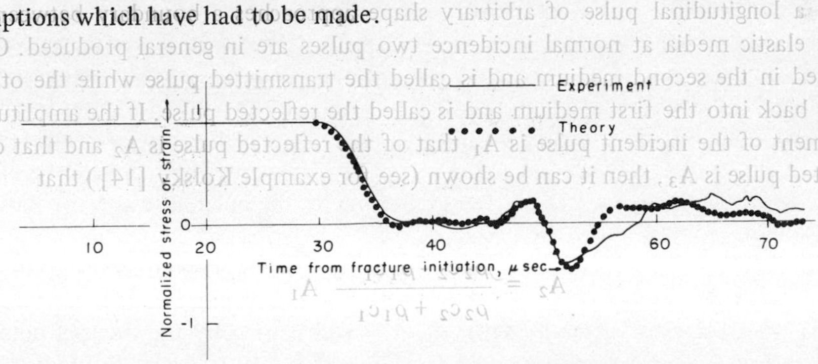

Fig. 8. Comparison between theoretical and experimental strain-time records for tensile fracture.

The more difficult problem is that of the stress pulses generated when a rod breaks in flexure. Here we can no longer use the assumption which worked so well for tensile fractures, namely, that the fracture grows so quickly that there is no time for the stress distribution to rearrange itself during the fracture process. If we do make this assumption we can only conclude that the fracture runs through the section of the specimen which is under tension up to the neutral axis, and then stays there. This is clearly physically quite unrealistic, since rods do break in flexure, and the neutral axis must clearly slowly move into the compression region once the fracture has begun to take place. Some early work by Phillips was misleading in that unloading waves from the ends of the specimen resulted in the fracture growing more rapidly than it otherwise would. Very recently Dr. James Lee working in the author's laboratory has shown that when very long rods are broken in flexure a sharp longitudinal pulse a few microseconds in duration is produced, but the tensile strain after falling to about two-thirds of its maximum value, suddenly starts to decrease very slowly, the total duration of the fracture being of more than 100 microseconds duration for a ½″ diameter rod. Presumably what happens is that the fracture process is identical to that described above for simple tensile fractures until the fracture surface reaches the neutral axis. At this point the fracture can no longer continue to propagate as before, and in the absence of reflections from the ends of the rod a comparatively slow diffraction process results in a slow movement of the neutral axis into the region which was formerly under compression, and the fracture follows this stress rearrangement through the newly created tensile region. This problem is at present being actively worked on at Brown and it is hoped to produce a more quantitative description of exactly what is occurring.

REFLECTION OF LONGITUDINAL PULSE AT NORMAL INCIDENCE AT A BOUNDARY BETWEEN TWO VISCOELASTIC SOLIDS WHEN 'CROSS-OVER EFFECT' OCCURS

When a longitudinal pulse of arbitrary shape approaches a boundary between two different elastic media at normal incidence two pulses are in general produced. One is propagated in the second medium and is called the transmitted pulse while the other is reflected back into the first medium and is called the reflected pulse. If the amplitude of displacement of the incident pulse is A_1 that of the reflected pulse is A_2 and that of the transmitted pulse is A_3, then it can be shown (see for example Kolsky [14]) that

$$A_2 = \frac{\rho_2 c_2 - \rho_1 c_1}{\rho_2 c_2 + \rho_1 c_1} A_1 \tag{1}$$

while

$$A_3 = \frac{2 \rho_1 c_1}{\rho_2 c_2 - \rho_1 c_1} A_1 \tag{2}$$

Thus if $\rho_2 c_2 = \rho_1 c_1$ then $A_2 = 0$ and the two media are said to have the same *acoustic impedance*. Acoustic impedance which is the product ρc is generally denoted by Z so that (1) can be written as $A_2 = A_1 (Z_2 - Z_1) / (Z_2 + Z_1)$. For elastic media Z is independent of frequency and the shapes of the incident reflected and transmitted pulses are all alike, although of course there is a change in sign of A_2 when $Z_1 > Z_2$. When one or both of the two media is viscoelastic, however, the problem becomes considerably more complicated since the acoustic impedances are now complex quantities and functions of the frequency for a sinusoidal wave. Thus for a pulse of arbitrary shape each Fourier component is reflected differently and a Fourier sum must be evaluated to determine the shapes of the reflected and transmitted pulses, these in general will differ from that of the shape of the incident pulse.

We will consider an infinite train of sinusoidal waves of angular frequency ω. Then we have that if the longitudinal displacements are u_1, u_2 and u_3,

$$u_1 = A_1 \exp i(\omega t - k_1 x)$$

$$u_2 = A_2 \exp i(\omega t + k_1 x)$$

and

$$u_3 = A_3 \exp i(\omega t - k_2 x)$$

the A's and the k's are here, in general, complex. Let us assume that the boundary is at $x = 0$, then the boundary conditions are $u_1 + u_2 = u_3$ so that

$$A_1 + A_2 = A_3$$

(continuity of displacement) and $\sigma_1 + \sigma_2 = \sigma_3$ where each σ corresponds to the normal stress produced by the stress pulse.

Now whereas in an elastic medium $k = \omega/c$ where c is the elastic wave velocity, for a viscoelastic medium $k = \omega/c - i\alpha$. c is here the phase velocity of the wave and is equal to $[E^*/\rho]^{1/2} \sec(\delta/2)$, E^* being the amplitude of the complex modulus and ρ the density, while δ is the angle by which the strain lags behind the stress. α is the attenuation coefficient and is given by $\alpha = \omega \tan(\delta/2)/c$. Similarly whereas for an elastic medium $\sigma = E \dfrac{\partial u}{\partial x}$ for a viscoelastic one E is complex and a function of frequency and may be written as $E = E_1 + iE_2$ (E_2/E_1 is here equal to $\tan \delta$ while $E^* = (E_1^2 + E_2^2)^{1/2}$). As in the case of an elastic solid we have for a viscoelastic solid $Z(\omega) = \sigma/v$ where v is the amplitude of the particle velocity associated with a wave train of frequency ω. Z is however now complex but is as before equal to Ek/ω where $E = E_1 + iE_2$. Now, whereas the reflection coefficient R which is the ratio A_2/A_1 is a simple real constant for elastic solids it is now also complex but we still have the same relation $R = [Z_2(\omega) - Z_1(\omega)] / [Z_2(\omega) + Z_1(\omega)]$. The formulation above is based on that of Moffett [15] who has recently carried out some experimental work on these phenomena, but essentially the same treatment was used by S. S. Lee and the author [16] in a theoretical study a few years ago where they were considering this problem.

In carrying out the computations Kolsky and Lee [16] used an approximate relations for E^* as a function of ω. This relation applies when δ is assumed to be constant, i.e., independent of frequency, the relation is

$$E(\omega) \cong E(\omega_o) \, [1+[2 \tan \delta/\pi] \ell n(\omega/\omega_o)]$$

this leads to

$$c(\omega) \cong c(\omega_o) \, [1+ [\tan \delta/\pi] \ell n(\omega/\omega_o)]$$

$E(\omega_o)$ and $c(\omega_o)$ are the values of E^* and c at some reference frequency ω_o. This relation is found to be very satisfactory for materials at temperatures remote from the glass-rubber transition.

In this work it was shown that when the values of the acoustic impedances of the two media are widely different, and the viscoelastic effects are comparatively small, the incident pulse, the reflected pulse and the transmitted pulse have all much the same shape. When however the values of Z lie close to each other and are changing with ω at vastly differing rates the shapes can differ widely. The most marked effect occurs when the values of Z cross over at some frequency ω_c, so that $Z_2 - Z_1$ has one sign on one side of this cross-over frequency and the opposite sign on the other side of ω_c; this has been called the *cross-over effect*. It can of course equally occur in reflections between an elastic and a viscoelastic medium if the values of Z are appropriate. Figure 9 shows how a d-c pulse travelling from an elastic solid (of appropriate elastic modulus) into a rod of polymethylmethacrylate is reflected as an s-shaped pulse as a result of the cross-over effect.

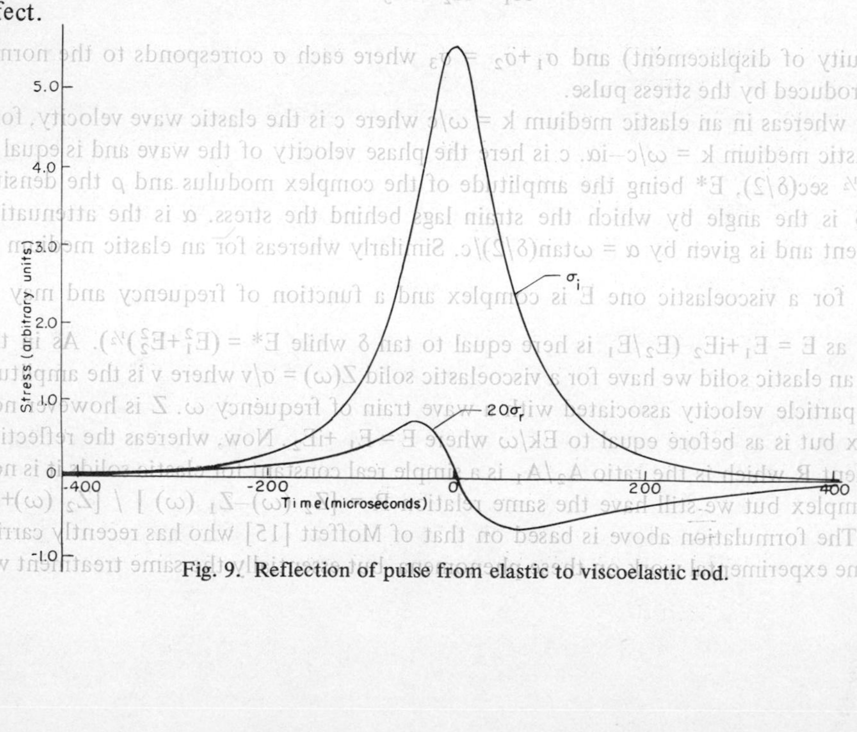

Fig. 9. Reflection of pulse from elastic to viscoelastic rod.

Figure 10 shows the complementary problem where a pulse which is travelling in a viscoelastic rod (p.m.m.) is reflected at the junction with a polystyrene rod [this is almost elastic]. In this figure two different values, slightly apart [(a) = 2700 m/sec and (b) = 2660 m/sec] were assumed for the 'elastic wave velocity' in polystyrene and this of course changed the value of the cross-over frequency. The resulting change, of the point at which the reflected pulse cut the stress axis can be seen in the figure. In comparing

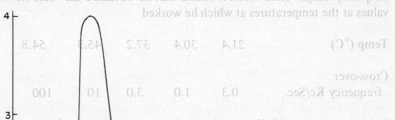

Fig. 10. Reflection of pulse from viscoelastic to elastic rod for two 'cross-over' frequencies.

Figs. 9 and 10 two points should be noted, first that whereas in Fig. 9 the tensile region is followed by a compressive one, in Fig. 10 the reverse is the case. Second that in Fig. 9, a symmetrical incident pulse has been assumed, while in Fig. 10 since the incident pulse was travelling in a viscoelastic medium it was considered more realistic to assume asymmetry, and the pulse is in fact the shape that a sharp pulse would assume after travelling along a 120 cm length of this material.

Recently Moffett has carried out some experiments to verify this effect experimentally. He used rods of polystyrene and unplasticised polyvinyl chloride and for these two materials a cross-over frequency occurs in the acoustic range at ordinary temperatures. By varying the temperature the value of the cross-over frequency can be made to span a wide frequency range. Thus Moffett found that he obtained the 'best fit' with the following values at the temperatures at which he worked

Temp (°C)	21.4	30.4	37.2	45.3	54.8
Cross-over frequency Kc/Sec.	0.3	1.0	3.0	10	100

In his experiments Moffett produced the incident pulse by firing a projectile in the form of a polystyrene cylindrical pellet (½″ diameter 5″ length) to hit a polystyrene rod of the same diameter normally. The other end of the polystyrene rod which was ground flat, was bonded to a rod of unplasticised PVC by means of a grease joint. Both rods and the junction were kept in a thermostatically controlled oven, the temperature of which could be varied. Only about 2 inches of the polystyrene rod was allowed to protrude from the oven. To prevent the generation of flexural pulses the end of the pellet was rounded to a 6″ radius.

In order to monitor the incident and reflected pulses strain gages (BLH C-7) were cemented onto the polystyrene rod. It was found that as the temperature increased the reflected strain pulse changed from a rectangular pulse of the same sign as the incident one to a pulse of opposite sign in a series of roughly s-shaped intermediaries as predicted in the curves shown in Figs. 9 and 10. The main differences lay in the fact that the initial pulse was rectangular in shape rather than u-shaped so that the intermediates were somewhat sharper than those shown here in the figures. A preliminary account of this work has been given at the recent meeting of the Acoustical Society of America [16] and Moffett has just completed a report [15] which it is hoped will soon be published.

ACKNOWLEDGMENTS

The author wants to express his thanks to Mrs. E. Fonseca for typing the manuscript, and to Miss E. Addison for preparing the drawings. It is also a pleasure to acknowledge the support given by the Army Research Office at Durham under Contract No. DA-31-124-ARO(D)-358 with Brown University.

REFERENCES

1. G. E. Duvall, *Appl. Mech. Rev.*, **15**, 849 (1962).
2. E. H. Lee and S. J. Tupper, *Appl. Mech.*, **21**, 63 (1954).

3. K. W. Hillier and H. Kolsky, *Proc. Phys. Soc.*, B **62**, 111 (1949).
4. K. W. Hillier, *Trans. Inst. Rubber Industry*, **26**, 64 (1950).
5. W. Ramberg and L. K. Irwin, *Proc. 9th Int. Congr. Appl. Mech.*, **8**, 480 (1957).
6. E. A. Ripperger and L. M. Yeakley, *Exp. Mech.*, **3**, 47 (1963).
7. H. Kolsky, *Nature*, **224**, 1301 (1969).
8. Y. M. Tsai and H. Kolsky, *J. Mech. Phys. Solids*, **15**, 263 (1967).
9. Y. M. Tsai, *Brown Univ. Report NSF-GP-2010/4*, (1966).
10. G. F. Miller and H. Pursey, *Proc. Roy. Socl*, A223, 521 (1954).
11. J. Phillips, *Int. J. Solids Structs*, **6**, 1403 (1970).
12. J. Miklowitz, *J. Appl. Mech.*, **20**, 122 (1953).
13. R. M. Davies, *Phil. Trans. Roy. Soc.*, **A240**, 375 (1948).
14. H. Kolsky, *Stress Waves in Solids,* Clarendon Press, Oxford, 1953; Dover reprint, 1964.
15. M. Moffett, *Brown Univ. Report AROD,* No. 18 (1971).
16. H. Kolsky and S. S. Lee, *Brown Univ. Report Nonr 562(30)/5,* (1962).
17. M. Moffett and H. Kolsky, Paper read to meeting of Acoustical Society of America, Washington (1971).

3. K. W. Hillier and H. Kolsky, Proc. Phys. Soc., B 62, 111 (1949).
4. K. W. Hillier, Trans. Inst. Rubber Industry, 26, 64 (1950).
5. W. Ramberg and L. K. Irwin, Proc. 9th Int. Congr. Appl. Mech., 8, 480 (1957).
6. E. A. Ripperger and L. M. Yeakley, Exp. Mech., 3, 47 (1963).
7. H. Kolsky, Nature, 224, 1301 (1969).
8. Y. M. Tsai and H. Kolsky, J. Mech. Phys. Solids, 15, 263 (1967).
9. Y. M. Tsai, Brown Univ. Report NSF-GP-2010/4, (1966).
10. G. F. Miller and H. Pursey, Proc. Roy. Soc., A223, 521 (1954).
11. J. Phillips, Int. J. Solids Struct., 6, 1403 (1970).
12. J. Miklowitz, J. Appl. Mech., 20, 122 (1953).
13. R. M. Davies, Phil. Trans. Roy. Soc., A240, 375 (1948).
14. H. Kolsky, Stress Waves in Solids, Clarendon Press, Oxford, 1953; Dover reprint, 1964.
15. M. Moffett, Brown Univ. Report AROD, No. 18 (1971).
16. H. Kolsky and S. S. Lee, Brown Univ. Report Nonr 562(30)/5, (1962).
17. M. Moffett and H. Kolsky, Paper read to meeting of Acoustical Society of America, Washington (1971).

DYNAMIC RESPONSE OF DISLOCATIONS IN SOLIDS

T. MURA

Department of Civil Engineering and Materials Research Center*
Northwestern University, Evanston, Illinois 60201

Abstract—In this paper solutions for the displacements, strains, stresses, energies, and generalized forces associated with uniformly moving dislocations, vibrating dislocations, and accelerating dislocations for given distributions and configurations will be reviewed and revised. Special emphasis will be placed upon the correlation between applied stresses and dynamic motion of dislocations.

This dynamic theory of dislocations may be the fundamental concept of all slip phenomena accompanying the inertia effect in metals as well as in noncrystal line materials such as soil and earth.

INTRODUCTION

Since Taylor [1], Orowan [2] and Polany [3] used Volterra's dislocation [4] to explain the plastic deformation of single crystals, dislocation theory has had considerable success in the quantitative prediction of many properties of metals, for instance, crystal growth and melt [5], grain boundaries [6], yielding, work-hardening, and creep [7-9], fracture [10-11], fatigue [12], internal friction [13], precipitation and aging [14], twin and martensite formation [15], radiation damage, diffusion, electric and magnetic resistivity, optical absorption and excitation, etc. [16].

Dislocation theory, however, has not been fully applied to such broad areas of mechanics as geology, biology, and structural engineering. Since dislocation theory is a study of slip phenomena, the theory can be extended to the study of faults and folds in the earth's crust [17-18]. Earthquakes are considered as dynamic motions of dislocations [19].

The effect of sonic and ultrasonic radiation on dislocations will improve workability of hard metals in forming, rolling, extruding, etc. [20-21]. The decrease in yield shearing stress caused by $2W/cm^2$ ultrasonic irradiation was found to be about the same order of magnitude (50-100 g/mm^2) for the plastic deformation of Al, Cd, and Zn crystals, the temperature being constant [22]. On the other hand, the structural elements in rockets or other high speed vehicles which are exposed to sonic and ultrasonic radiation of high intensity will lose their material strength [23].

Another important area with respect to the future application of the dynamic theory of dislocations is acoustic emission. Acoustic emission is the pressure wave produced in metals by the energy released as the metal deforms and fractures. Acoustic emission can be used as a nondestructive tool for detecting the nucleation and growth of cracks [24-29].

---

*This research was supported by the Advanced Research Projects Agency of the Department of Defense through the Northwestern University Materials Research Center.

345

The most important feature of dislocation theory is, according to the present author's opinion, to give explanations of inelastic properties of materials by means of elasticity theory. This can be seen most effectively in the earliest dislocation paper by Taylor [1]. The parabolic stress strain curves of f.c.c. single crystals were derived from the elastic strain energy caused by a pair of positive and negative dislocations and the assumption that plastic strain is proportional to dislocation density.

In this paper the emphasis will be placed upon the result that inelastic dynamic properties of materials (energy dissipation, viscosity of materials, etc.) can be explained by the elasticity theory when dynamic response of dislocations is taken into account in the analysis.

GENERAL THEORY OF DISLOCATION MOTION

In this section the fundamental concepts of dislocation theory and the general method of obtaining the related elastic fields will be introduced.

The dislocation loop ∂S is defined as the boundary of a slip surface S. The slip vector **b** is called the Burgers vector which is a multiple value of the displacement on S. When b_i is divided by the thickness of S (which is zero), plastic distortion may be defined. Although this plastic distortion is infinite, its product with the volume element, $dx = dx_1 dx_2 dx_3$, remains finite and is proportional to the product of b_i and surface element dS_j which is included in dx. Thus the plastic distortion tensor β_{ji}^* is defined by

$$\beta_{ji}^* \, dx = -b_i \, dS_j \tag{1}$$

This β_{ji}^* behaves as the product of a one-dimensional Dirac delta function in the direction of the thickness of S, and a two-dimensional Heaviside step function which takes 1 on S and zero elsewhere. Since these functions can be treated as regular functions by Fourier integrals, β_{ji}^* may be expressed as

$$\beta_{ji}^* \, (x, t) = \int_{-\infty}^{\infty} \int_{-\infty}^{\infty} \bar{\beta}_{ji}^* \, (\boldsymbol{\xi}, \omega) \exp \left\{ i \, (\boldsymbol{\xi} \cdot x + \omega t) \right\} d\boldsymbol{\zeta} \, d\omega \tag{2}$$

where $d\boldsymbol{\xi} = d\xi_1 \, d\xi_2 \, d\xi_3$, $\boldsymbol{\xi} \cdot x = \xi_i x_i$, and t denotes the time variable. $\bar{\beta}_{ji}^*$ is the Fourier transform of β_{ji}^* or

$$\bar{\beta}_{ji}^* \, (\boldsymbol{\xi}, \omega) = (2\pi)^{-4} \int_{-\infty}^{\infty} \int_{-\infty}^{\infty} \beta_{ji}^* \, (x, t) \, (\exp \left\{ -i \, (\boldsymbol{\xi} \cdot x + \omega t) \right\} dx \cdot dt \tag{3}$$

which becomes from (1)

$$\bar{\beta}_{ji}^* (\xi, \omega) = -(2\pi)^{-4} \, b_i \int_{-\infty}^{\infty} dt \int_{S(t)} \exp\left\{-i \, (\xi \cdot x + \omega t)\right\} dS_j (x) \cdot \qquad (4)$$

Most elastic problems in dislocation theory are concerned with finding the elastic field when $S(t)$ is prescribed. The displacement u_i is obtained as follows. The equation of motion is

$$\sigma_{pq,q} = \rho \, \ddot{u}_p \qquad (5)$$

where $f_{,j}$ denotes $\partial f / \partial x_j$ and the summation convention is employed for the repeated indices, ρ is the density, and $\ddot{u}_p$ the second derivative of u_p with respect to time. The total distortion $u_{i,j}$ is the sum of the elastic distortion β_{ji} and the plastic distortion β_{ji}^*:

$$u_{i,j} = \beta_{ji} + \beta_{ji}^* \qquad (6)$$

and β_{ji} is related to stress through Hooke's law:

$$\sigma_{ij} = C_{ijkl} \beta_{lk} \qquad (7)$$

Combination of the above three equations leads to the relation

$$C_{pqmn} (u_{m,nq} - \beta_{nm,q}^*) = \rho \, \ddot{u}_p \qquad (8)$$

When the material is infinitely extended, and if the displacement is assumed to have the form

$$u_m (x, t) = \int_{-\infty}^{\infty} \int_{-\infty}^{\infty} \bar{u}_m (\xi, \omega) \exp\left\{ i \, (\xi \cdot x + \omega t) \right\} d\xi \, d\omega, \qquad (9)$$

$\bar{u}_m$ can be obtained by substituting (9) and (2) into (8), resulting in the expression

$$\bar{u}_m (\xi, \omega) = -i \, C_{klij} \; \xi_l \; \frac{N_{mk} (\xi, \omega)}{D (\xi, \omega)} \; \bar{\beta}_{ji}^* (\xi, \omega) \qquad (10)$$

Here D is the determinant and N_{mk} is the cofactor of the matrix whose mk element is $C_{mikj} \xi_i \xi_i - \rho \omega^2 \, \delta_{mk}$. N_{mk}/D is the Fourier transform of the Green's function $G_{mk} (x, t)$:

$$G_{mk} (x, t) = (2\pi)^{-4} \int_{-\infty}^{\infty} \int_{-\infty}^{\infty} \frac{N_{mk} (\xi, \omega)}{D (\xi, \omega)} \exp\left\{ i \, (\xi \cdot x + \omega t) \right\} d\xi \, d\omega \qquad (11)$$

Generally, N_{mk} and D are homogeneous polynomials of ξ and ω of degree 4 and 6 respectively, and the above integral can be expressed in a series form. N_{mk}/D has a simple form for isotropic materials as

$$\frac{N_{mk}(\xi,\omega)}{D(\xi,\omega)} = \frac{\delta_{mk}\left\{(\lambda+2\mu)\,\xi^2 - \rho\omega^2\right\} - \xi_m\,\xi_k\,(\lambda+\mu)}{(\mu\xi^2 - \rho\omega^2)\left\{(\lambda+2\mu)\,\xi^2 - \rho\omega^2\right\}} \tag{12}$$

where $\xi^2 = \xi_i\,\xi_i$, and λ, μ are Lamé's constants.

For a prescribed $\mathbf{S}$ (t), $\bar{\beta}_{ji}^{*}$ (ξ, ω) is calculated from (4) and the displacement is obtained from (9) and (10), resulting in the expression

$$u_m(\mathbf{x},t) = i\,(2\pi)^{-4}\,b_i \int_{-\infty}^{\infty}\int_{-\infty}^{\infty} d\xi\, d\omega \int_{-\infty}^{\infty} dt' \int_{S(t')} C_{klij}\,\xi_l\,\frac{N_{mk}(\xi,\omega)}{D(\xi,\omega)}$$

$$\times \exp\left\{\,i\,[\,\xi\cdot(\mathbf{x}-\mathbf{x}') + \omega\,(t-t')\,]\,\right\}\,dS_j(\mathbf{x}') \tag{13}$$

or by using (11),

$$u_m(\mathbf{x},t) = b_i \int_{-\infty}^{\infty} dt' \int_{S(t')} C_{klij}\,G_{mk,l}(\mathbf{x}-\mathbf{x}', t-t')\,dS_j(\mathbf{x}') \cdot \tag{14}$$

Nabarro [30] first pointed out the corresponding expression of (14) for isotropic materials and infinitesimal $\mathbf{S}$. Mura [31] showed that the integration in (14) with respect to dS_j can generally be performed when u_m and $u_{m,n}$ are considered. The result is

$$\dot{u}_m(\mathbf{x},t) = b_i\,\epsilon_{jnh} \int_{-\infty}^{\infty} dt' \oint_{\partial S(t')}\left\{C_{klij}\,G_{mk,l}(\mathbf{x}-\mathbf{x}')\,t-t'\right\} \tag{15}$$

$$\times\,V_n(\mathbf{x}',t')\,dl_h(\mathbf{x}')$$

$$u_{m,n}(\mathbf{x},t) = -b_i\,\epsilon_{jnh} \int_{-\infty}^{\infty} dt' \oint_{\partial S(t')}\left\{C_{klij}\,G_{mk,l}(\mathbf{x}-\mathbf{x}', t-t')\right.$$

$$\left. +\rho\,\dot{G}_{im}(\mathbf{x}-\mathbf{x}', t-t')\,V_j(\mathbf{x}',t')\right\}\,dl_h(\mathbf{x}') + \beta_{nm}^{*}(\mathbf{x},t) \tag{16}$$

where V is the velocity of the dislocation line element $d\ell$. The key relation used for the derivation of the above result is

$$dS_n = \epsilon_{njh} \, V_j \, d\ell_h \, . \tag{17}$$

It is often convenient to express (15) and (16) in the Fourier integral forms [32-33]. Then

$$\dot{u}_m(x, t) = i(2\pi)^{-4} \, b_i \, \epsilon_{jnh} \int_{-\infty}^{\infty} \int_{-\infty}^{\infty} d\xi \, d\omega \int_{-\infty}^{\infty} dt' \oint_{\partial S(t')} C_{klij} \, \xi_l \, L_{mk}(\xi, \omega) \, V_n(x't')$$

$$\times \exp\left\{ i\left[\xi \cdot (x-x') + \omega(t-t') \right] \right\} \, d\ell_h(x') \tag{18}$$

$$u_{m,n}(x, t) = -i(2\pi)^{-4} \, b_i \, \epsilon_{jnh} \int_{-\infty}^{\infty} \int d\xi \, d\omega \int_{-\infty}^{\infty} dt' \oint_{\partial S(t')} [C_{klij} \, \xi_l \, L_{mk}(\xi, \omega)$$

$$+ \rho\omega \, L_{im}(\xi, \omega) \, V_j(x', t')] \exp\left\{ i\left[\xi \cdot (x-x') + \omega(t-t') \right] \right\} \, d\ell_h(x')$$

$$+ \beta^*_{nm}(x, t) \tag{19}$$

where

$$L_{mk}(\xi, \omega) = N_{mk}(\xi, \omega) / D(\xi, \omega) \, . \tag{20}$$

Three special cases of the above relations will now be considered:

1) For stationary dislocations we have $V = 0$, $\omega = 0$, $(2\pi)^{-4}$ changes to $(2\pi)^{-3}$, and the integrations with respect to ω, t' must be dropped. In this case Willis [34] evaluated analytically the integrals in (19) for a dislocation segment in an anisotropic medium.

2) For a dislocation loop or segment moving with a constant velocity V, $\omega = -\xi \cdot V$, $t' = 0$, $(2\pi)^4$ changes to $(2\pi)^{-3}$, and the integrations with respect to ω, t' must be dropped. The integration method of Willis was applied by Mura [35] for this case.

3) For an oscillating dislocation, $t' = 0$, $V_n = i\omega A_n$ (A_n: constant amplitude of the dislocation line), $(2\pi)^{-4}$ changes to $(2\pi)^{-3}$, and the integrations with respect to ω and t' must be dropped.

For uniformly moving dislocations in an anisotropic medium the elastic field is essentially static. Sáenz [36] and Stroh [37] studied this case and the condition of dislocation velocities under which the equations of motion are of the elliptic type [38]. Three critical velocities are then obtained from $D\,(\underset{\sim}{\xi},-\underset{\sim}{\xi}\cdot \mathbf{V})=0$ which vary with the direction of motion. For isotropic materials, the three critical velocities degenerate in shearing wave velocity $c=(\mu/\rho)^{\frac{1}{2}}$ and dilatational wave velocity $a=\left\{\,(\lambda+2\mu)/\rho\,\right\}^{\frac{1}{2}}$.

Motion of dislocations is primarily caused by an applied stress. The correlation between the applied stress and the motion of a dislocation loop is the most important part of dislocation theory. Let's denote the applied stress by σ_{ij}^{A}, the dislocation stress by σ_{ij}, and the total displacement by u_i. Then, the associate fundamental equations for an elastic continuum D are:

$$\sigma_{ij,j}+\sigma_{ij,j}^{A}=\rho\,\ddot{u}_i$$

$$u_{i,j}=\beta_{ji}+\beta_{ji}^{*}$$

$$\sigma_{ij}+\sigma_{ij}^{A}={}^{\cdot}C_{ijkl}\,\beta_{lk}$$

$$(\sigma_{ij}+\sigma_{ij}^{A})\,n_j=X_i \text{ on } \partial D$$

(21)

where ∂D is the boundary of D where the applied traction $\mathbf{X}$ is given. The Lagrangian L of the system becomes

$$\int_{t_0}^{t_1} L dt=\int_{t_0}^{t_1} dt \int_D \frac{1}{2}\rho\,\dot{u}_i\,\dot{u}_i\,dx-\int_{t_0}^{t_1} dt \int_D \frac{1}{2}(\sigma_{ij}+\sigma_{ij}^{A})\,\beta_{ji}\,dx$$

$$+\int_{t_0}^{t_1} dt \int_{\partial D} X_i\,u_i\,dS\,.$$

(22)

The variation of (22) with respect to δu_i and $\delta\beta_{ji}^{*}$ under fixed $\mathbf{X}$ leads to

$$\delta\int_{t_0}^{t_1} L dt=\int_{t_0}^{t_1} dt \int_D (-\rho\,\ddot{u}_i+\sigma_{ij,j}+\sigma_{ij,j}^{A})\,\delta u_i\,dx$$

$$+\int_{t_0}^{t_1} dt \int_{\partial D} (X_i-\sigma_{ij}\,n_j-\sigma_{ij}^{A}\,n_j)\,\delta u_i\,dS$$

(23)

$$+\int_{t_0}^{t_1} dt \int_D (\sigma_{ij}+\sigma_{ij}^{A})\,\delta\beta_{ji}^{*}\,dx\,.$$

where $\delta\beta_{ji} = \delta u_{i,j} - \delta\beta_{ji}^*$ has been used. The first two integrals in (23) become zero when the field quantities satisfy the conditions (21). Moreover, from (1) we have

$$\delta\beta_{ji}^* \, dx = -b_i \, \delta \, dS_j \tag{24}$$

where $\delta \, dS_j$ is the variation of the slip plane due to variation of the dislocation displacement $\delta\zeta$. Thus, we have

$$\delta \, dS_j = \epsilon_{jlh} \, \delta\zeta_l \, d\ell_h \equiv n_j \, \delta S \equiv \delta S_j \tag{25}$$

and

$$\delta \, \beta_{ji}^* \, dx = -b_i \, \epsilon_{jlh} \, \delta\zeta_l \, d\ell_h \, . \tag{26}$$

Equation (23) now becomes

$$\delta \int_{t_o}^{t_1} L dt = -\int_{t_o}^{t_1} dt \oint_{\partial S} (\sigma_{ij} + \sigma_{ij}^A) \, \epsilon_{jlh} \, b_i \, \delta\xi_l \, \nu_h \, d\ell \tag{27}$$

where $\nu_h \, d\ell = d\ell_h$. The Peach-Koehler force [39] is defined by the coefficient of $\delta\zeta_l$ in the integrand of (27), that is

$$f_l = \epsilon_{ljh} (\sigma_{ij} + \sigma_{ij}^A) \, b_i \, \nu_h \, . \tag{28}$$

When no other forces are considered, the equilibrium condition of the dislocation loop is defined as

$$f_l = 0 \text{ along } \partial S \, . \tag{29}$$

The stress components involved in (28) are $\sigma_{12} + \sigma_{12}^A$ for the straight edge dislocation and $\sigma_{23} + \sigma_{23}^A$ for the straight screw dislocation when both of them are parallel to the x_3 axis. It is well known that these stress components vanish along the dislocation lines when they are moving with a constant subsonic velocity V in the x_1 direction, the stress being evaluated at $x_1 = Vt$, $y = +\epsilon \approx 0$.* This means that the uniform motion of a straight dislocation (a discrete dislocation) can exist in the absence of an applied stress ($\sigma_{ij}^A = 0$) when there is no friction.

---

*Nabarro [79] suggested $\epsilon = b/2$ in view of the Peierls dislocation.

Peierls [40-41] introduced in (29) an extra term of atomic force caused by slip [w] on S: $- (\mu/2\pi) \sin (2\pi$ [w] $/b)$. In this case, the dislocation is distributed along the x_1 -axis rather than isolated discretely at $\overline{x}_1 = y = 0$. This dislocation is called the Peierls dislocation and [w] $= \dfrac{b}{2} (1 + \dfrac{2}{\pi} \tan^{-1} \dfrac{x_1}{\overline{\zeta}})$. The Peierls dislocation is a distributed dislocation of strength (d [w] $/dx_1$). The strength becomes $(b/\pi) \overline{\zeta} / (\overline{x}_1{}^2 + \zeta^2)$, where $\overline{x}_1$ $= x_1 - Vt$. The parameter $2\overline{\zeta}$ is called the dislocation width. The equilibrium condition must hold at every point where the spread dislocation is distributed. The uniform motion of a Peierls dislocation also can exist without an applied stress σ_{ij}^A when the dislocation width $\overline{\zeta}$ is chosen suitably. A frictional force (drag force) can also be added to the equilibrium equation (29). The accelerating motion of dislocations generally requires the application of external stress fields.

Consider now the case of an oscillating dislocation in an infinitely extended medium. When the atomic force and the frictional force are neglected, the equilibrium condition (29) must be satisfied on the dislocation line. The rate of work done by the applied stress is

$$\frac{dW}{dt} = \int\limits_D \sigma_{ij}^A \, \dot{\beta}_{ji}^* \, dx = \int\limits_{\partial S} \sigma_{ij}^A \, b_i \, \epsilon_{ljh} \, V_l \, d\ell_h \qquad (30)$$

where from (26)

$$\dot{\beta}_{ji}^* \, dx = \frac{\partial \beta_{ji}^*}{\partial t} \, dx = b_i \, \epsilon_{ljh} \, \frac{\delta \zeta_l}{\partial t} \cdot \nu_h \, dl, \quad V_l = \frac{\partial \zeta_l}{\partial t} \qquad (31)$$

The equilibrium equation (29) can be written from (28) and (31) as

$$(\sigma_{ij} + \sigma_{ij}^A) \, \dot{\beta}_{ji}^* = 0 . \qquad (32)$$

It will be shown that $\int_o^{2\pi/\omega} \dfrac{dW}{dt} dt$ is the energy flux radiated from the oscillating dislocation with a period $2\pi/\omega$. By assigning a mode of dislocation oscillation, $u_m, \dot{u}_m,$ and $u_{m,n}$ are obtained from (13), (18), and (19), respectively (excluding the applied displacement). For any domain D_1 which includes the dislocation, Gauss' theorem leads to

$$\int_{\partial D_1} \sigma_{ij}\, n_j\, \dot{u}_i\, dS = \int_{D_1} \sigma_{ij,j}\, \dot{u}_i\, dx + \int_{D_1} \sigma_{ij}\, \dot{u}_{i,j}\, dx$$

$$= \int_{D_1} \rho\, \ddot{u}_i\, \dot{u}_i\, dx \int_{D_1} \sigma_{ij}\, (\dot{\beta}_{ji} + \dot{\beta}_{ji}^{*})\, dx$$

$$= \frac{d}{dt} \left(\int_{D_1} \frac{1}{2}\, \rho\, \dot{u}_i\, \dot{u}_i\, dx + \int_{D_1} \frac{1}{2}\, \sigma_{ij}\, \beta_{ji}\, dx \right) + \int_{D_1} \sigma_{ij}\, \dot{\beta}_{ji}^{*}\, dx \qquad (33)$$

When (33) is integrated with respect to t from 0 to $2\pi/\omega$, the first two terms in the right hand side vanish. Then from (32)

$$-\int_{o}^{2\pi/\omega} \left(\int_{\partial D_1} \sigma_{ij}\, n_j\, \dot{u}_i\, dS \right) dt = \int_{o}^{2\pi/\omega} \left(\int_{D_1} \sigma_{ij}^{A}\, \dot{\beta}_{ji}^{*}\, dx \right) = \int_{o}^{2\pi/\omega} \frac{dW}{dt}\, dt \qquad (34)$$

since $\int_{D_1} \sigma_{ij}^{A}\, \dot{\beta}_{ji}^{*}\, dx = \int_{D} \sigma_{ij}^{A}\, \dot{\beta}_{ji}^{*}\, dx$; ($\dot{\beta}_{ji}^{*}$ is zero everywhere except on the dislocation). It should be noted that the energy flux is independent of the size of D_1 as far as D_1 includes the dislocation. The left side in (34) is the energy flux radiated from the oscillating dislocation with period $2\pi/\omega$. The free oscillation of the dislocation where σ_{ij}^{A} = 0 has no energy flux.

The general theory of moving dislocations similar to that described in this section has been developed by Kosevich [42-43], Mura [31], [44], and Bross [45]. Günther [46] has investigated the geometry and field equations of moving dislocations considering some nonlinear effects. The earliest work on field equations has been done by Höllander [47-48] where Poisson's ratio has been assumed to be zero. There are more papers [49-54] on the general theory emphasizing various aspects of moving dislocations.

UNIFORMLY MOVING DISLOCATIONS

The elastic solution of a uniformly moving straight dislocation has been obtained by Frank [55] and Eshelby [56]. Frank has shown that when a screw dislocation is moving with a constant velocity V, the total energy (the sum of elastic strain energy and kinetic energy) is given by $E_o/ (1-V_2/c_2)^{1/2}$, where E_o is the elastic strain energy of the dislocation at rest, and c the shear wave velocity $(\mu/\rho)^{1/2}$. His calculation was stimulated by the work done by Kontorova and Frenkel [57] who studied a one-dimensional dislocation model and found a strikingly analogous form to that of a particle in special relativity. Frank's calculation was extended to an edge dislocation by Eshelby [56]. In this case no relativistic relation is found since the dilatational wave velocity is also involved.

As mentioned in the last section, the uniform motion of a straight dislocation is possible in the absence of an applied stress. It has not been clear yet how the dislocation arrives at the constant velocity. Eshelby also extended his analysis to the Peierls dislocation for this case. No applied stress is necessary for uniform motion if the width of the dislocation is $2\overline{\zeta} = \left\{ b/(1-\nu) \right\}$ D (V)/D (0) for an edge dislocation or $2\overline{\zeta} = b\beta$ for a screw dislocation, where D (V) = $- 2\mu$ $(2c^2/V^2)$ $(\gamma-\alpha^4/\beta)$,$\gamma = (1-V^2/a^2)^{1/2}$, $\beta = (1-V^2/c^2)^{1/2}$, $\alpha = (1-V^2/2c^2)^{1/2}$, $c = (\mu/\rho)^{1/2}$, $a = \left\{ (\lambda+2\mu)/\rho \right\}^{1/2}$. It is interesting to note that the width of the edge dislocation vanishes when D (V) = 0, the solution of which is the Rayleigh wave velocity. The width of the screw dislocation vanishes at V = c. Eshelby's calculation was extended by Leibfried and Dietze [58] to a dislocation moving in the middle of a plate (direction of dislocation parallel to the plate surface). Weertman [59-61] completed Eshelby's analysis [56] by calculating the elastic strain energy and the kinetic energy of the edge dislocation. He found that the shear stress on the slip plane (except at the dislocation position) decreases with increasing dislocation velocity, vanishes at the Rayleigh velocity, and changes sign at velocities higher than the Rayleigh velocity. Because of this, edge dislocations of like sign attract each other contrary to the usual situation) when the dislocations are traveling at velocities above the Rayleigh velocity. Such a critical velocity of a dislocation is called the threshold velocity. The threshold velocities have been calculated for various crystals by Teutonico [62-66], Weertman [67-68] and his coworkers [69-70]. They started their stress analysis from the general solution obtained by Bullough and Bilby [71] for a uniformly moving straight dislocation in an anisotropic media. If a dislocation of general shape is considered, the formula in [35] is recommended for use.

The effect of free surfaces and the inhomogeneity of materials on dislocation motion has not been fully investigated. Weertman [72] considered a dislocation moving on the interface between two isotropic media of different elastic constants and densities. The case when a dislocation is moving parallel to the interface has been solved recently by Lee and Dundurs [73].

Although the energy of a moving dislocation becomes infinite as its velocity approaches the sound velocity, there are dislocation solutions of the elastic equations for greater dislocation velocities than c. This dislocation-like solution was first demonstrated by Eshelby [74]. He suggested its possible application to the propagation of diffusionless transformations by dislocations; however, no successful application has been reported.

When a straight screw dislocation parallel to the x_3 -axis is moving along the x_1 -axis with a supersonic velocity V $>$ c, the displacement components are obtained from (13) as

$$u_1 = u_2 = 0$$

$$u_3 = (1/2) bH (Vt - x_1 - \beta | x_2 |) \text{ sgn } x_2$$

(35)

Here $\beta = (V^2/c^2 - 1)^{1/2}$ and H and sgn are the Heaviside and the signature functions, respectively.

For an edge dislocation moving with a supersonic velocity c $<$ a $<$ V, we have

$$u_1 = (b/4)\,(c^2/V^2)\,\Big\{ -2H\,(x_2)\,[H\,(x_1-Vt+\gamma x_2)-H\,(-x_1+Vt-\gamma x_2)]$$

$$-(V^2/c^2-2)\,H\,(x_2)\,[H\,(x_1-Vt+\beta x_2)-H\,(-x_1+Vt-\beta x_2)]$$

$$+2H\,(-x_2)\,[H\,(x_1-Vt-\gamma x_2)-H\,(-x_1+Vt+\gamma x_2)]$$

$$+(V^2/c^2-2)\,H\,(-x_2)\,[H\,(x_1-Vt-\beta x_2)-H\,(-x_1+Vt+\beta x_2)]\Big\}$$

$$+(b/4)\,[H\,(x_2)-H\,(-x_2)]$$

$$u_2 = (b/2)\,(c^2/V^2)\,\Big\{ -(1/\beta)\,(V^2/c^2-2)\,[H\,(x_2)\,H\,(-x_1+Vt-\beta x_2)$$

$$+H\,(-x_2)\,H\,(-x_1+Vt+\beta x_2)]\ +2\gamma\,[H\,(x_2)\,H\,(-x_1+Vt-\gamma x_2)$$

$$+H\,(-x_2)\,H\,(-x_1+Vt+\gamma x_2)]\Big\}$$

$$u_3 = 0 \tag{36}$$

where

$$\beta^2 = V^2/c^2 - 1$$
$$\gamma^2 = V^2/a^2 - 1$$

For an edge dislocation moving with a transonic velocity $c < V < a$, we have

$$u_1 = -(b/4)\,(c^2/V^2)\,(V^2/c^2-2)\,\Big\{ H\,(x_2)\,[H\,(x_1-Vt+\beta x_2)-H\,(-x_1+Vt-\beta x_2)]$$

$$-H\,(-x_2)\,[H\,(x_1-Vt-\beta x_2)-H\,(-x_1+Vt+\beta x_2)]\Big\}$$

$$-(b/2\pi)\,(c^2/V^2)\,\Big\{ H\,(x_2)\,\Big[-2\tan^{-1}\frac{\gamma x_2}{x_1-Vt}+\pi H\,(x_1-Vt)-\pi H\,(-x_1+Vt)\Big]$$

$$-H\,(-x_2)\,\Big[-2\tan^{-1}\frac{\gamma x_2}{x_1-Vt}+\pi H\,(x_1-Vt)-\pi H\,(-x_1+Vt)\Big]\Big\}$$

$$+(b/4)\,[H\,(x_2)-H\,(-x_2)]$$

$$u_2 = -(b/4)\,(c^2/V^2)\,(2-V^2/c^2)\,(1/\beta)\,\Big\{ H\,(x_2)\,[H\,(x_1-Vt+\beta x_2)-H\,(-x_1+Vt-\beta x_2)]$$

$$+H\,(-x_2)\,[H\,(x_1-Vt-\beta x_2)-H\,(-x_1+Vt+\beta x_2)]\Big\}$$

$$+(b/2\pi)\,(c^2/V^2)\,\gamma\log\Big\{ (x_1-Vt)^2+\gamma^2\,x_2^{\,2}\Big\}$$

$$u_3 = 0. \tag{37}$$

where

$$\beta^2 = V^2/c^2 - 1$$
$$\gamma^2 = 1 - V^2/a^2$$

and

$$\pi/2 \geqslant \tan^{-1} \frac{\gamma x_2}{x_1 - Vt} \geqslant -\pi/2 \cdot$$

The stress components at $x_1 = Vt$ and $x_2 = +\epsilon \approx 0$ for the Peach-Koehler force are obtained from the above displacements. Then, we have $\sigma_{32} = \mu u_{3,2} = -(\mu b \beta/2)\, \delta\, (Vt - x_1)$ for the screw dislocation and $\sigma_{12} = \mu\, (u_{1,2} + u_{2,1}) = -2\mu b\, [\gamma + (\beta^2 - 1)^2/4\beta]\, (\beta^2 + 1)^{-1}\, \delta\, (Vt - x_1)$ for the supersonic edge dislocation, where δ is the Dirac delta function. These stresses are infinite at $x_1 = Vt$. This difficulty is avoided by considering a spread dislocation with a distribution function $B\, (x_1 - Vt)$. Then, the factor $b\delta\, (Vt - x_1)$ in the above stress components is replaced by $B\, (x_1 - Vt)$ due to the characteristic of the Dirac delta function. If $B\, (x_1)$ is constant in the region $-A < \overline{x_1} < A$ and zero elsewhere, the equilibrium condition (29) is satisfied for a uniformly applied stress since, for this case, $\sigma_{32} + \sigma_{32}^A = 0$ in $-A < \overline{x_1} < A$ and $x_2 = +\epsilon \approx 0$ for the screw dislocation while $\sigma_{12} + \sigma_{12}^A = 0$ in the same region for the edge dislocation. Weertman [75-76] discussed atomic force laws under which the equilibrium condition is satisfied without applied stresses, where the distribution function $B\, (\overline{x_1})$ has been taken as $(b/\pi)\, \overline{\zeta}\, /\, (\overline{x_1}^2 + \overline{\zeta}^2)$.

Almost all calculations for moving dislocations have been limited to straight infinite dislocations. Very few calculations have been done for circular dislocation loops. Günther [77] obtained the elastic solution for a circular edge dislocation moving uniformly in the direction of the normal to the circular plane. In the region $V < c < a$ the elastic field shows double Lorentz contractions and in the region $C < a < V$ there arise two Mach cones radiated from the dislocation loop with angles $\pm \sin^{-1}\, (V/c)$ and $\pm \sin^{-1}\, (V/a)$ with respect to the direction of the motion. No disturbance is seen outside the domain bounded by the cones and the circular plane.

VIBRATING DISLOCATIONS

Unlike the uniform motion of a straight dislocation with a subsonic velocity, a vibrating straight dislocation generally requires the application of an oscillating stress (forced vibration). Free vibrations, however, are possible under special circumstances. By using Eshelby's elastic solution [78] for an oscillating* screw dislocation and the equilibrium condition (29), Nabarró [79] obtained the amplitude and phase-difference

---

*Oscillation is referred to a rigid string-like motion and vibration is referred to an elastic string-like motion.

for the position of the oscillating dislocation as functions of the amplitude and wave number of an incident sound wave. He also calculated the scattering cross-section (the ratio of the rate of radiation to the incident energy flux). Same calculation was done for an edge dislocation by Kiusalaas and Mura [80]. On the other hand, Leibfried [81] first pointed out that a dislocation moving through a flux of sound waves (which are a natural consequence of the thermal energy of the crystal) will experience a retarding force which is proportional to the dislocation velocity. The numerical value of his result, however, was criticized by Nabarro [79], Lothe [82], and Eshelby [83]. Eshelby used a kink model for mathematical simplicity and found that a kink moving through an isotropic flux of elastic waves has a scattering cross-section proportional to the square of its width, and experiences a retarding force proportional to the product of its velocity and the energy density of the waves. Pegel [84], Laub and Eshelby [85], and Ninomiya and Ishioka [86] discussed dispersion curves for the free vibration of a straight dislocation. They found that free vibrations are possible for a certain range of frequencies and wavelengths, but that in general a suitable applied stress is required to maintain the vibration. In order to obtain quantitative relations between the applied stress and the dislocation vibration, Kuo and Mura [87] extended the approach of [79] to this case of an elastic string-like vibration, as shown below.

Consider a screw dislocation with Burgers vector $(0, 0, b)$ lying along the x_3 direction and vibrating in the $x_1 x_3$ plane. The position of dislocation line is described by

$$x_1 = A \sin (\omega t + k_3 x_3 + \alpha), \quad x_2 = 0, \tag{38}$$

where A, k_3 and α are unknown constants. In order to maintain the motion (38), the following elastic shear wave must be applied:

$$u_2^A = u_o \sin \varphi \cos (\omega t + \varkappa \sin \varphi x_3 - \varkappa \cos \varphi x_2),$$

$$u_3^A = u_o \cos \varphi \cos (\omega t + \varkappa \sin \varphi x_3 - \varkappa \cos \varphi x_2), \tag{39}$$

$$\varkappa = \omega/c, \quad c^2 = \mu/\rho,$$

where φ is the angle between the wave front and the x_3-axis. The equilibrium condition (29) must be satisfied along the dislocation line. σ_{23} is calculated from (9), (10), (4) by using

$$\beta_{23}^* = b\delta (x_2) H \ [A \sin (\omega t + k_3 x_3 + \alpha) - x_1 \ ] , \tag{40}$$

and σ_{23}^A is $\mu (u_{2,3}^A + u_{3,2}^A)$. The expression (40) is obvious since the plastic distortion is constrained on the plane $x_2 = 0$ and $x_1 \leqslant A \sin (k_3 x_3 + \omega t + \alpha)$. The dislocation stress σ_{23} becomes an infinite double series containing high derivatives of the Bessell function

K_o. It takes a simple form, however, when the terms have coefficients of order greater than A^2 are neglected. This stress is singular at the position of the dislocation; but terms having a higher order of singularity than logarithmic will disappear when the stress is averaged over a small area of the dislocation core. Under these constraint conditions the equilibrium condition ($\sigma_{23} + \sigma_{23}^A = 0$ at the dislocation line) leads to the determination of k_3, α, and A as functions of $\varkappa$, φ, and u_o, where $\varkappa b \leqslant 1$ has been assumed. This assumption is automatically satisfied for most cases of practical concern. The result is

$$k_3 = \varkappa \sin \varphi$$

$$\alpha = - \tan^{-1} (H/F) \tag{41}$$

$$A = \frac{4u_o \varkappa \cos 2\varphi}{b (F^2 + H^2)^{1/2}}$$

where

(i) for $\omega/a < k_3 < \omega/c$ or $\dfrac{\mu}{\lambda + 2\mu} < \sin^2 \varphi < 1$

$$F = -\frac{\varkappa^2}{2} (\sin^2 \varphi + \cos^2 2\varphi) \, Y_o \, (\frac{\varkappa b}{2} \cos \varphi)$$

$$-\frac{4}{\pi} \varkappa^2 \sin^2\varphi \, (\sin^2\varphi - c^2/a^2) \, K_o \, [\, \varkappa \, (\sin^2\varphi - c^2/a^2)^{1/2} \, b/2 \,]$$

$$H = -\frac{\varkappa^2}{2} (\sin^2\varphi + \cos^2 2\varphi) \, J_o \, (\frac{\varkappa b}{2} \cos \varphi) \tag{42}$$

(ii) for $k_3 < \omega/a$ or $\sin^2 \varphi < \dfrac{\mu}{\lambda + 2\mu}$

$$F = -\frac{\varkappa^2}{2} (\sin^2\varphi + \cos^2 2\varphi) \, Y_o \, (\frac{\varkappa b}{2} \cos \varphi)$$

$$- 2 \varkappa^2 \sin^2 \varphi \, (c^2/a^2 - \sin^2 \varphi) \, Y_o \, [\, \varkappa \, (c^2/a^2 - \sin^2 \varphi)^{1/2} \, b/2 \,]$$

$$H = -\frac{\varkappa^2}{2} (\sin^2 \varphi + \cos^2 2\varphi) \, J_o \, (\frac{\varkappa b}{2} \cos \varphi)$$

$$- 2\varkappa^2 \sin^2 \varphi \, (c^2/a^2 - \sin^2 \varphi) \, J_o \, [\, \varkappa \, (c^2/a^2 - \sin^2 \varphi)^{1/2} \, b/2 \,] \tag{43}$$

and

$$a^2 = (\lambda + 2\mu)/\rho \; .$$

The energy dissipated per cycle per unit length of dislocation e is calculated by (34):

$$e = \frac{k_3}{2\pi} \int_0^{2\pi/\omega} dt \int_0^{2\pi/k_3} dx_3 \int_{-\infty}^{\infty} \int \sigma_{23}^A \frac{d}{dt} \beta_{23}^* dx_1 dx_2$$

$$= 4\pi\mu u_0^2 \kappa^2 \cos^2 2\varphi \frac{-H}{F^2 + H^2}, \tag{44}$$

where F and H are defined by (42) ∼ (43). Nabarro's calculation [79] corresponds to $\varphi = 0$ (and therefore $k_3 = 0$) and the dislocation is oscillating as a rigid string. The order of magnitude of the quantities α, A/u_0, $e/\mu u_0^2$, however, is the same as his result. The most prominent effect of φ is that at about $\varphi = 89°40' \sim 90°$, $e/\mu u_0$ attains its maximum value of 4π as shown in Fig. 1 for aluminum (b = 3×10^{-8} cm, $\mu = 2.6 \times 10^{11}$ dyne/cm^2, $\lambda = 5.6 \times 10^{11}$ dyne/cm^2, and $\rho = 2.7$ gr/cm^3).

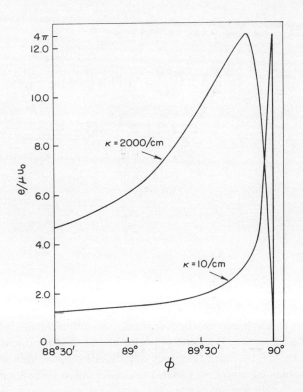

Fig. 1. The energy dissipated per cycle per unit length of vibrating screw dislocation e is divided by μu_0^2 where μ is the shear modulus and u_0 the amplitude of applied elastic shear wave. φ is the angle between the wave front and the dislocation line.

Due to the energy dissipation, the vibrating dislocation is analogous to a vibrating elastic string with damping. The analogy will be established if the equilibrium condition $(\sigma_{23} + \sigma_2\,{}^A_3 = 0$ at the dislocation line) can be written as

$$m\ddot{\zeta} + R\dot{\zeta} - \tau \, \frac{\partial^2 \zeta}{\partial x_3{}^2} = f_o \sin(\omega t + \varkappa \, \sin \varphi \, x_3) \tag{45}$$

Where ζ is the x_1 coordinate of the dislocation displacement. As a matter of fact, the equilibrium condition is written as (45) if the following m (effective mass of dislocation), τ (line tension of dislocation), R (damping constant), and f_o (external force amplitude) are chosen for the dislocation displacement (38), namely

$$\zeta = A \sin(\omega t + k_3 x_3 + \alpha)$$

$$f_o = b\mu\, u_o \, \varkappa \cos 2\varphi \qquad\qquad (\varkappa = \omega/c)$$

$$m = \frac{\rho b^2}{4\pi} \left[-\log \frac{b}{4} \, | k_3^2 - \frac{\omega^2}{c^2} |^{\frac12} - \gamma \, \right]$$

$$\tau = \frac{\mu b^2}{2\pi} \left\{ \left[-\frac{3}{2} + 2 \left(\frac{k_3 c}{\omega} \right)^2 \right] \left[\log \frac{b}{4} \, | k_3^2 - \frac{\omega^2}{c^2} |^{\frac12} + \gamma \, \right] \right.$$

$$\left. - 2 \left[\left(\frac{k_3 c}{\omega} \right)^2 - \left(\frac{c}{a} \right)^2 \right] \left[\log \frac{b}{4} \, | k_3^2 - \frac{\omega^2}{a^2} |^{\frac12} + \gamma \, \right] \right.$$

$$R = \frac{\mu b^2}{8\omega} \left(-3k_3^2 + \frac{\omega^2}{c^2} + 4 \, \frac{c^2 k_3^4}{\omega^2} \right) \quad \text{for } \omega/a < k_3 < \omega/c$$

$$R = \frac{\mu b^2}{8\omega} \left(-3k_3^2 + \frac{\omega^2}{c^2} + 4 \, \frac{c^2 k_3^2}{a^2} \right) \quad \text{for } k_3 < \omega/a$$

$$\varkappa \, \sin \varphi = k_3 \tag{46}$$

In the above calculation, $K_o(z) = -(\log z/2 + \gamma)$, $\gamma = 0.5772$, have been used due to the small argument of the Bessel function. It is easily seen that the damping coefficient is proportional to the frequency. Equation (45) also can be used for discussion on the free vibration of a dislocation, the eigen frequency of a dislocation, and dispersion curves. The eigen frequency is obtained from $\omega = (\tau\, k_3^2/m)^{\frac12}$. Since τ and m are functions of ω and k_3, $\omega^2 = \tau k^2/m$ gives the dispersion curves. The same discussion can be done for an edge dislocation.

The treatment of many oscillating dislocations by considering their interaction has not been reported, except for the free oscillation of two parallel dislocations [88].

When an infinite number of edge dislocations of strength b parallel to the x_3 axis is distributed uniformly on plane $x_2 = 0$, the treatment is rather simple as follows. If this continuous distribution of dislocations oscillates rigidly in the $\pm\, x_1$ direction, with amplitude A and frequency ω, the dislocation stress component σ_{12} is obtained from (19):

$$\sigma_{12} = \frac{b\mu\omega A}{2c} \sin\left(\frac{\omega}{c}\,|x_2|-\omega t\right)\cdot \tag{47}$$

The applied elastic shear wave necessary to preserve the oscillation is

$$u_1^A = u_o \cos\left(\frac{\omega}{c} x_2 - \omega t\right)$$

or

$$\sigma_{12}^A = -\mu\, u_o\, \frac{\omega}{c}\, \sin\left(\frac{\omega}{c} x_2 - \omega t\right). \tag{48}$$

The equilibrium condition $(\sigma_{12} + \sigma_{12}^A = 0$, at $x_2 = 0)$ leads to

$$A/u_o = 2/b\cdot \tag{49}$$

The energy dissipated per unit cycle per unit length of distribution of dislocation is

$$e = \int_0^{2\pi/\omega} (\sigma_{12}^A)_{x_2=0}\ bA\omega \sin\omega t = 2\pi\,\mu\,u_o^2\ \omega/c \tag{50}$$

It is interesting to note that (50) is independent of the strength of the dislocation distribution. Also it can be seen that the factor ω/c in (50) is attributed to the interaction between dislocations, because $e/\mu u_o^2$ for a single dislocation attains a maximum value of 4π as mentioned before.

Almost no research has been done on the effects of free surfaces. Let us consider a torsion bar with a uniform circular section. Mura [89] showed that if screw dislocations of strength b parallel to the axis of the bar are distributed uniformly along a circle of radius r_o (see Fig. 2), all the stress components disappear except $\sigma_{\theta z}$ which becomes -$\mu b r_o/r$ for $r_o < r < r_1$ and zero for $0 < r < r_o$. These distributed dislocations will oscillate in the radial direction according to A exp (iωt) if an oscillating torque M_o exp (iωt) is applied at the end of the bar. The dynamic dislocation stress field is calculated as follows.

The stress at point (r, θ) caused by a dislocation at point (r_o, φ) oscillating in the radial direction is

$$\sigma_{Rz} = -\frac{\mu b\omega^2 A}{4c^2}\ \sin\,\alpha\ H_1^{(2)'}\left(\frac{\omega R}{c}\right)\exp\,(i\omega t)$$

$$\sigma_{\alpha z} = -\frac{\mu b\omega A}{4cR}\ \cos\,\alpha\ H_1^{(2)}\left(\frac{\omega R}{c}\right)\ \exp\,(i\omega t), \tag{51}$$

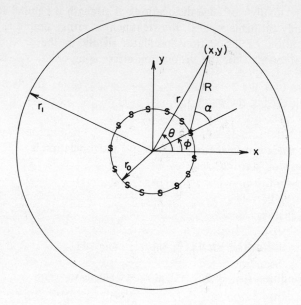

Fig. 2. A solid torsion bar with radius r_1 has a continuous distribution of screw dislocations at $r = r_0$. The dislocations are parallel to the axis of the bar.

where $H_1^{(2)'}(z) = \frac{d}{dz} H_1^{(2)}(z)$, and $H_1^{(2)}$ is the Hankel function. Transforming (51) into (r, θ) components σ_{rz}, $\sigma_{\theta z}$ and then integrating them with respect to φ, we have

$$\sigma_{rz} = 0$$

$$\sigma_{\theta z} = -\frac{\mu b \omega A r_0}{2c} \exp(i\omega t) \int_o^\pi \left\{ \frac{d}{d\beta} \left[H_1^{(2)}\left(\frac{\omega R}{c}\right) \right] \frac{\sin \beta}{R} \right.$$

$$\left. + H_1^{(2)}\left(\frac{\omega R}{c}\right) \frac{(r \cos \beta - r_0)(r - r_0 \cos \beta)}{R^3} \right\} \, d\beta = 0 \qquad (52)$$

where $\theta - \varphi = \beta$, $R^2 = r^2 + r_0^2 - 2rr_0 \cos(\theta - \varphi)$.

The above result (52) means that the dynamic dislocation stress vanishes everywhere although the dislocations are oscillating. This is a puzzle. At this moment, no explanation can be given.

OTHER PROBLEMS

In this section, we will discuss more general motions of dislocations as well as several other topics which will require further investigations in the future.

Eshelby [90] considered several cases of the accelerated motion of a screw dislocation. The dislocation is at rest for negative t and thereafter moved 1) with a velocity $c^2 t / (x_0^2 + c^2 t^2)^{1/2}$ and acceleration $c^2 x_0^2 / (x_0^2 + c^2 t^2)^{3/2}$, 2) with a constant acceleration, 3) with a constant velocity. He calculated the uniform applied stresses required for these motions. He also calculated the velocity of the dislocation when 1) a constant stress is applied at t = 0 and t = t_1 so that the total impulse is a constant value. All these calculations are nothing more than solving the equilibrium condition (29) using (28). Like Nabarro [79], Eshelby also took account of the Peierls stress law. According to the present author's opinion, no such attention is necessary as long as the equilibrium condition is being considered at $\bar{x}_1 = 0$, $x_2 = b/2$. Some details of the elastic field caused by the uniform motion of a dislocation starting from rest are seen in [91] for an edge dislocation and in [92] for a screw dislocation.

The interaction between a varying stress field and a moving screw dislocation has an electro-magnetic analogy as shown by Eshelby [90] and Nabarro [79]. From this analogy, Nabarro [79] predicted that a moving dislocation will receive a Lorentz force (in addition to the Peach-Koehler force) similar to a moving line density of charge. Since the Lorentz force has been predicted strictly from this analogy, its existence has been criticized by several authors [93-96]. Although in some papers [43, 47, 49, 51] the Lorentz force is derived from a Lagrangian formalism, either the mathematics or the physics involved in these derivations are quite questionable.

Let us consider the following Lagrangian functional for a moving dislocation loop in an infinitely extended material:

$$\int_{t_0}^{t_1} L \, dt = \int_{t_0}^{t_1} dt \int_D (\frac{1}{2} \rho \, \dot{u}_i \, \dot{u}_i - \frac{1}{2} \sigma_{ij} \beta_{ji}) \, dx \qquad (53)$$

where

$$\beta_{ji} = u_{i,j}$$
$$\sigma_{ij,j} = \rho \, \ddot{u}_i \qquad (54)$$
$$\sigma_{ij} = C_{ijkl} \beta_{lk}$$

The above formulae are defined everywhere except on the slip surface S. The boundary of S (which is the dislocation line) is moving with velocity ζ. The variation of (53) becomes

$$\delta \int_{t_0}^{t_1} L \, dt = \int_{t_0}^{t_1} dt \int_D (\rho \, \dot{u}_i \, \delta \, \dot{u}_i - \sigma_{ij} \, \delta \, \beta_{ji}) \, dx$$

$$= \int_{t_0}^{t_1} dt \int_D [\rho u_i \frac{\partial}{\partial t} (\delta u_i) - \sigma_{ij} \frac{\partial}{\partial x_j} (\delta u_i)] \, dx \qquad (55)$$

Applying the Gauss theorem, we have

$$\int_D \sigma_{ij} \frac{\partial}{\partial x_j} (\delta u_i) \, dx = \int_{S+\delta S} \sigma_{ij} \, n_j \, [\delta u_i] \, dS - \int_D \sigma_{ij,j} \, \delta u_i \, dx$$

$$= \sigma_{ij} \, n_j \, b_i \, \delta S - \int_D \sigma_{ij,j} \, \delta u_i \, dx \qquad (56)$$

where δS is the variation of slip surface S, and is expressed by the vector product of a virtual displacement of the dislocation position $\delta \zeta$ and the dislocation line element $\nu \, d\ell$ [see (25)] :

$$n_j \, \delta S = \epsilon_{jln} \, \delta \zeta_l \, \nu_h \, d\ell \qquad (57)$$

In the above derivation, $[u_i]$ is the difference of u_i evaluated at the upper and lower surfaces of the slip plane. $[u_i] = b_i$ on S while $[u_i] = b_i$ on $S + \delta S$ after giving a vertical displacement $\delta \, \zeta_l$ to the position of the dislocation. Then the variation $[\delta u_i] = \delta \, [u_i]$ is zero on S and b_i on δS.

Furthermore, we have

$$\int_{t_o}^{t_1} \rho \, \dot{u}_i \, \frac{\partial}{\partial t} (\delta u_i) \, dt \, dx = - \rho \, \dot{u}_i \, b_i \, \dot{\zeta}_j \, n_j \, \delta S - \int_{t_o}^{t_1} \rho \, \ddot{u}_i \, \delta u_i \, dt \, dx \, \cdot \qquad (58)$$

The first term in the right side of (58) is zero from (57) when $\dot{\zeta}_j \, \delta t = \delta \, \zeta_j$ or the surface δS is parallel to the velocity of the dislocation. When the surface is not parallel to ζ, the volume $\zeta_j \, n_j \, \delta S$ swept by δS during its motion by ζ will be subjected to an impulsive material displacement due to b_i. Namely, $\partial/\partial t \, (\delta u_i)$ is the derivative of a step function in the small domain $dx = \zeta_j \, n_j \, \delta S$ and $\partial/\partial t \, (\delta u_i) dt$ is b_i in the domain. Substituting (56), (58) into (55) leads to

$$\delta \int_{t_o}^{t_1} L \, dt = - \int_{t_o}^{t_1} dt \oint_{\partial S} (\rho \, \dot{u}_i \, \dot{\zeta}_j + \sigma_{ij}) \, b_i \, \epsilon_{jlh} \, \delta \, \zeta_l \, \nu_h \, d\ell \, \cdot \qquad (59)$$

The term $\rho \, \dot{u}_i \, \dot{\zeta}_j \, b_i \, \epsilon_{jlh} \, \nu_h$ is called the Lorentz force.

The scattering of elastic waves by a dislocation discussed in the last section is only one of the many energy dissipative mechanisms which have been suggested by a number of investigators. Eshelby [78] has shown that the thermoelastic effect around a moving edge dislocation produces an irreversible heat flow causing energy dissipation. A more rigorous analysis of this thermoelastic dissipation due to an edge dislocation moving at an arbitrary speed has been performed by Weiner [97]. In order to produce a similar thermal effect for a moving screw dislocation, Mason [98] proposed a shear wave effect which would instantaneously raise the temperature of those phonons which have components along the compressed direction, while lowering the temperature of those which have components along the extended direction.

As an element of the dislocation line moves through the lattice, its potential energy varies in a roughly sinusoidal manner [99]. Therefore, the traveling dislocation experiences an oscillatory force. The energy is dissipated through this oscillation by a mechanism similar to that mentioned in the last section. Hart [100] calculated the magnitude of this dissipation. Various other dissipative mechanisms have been studied by Lothe [101].

CONCLUSIONS

As seen in this paper, the motion of a dislocation line is similar to that of one-dimensional line substance, except that its mass, line tension, and damping constant are dependent on motion. It moves like a rigid string (bar) with a constant subsonic velocity without any applied force. When it is accelerated by an applied force, Newton's first law holds by taking a mass which is a functional of history of motion. The dislocation line vibrates like an elastic string but its mass, line tension, and damping constant are functions of a frequency. This analogy will be extended to a helical dislocation under applied stress σ_{zz}, as shown in Fig. 3 (a). The vibration of the dislocation will be analogous to a dashpot-spring-mass model as shown in Fig. 3 (b). The analogy is established by solving our fundamental equation (29) for this helical dislocation. The damping of the vibrations of a coiled spring due to relaxation or creep will be discussed in a similar manner as in the paper by Hoff [102].

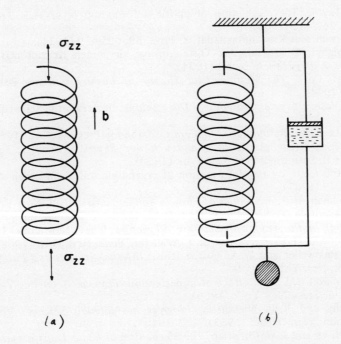

Fig. 3. Analogy between a helical dislocation and viscoelastic spring model.

In the present paper no comparison is made between the theory and experiments. Dislocation configurations treated here are so simplified that physical reality seems to be remote. In order to compare, for instance, the theory with internal friction experiments, we have to take into account the dislocation pinning by impurity particles (finite length of dislocation between pinning points), the distribution function of dislocation lengths, the breakaway model of Koehler [103], etc. These considerations have been done most successfully by Granato and Lücke [104-105] by assuming that the mass, line tension, and damping constant of a dislocation line are independent of motion (constant). According to our solution shown in Fig. 1, the maximum decrement (the energy lost per cycle divided by the total vibration energy of the specimen) Δ becomes $\Delta = ne/\mu\, u_0^2\, \kappa^2 = n\, 4\,\pi/\kappa^2$ where n is the dislocation density. Experimental observations [103, 105] give the order of Δ to be 10^{-3} for $\kappa \approx 1$. In order to reach agreement between our theoretical result and the experiments, the magnitude of order of n is to be taken as 1. The value is too small. The numerical value of R in (46) also seems too small compared with experimental values of Mason [106]. These numerical difficulties may be based upon the assumption that the whole length of dislocation line is moving. The dislocation lines are constrained by pinning of impurities so that only fractions of dislocation lines are able to move.

REFERENCES

1. G. I. Taylor, The mechanism of plastic deformation of crystals. *Proc. Roy. Soc.* A**145**, 362-415 (1934).
2. E. Orowan, Zur Kristallplastizilät. *Z. Phys.* **89**, 605-639 (1934).
3. M. Polanyi, Über eine Art Gitterstörung, die einen Kristall plastisch machen könnte. *Z. Phys.* **89**, 660-664 (1934).
4. A. E. H. Love, *The Mathematical Theory of Elasticity.* Dover Publication New York (1944).
5. A. R. Verma, *Crystal Growth and Dislocations.* Butterworths Scientific Publication (1953).
6. D. McLean, *Grain Boundaries in Metals.* Clarendon Press, Oxford (1957).
7. A. Seeger, Kristallplastizität. *Handbuch der Physik.* Vol. VII, Part 2, Crystal Physics II. Springer-Verlag New York (1958).
8. Proc. Int. Conf. on the deformation of crystalline solids, *Canadian J. of Phys.* **45**, 453-1249 (1967).
9. J. J. Gilman, *Micromechanics of Flow in Solids.* McGraw-Hill, New York (1969).
10. B. L. Averbach, D. K. Felbeck, G. T. Hahn and D. A. Thomas, eds., Fracture, *Proc. Int. Conf. on the Atomic Mechanism of Fracture.* Wiley, New York (1959).
11. H. Leibowitz (Ed.) *Fracture,* Vol. I, Academic Press, New York (1968).
12. G. M. Rassweiler and W. L. Grube (Eds.) *Internal Stresses and Fatigue in Metals.* Elseiver, New York (1959).
13. W. P. Mason (Ed.) The effect of imperfection. *Physical Acoustics* Vol. III, Part A. Academic Press, New York (1966).
14. A. Kelly and R. B. Nicholson, *Progress in Materials Science,* Vol. X, No. 3. Pergamon Press, Oxford (1963).
15. B. A. Bilby and J. W. Christian, *The Mechanism of Phase Transformation in Metals.* The Inst. of Metals (1956).

16. F. R. N. Nabarro, *Theory of Crystal Dislocations.* Clarendon Press, Oxford (1967).
17. A. E. Scheidegger, *Principles of Geodynamics.* Springer-Verlag, New York (1958).
18. M. A. Chinnery, The deformation of the ground around surface fault. *Bull. Seism. Soc. Amer.* **51**, 355-372 (1961), **53**, 921-932 (1963).
19. J. Weertman, Continuum Distribution of Dislocations on Faults with Finite Friction. *Bull. Seism. Soc. Amer.* **54**, 1035-1058 (1964).
20. B. Langenecker, C. W. Fountain and V. O. Jones, Ultrasonics: An aid to metal forming? *Metal Progress.* **85**, 97-101 (1964).
21. B. Langenecker, Effects of ultrasound on deformation characteristics of metals. *IEEE Trans. on Sonic and Ultrasonics.* **13**, 1-8 (1966).
22. F. Blaha and B. Langenecker, Plastizitätsuntersuchungen von metallkristallen in ultraschallfeld. *Acta Met.* **7**, 93-100 (1959).
23. B. Langenecker, effect of sonic and ultrasonic radiation on safety factors of rockets and missiles. *AIAA J.* **1**, 80-83 (1963).
24. R. M. Fisher and J. S. Lally, Microplasticity detected by an acoustic technique. *Canadian J. Phys.* **45**, 1147-1159 (1967).
25. P. H. Hutton, Acoustic emission in metals as an NDT tool. *Materials Evaluation.* **26**, 125-129 (1968).
26. H. L. Dunegan, D. O. Harris and C. A. Tatro, Fracture analysis by use of acoustic emission. *Engr. Fracture Mech.* **1**, 105-122 (1968).
27. A. T. Green, Detection of incipient failures in pressure vessels by stress-wave emissions. *Nuclear Safety.* **10**, 4-18 (1969).
28. J. R. Frederick, Acoustic emission as a technique for non-destructive testing. *Materials Evaluation.* **28**, 43-47 (1970).
29. J. R. Frederick, *Ultrasonic Engineering.* Wiley, New York (1965).
30. F. R. N. Nabarro, The synthesis of elastic dislocation fields. *Phil. Mag.* **42**, 1224-1231 (1951).
31. T. Mura, Continuous distribution of moving dislocations. *Phil. Mag.* **8**, 843-857 (1963).
32. T. Mura, Periodic distribution of dislocations. *Proc. Roy. Soc.* A280, 528-544 (1964).
33. T. Mura, The elastic field of moving dislocations and disclinations. *Fundamental Aspect of Dislocation Theory,* Vol. I, II, eds., J. A. Simmons, R. DeWit, and R. Bullough, NBS Special Pub. 317 (1971).
34. J. R. Willis, Stress fields produced by dislocations in anisotropic media. *Phil. Mag.* **21**, 931-949 (1970).
35. T. Mura, Stress and velocity field produced by uniformly moving dislocations in anisotropic media. *Phil. Mag.* **23**, 235-237 (1971).
36. A. W. Sáenz, Uniformly moving dislocations in anisotropic media. *J. Rat. Mech. Anal.* **2**, 83-98 (1953).
37. A. N. Stroh, Steady state problems in anisotropic elasticity. *J. Math. Phys.* **41**, 77-103 (1962).
38. F. John, *Plane Waves and Spherical Means Applied to Partial Differential Equations,* Interscience, New York (1955).
39. M. Peach and J. S. Koehler, The forces exerted on dislocations and the stress fields produced by them. *Phys. Rev.* **80**, 436-439, (1950).
40. R. Peierls, The size of a dislocation. *Proc. Phys. Soc.* **52**, 34-37 (1940).
41. F. R. N. Nabarro, Dislocations in a simple cubic lattice. *Proc. Phys. Soc.* **59**, 256-272 (1947).
42. A. M. Kosevich, The deformation field in an isotropic elastic medium containing moving dislocations. *Soviet Physics JETP.* **15** 108-115 (1962).
43. A. M. Kosevich, The equation of motion of a dislocation. *Soviet Physics JETP.* **16**, 455-462 (1963).

44. T. Mura, On dynamic problems of continuous distribution of dislocations. *Int. J. Engng. Sci.* **1**, 371-381 (1963).
45. H. Bross, Zur theorie bewegter Versetzungen. *phy. stat. sol.* **5**, 329-342 (1964).
46. H. Günther, Zur nichtlinearen Kontinuumstheorie bewegter Versetzungen. *Schriftenreihe der Institute für Mathematik, bei der Deutschen Akademie der Wissenschaften zu Berlin.* Akademie-Verlag (1967).
47. E. F. Höllander, The basic equations of the dynamics of the continuous distribution of dislocations. *Czech. J. Phys.* **10**, 409-418 (1960). **10** 479-487 (1960). **10**, 551-560 (1960).
48. E. F. Höllander, The geometric equations of dislocation dynamics. *Czech. J. Phys.* 35-47 (1962).
49. H. Zorski, Theory of Discrete Defects. *Arch. Mech. Stos.* **18**, 301-372 (1966).
50. R. J. Beltz, T. L. Davis and K. Malén, Some unifying relations for moving dislocations. *phys. stat. sol.* **26**, 621-636 (1968).
51. E. Kossecka, Theory of dislocation lines in a continuous medium. *Arch. Mech. Stos.* **21**, 167-190 (1969).
52. G. Stenzel, Lineare Kontinuumstheorie bewegter Versetzungen. *phy. stat. sol.* **34**, 351-364 (1969).
53. G. Stenzel, Zum Peierls – Modell bewegter Versetzungen. *phy. stat. sol.* **34**, 495-500 (1969).
54. G. Kluge, Zur Dynamik der allgemeinen Versetzungstheorie bei Berücksichtigung von Momentenspannungen. *Int. J. Engng. Sci.* **7**, 169-182 (1969).
55. F. C. Frank, On the equations of motion of crystal dislocations. *Proc. Phys. Soc.* A62, 131-134 (1949).
56. J. D. Eshelby, Uniformly moving dislocations. *Proc. Phys. Soc.* A62, 307-314 (1949).
57. T. Kontorova and Y. I. Frenkel, Theory of plastic deformation and twining. *Zh. éksp. teor. Fiz.* **8**, 89-95 (1938).
58. G. Leibfried and H. D. Dietze, Zur theorie der Schraubensetzung. *Z. Phys.* **126**, 790-808 (1949).
59. J. Weertman, High velocity dislocations. *Response of Metals to High Velocity Deformation*, eds., P. G. Shewmon and V. F. Zackay, Interscience, New York (1961).
60. J. Weertman, Possibility of the existence of attractive forces between dislocations of like sign. *Phys. Rev.* **119**, 1871-1872 (1960).
61. J. Weertman, Coalescence energy of two moving dislocations of like sign. *J. Appl. Phys.* **37**, 4925-4927 (1966).
62. L. J. Teutonico, Dynamic behavior of dislocations in anisotropic media. *Phys. Rev.* **124**, 1039-1045 (1961).
63. L. J. Teutonico, Moving edge dislocations in cubic and hexagonal materials. *Phys. Rev.* **125**, 1530-1533 (1962).
64. L. J. Teutonico, Uniformly moving dislocations of arbitrary orientation in anisotropic media. *Phys. Rev.* **127**, 413-418 (1962).
65. L. J. Teutonico, Dislocations in the primary slip systems of face-centered cubic crystals. *Acta Met.* **11**, 391-397 (1963).
66. L. J. Teutonico, Dynamic behavior of dislocations in the primary slip systems of body-centered cubic crystals. *J. Appl. Phys.* **34**, 950-955 (1963).
67. J. Weertman, Fast moving edge dislocations on the (110) plane in anisotropic body-centered cubic crystals. *Phil. Mag.* **7**, 617-631 (1962).
68. J. Weertman, Fast moving edge dislocations on the (111) plane in anisotropic face-centered cubic crystals. *J. Appl. Phys.* **33**, 1631-1635 (1962).
69. A. V. Hull and J. Weertman, Fast moving dislocations in various face-centered cubic metals. *J. Appl. Phys.* **33**, 1636-1637 (1962).

70. J. Cotner and J. Weertman, Fast moving edge dislocations in alpha iron. *Acta Met.* **10**, 515-517 (1962).
71. R. Bullough and B. A. Bilby, Uniformly moving dislocations in anisotropic media. *Proc. Phys. Soc.* B67, 615-624 (1954).
72. J. Weertman, Dislocations moving uniformly on the interface between isotropic media of different elastic properties. *J. Mech. Phys. Solids.* **11**, 197-204 (1963).
73. M. S. Lee and J. Dundurs, Uniformly Moving Screw (or edge) Dislocation in a Bimetallic Material. To be Published.
74. J. D. Eshelby, Supersonic dislocations and dislocations in dispersive media. *Proc. Phys. Soc.* B69, 1013-1019 (1956).
75. J. Weertman, Uniformly moving transonic and supersonic dislocations. *J. Appl. Phys.* **38**, 5293-5301 (1967).
76. J. Weertman, Dislocations in uniform motion on slip or climb planes having periodic force laws. *Mathematical Theory of Dislocations.* ed., T. Mura. A.S.M.E. (1969).
77. H. Günther, Überschallbewegung von Eigenspannungsquellen in der Kontinuums-theorie. *Ann. Phys.* **21**, 93-105 (1968).
78. J. D. Eshelby, Dislocations as a cause of mechanical damping in metals. *Proc. Roy. Soc.* A197, 396-416 (1949).
79. F. R. N. Nabarro, The interaction of screw dislocations and sound waves. *Proc. Roy. Soc.* A209. 278-290 (1951).
80. J. Kiusalaas and T. Mura, On the elastic field around an edge dislocation with application to dislocation vibration. *Phil. Mag.* **9**, 1-7 (1964).
81. G. Leibfried, Über den Einfluss thermisch angeregter Schallwellen auf die plastische Deformation. *Z. Phys.* **127**, 344-356 (1950).
82. J. Lothe, Aspects of the theories of dislocation mobility and internal friction. *Phys. Rev.* **117**, 740-709 (1960).
83. J. D. Eshelby, The interaction of kinks and elastic waves. *Proc. Roy. Soc.* A266, 222-246 (1962).
84. B. Pegel, Strahlungslose Eigenschwingungen von Versetzungen. *phy. stat. sol.* **14**, K165-167 (1966).
85. T. Laub and J. D. Eshelby, The velocity of a wave along a dislocation. *Phil. Mag.* **11**, 1285-1293 (1966).
86. T. Ninomiya and S. Ishioka, Dislocation vibration: Effective mass and line tension. *J. Phy. Soc. Japan.* **23**, 361-372 (1967).
87. H. H. Kuo and T. Mura, Dislocation vibration and energy dissipation. To be published in *J. Appl. Phys.*
88. B. Pegel, Self-vibration of parallel edge dislocations. *phys. stat. sol.* **29**, K133-136 (1968).
89. T. Mura, Individual dislocations and continuum mechanics. *Inelastic Behavior of Solids.* eds., M. F. Kanninen, W. F. Adler, A. R. Rosenfield, and R. I. Jaffee, McGraw-Hill, New York (1970).
90. J. D. Eshelby, The equation of motion of a dislocation. *Phys. Rev.* **90**, 248-255 (1953).
91. D. D. Ang and M. L. Williams, The dynamic stress field due to an extensional dislocation. *4th Midwestern Conference on Solid Mechanics.* The Univ. of Texas (1959).
92. J. Kiusalaas and T. Mura, On the motion of a screw dislocation. *Recent Advances in Engineering Science* Vol. I, ed. by A. C. Eringen, Gordon & Breach, New York (1964).
93. J. Lothe, Lorentz Force on Screw Dislocations and Related Problems. *Phys. Rev.* **122**, 78-82 (1961).

94. H. A. Bahr, W. Pompe and H. G. Schöpf, On the dynamic dislocations. *phys. stat. sol.* **28**, K23-26 (1968).

95. K. Malén, On the analysis of the force on a moving dislocation. *phys. stat. sol.* **37**, 267-274 (1970).

96. H. A. Bahr and H. G. Schöpf, *phys. stat. sol.* **40**, K21-K24 (1970).

97. J. H. Weiner, Thermoelastic dissipation due to high-speed dislocations. *J. Appl. Phys.* **29**, 1305-1307 (1958).

98. W. P. Mason, Phonon viscosity and its effect on acoustic wave attenuation and dislocation motion. *J. Acoust. Soc. Am.* **32**, 458-472 (1960).

99. E. Orowan, Problems of plastic gliding. *Proc. Phys. Soc.* **52**, 8-22 (1940).

100. E. W. Hart, Lattice resistance to dislocation motion at high velocity. *Phys. Rev.* **98**, 1775-1776 (1955).

101. J. Lothe, Theory of Dislocation Mobility in Pure Slip. *J. Appl. Phys.* **33**, 2116-2125 (1962).

102. N. J. Hoff, Damping of the vibrations of a coiled spring due to creep. *Creep in Structures,* ed. by N. J. Hoff, IUTAM Colloquium Stanford, Springer-Verlag, New York 355-373 (1962).

103. J. S. Koehler, The influence of dislocations and impurities on the damping and the elastic constants of metal single crystals. *Imperfections in Nearly Perfect Crystals,* ed. by W. Shockley, J. H. Hollomon, R. Maurer, and F. Seitz, Wiley, New York 197-216 (1952).

104. A. Granato and K. Lücke, Theory of mechanical damping due to dislocations. *J. Appl. Phys.* **27**, 583-593 (1956).

105. A. Granato and K. Lücke, Application of dislocation theory to internal friction phenomena at high frequencies. *J. Appl. Phys.* **27**, 789-804 (1956).

106. W. P. Mason, Dislocation drag mechanisms and their effects on dislocation velocities. *Dislocation Dynamics,* ed. by A. R. Rosenfield, G. T. Hahn, A. L. Bement, Jr., and R. I. Jaffee, McGraw-Hill, New York, 487-505 (1968).